NITRIC OXIDE

NATO Science Series

A Series presenting the results of activities sponsored by the NATO Science Committee.
The Series is published by IOS Press and Kluwer Academic Publishers in conjunction with the NATO Scientific Affairs Division.

General Sub-Series

A Life Sciences	IOS Press
B Physics	Kluwer Academic Publishers
C Mathematical and Physical Sciences	Kluwer Academic Publishers
D Behavioural and Social Sciences	Kluwer Academic Publishers
E Applied Sciences	Kluwer Academic Publishers
F Computer and Systems Sciences	IOS Press

Partnership Sub-Series

1 Disarmament Technologies	Kluwer Academic Publishers
2 Environmental Security	Kluwer Academic Publishers
3 High Technology	Kluwer Academic Publishers
4 Science and Technology Policy	IOS Press
5 Computer Networking	IOS Press

The Partnership Sub-Series incorporates activities undertaken in collaboration with NATO's Partners in the Euro-Atlantic Partnership Council - countries of the CIS and Central and Eastern Europe - in Priority Areas of concern to those countries.

NATO-PCO-DATA BASE

The NATO Science Series continues the series of books published formerly in the NATO ASI Series. An electronic index to the NATO ASI Series provides full bibliographical references (with keywords and/or abstracts) to more than 50 000 contributions from international scientists published in all sections of the NATO ASI Series.
Access to the NATO-PCO-DATA BASE is possible in two ways:
- via online FILE 128 (NATO-PCO-DATA BASE) hosted by ESRIN, Via Galileo Galilei, I-00044 Frascati, Italy;
- via CD-ROM "NATO-PCO-DATA BASE" with user-friendly retrieval software in English, French and German (© WTV GmbH and DATAWARE Technologies Inc., 1989).

The CD-ROM of the NATO ASI Series can be ordered from PCO, Overijse, Belgium.

Series A: Life Sciences - Vol. 317

ISSN: 1387-6686

Nitric Oxide

Basic Research and Clinical Applications

Edited by

Ryszard J. Gryglewski

Department of Pharmacology, Jagiellonian University Medical College, Cracow, Poland

and

Pietro Minuz

Department of Biomedical and Surgical Sciences, University of Verona, Verona, Italy

IOS Press

Ohmsha

Amsterdam • Berlin • Oxford • Tokyo • Washington, DC

Published in cooperation with NATO Scientific Affairs Division

Proceedings of the NATO Advanced Study Institute on
Nitric Oxide: Basic Research and Clinical Applications
7–17 September 2000
Erice, Italy

ISBN 1 58603 164 3 (IOS Press)
ISBN 4 274 90433 4 C3045 (Ohmsha)
Library of Congress Catalog Card Number: 2001087802

Publisher
IOS Press
Nieuwe Hemweg 6B
1013 BG Amsterdam
Netherlands
fax: +31 20 620 3419
e-mail: order@iospress.nl

Distributor in the UK and Ireland
IOS Press/Lavis Marketing
73 Lime Walk
Headington
Oxford OX3 7AD
England
fax: +44 1865 75 0079

Distributor in the USA and Canada
IOS Press, Inc.
5795-G Burke Centre Parkway
Burke, VA 22015
USA
fax: +1 703 323 3668
e-mail: iosbooks@iospress.com

Distributor in Germany, Austria and Switzerland
IOS Press/LSL.de
Gerichtsweg 28
D-04103 Leipzig
Germany
fax: +49 341 995 4255

Distributor in Japan
Ohmsha, Ltd.
3-1 Kanda Nishiki-cho
Chiyoda-ku, Tokyo 101
Japan
fax: +81 3 3233 2426

PRINTED IN THE NETHERLANDS

Preface

In 1998 Robert Furchgott, Louis Ignarro and Ferid Murad received the Nobel Prize in Physiology and Medicine for their discovery of biological significance and mechanisms of action of nitric oxide (NO). It was the most unusual discovery in biological sciences. The message passed around that toxic gas, composed of two atoms of two major air components is actually generated not only during tempest by lightnings or in working engines of cars, but also by cells of living matter. There, in disguise of a free radical NO had been serving for ages as a regulator of life in bodies of fruit fly, crayfish and humans. This incredible information quickly spread from laboratory benches to the clinics. No wonder – firstly, many drugs including old nitroglycerine act upon the NO principle, secondly, mechanisms of development of atherosclerosis, diabetes or septic shock become more clear with NO as a major player, and finally, many physiological functions of human body like memory, penile erection or free breathing depend on presence of NO at the right place and at the right time.

In the exclusive and yet friendly atmosphere of Centro Ettore Majorana in Erice during the NATO ASI Course entitled: "Nitric Oxide – Basic Research and Clinical Applications", Robert Furchgott and Ferid Murad disputed the subject with their colleagues and with young researchers. A very special feature of this course was the geographical mosaic of its participants' origin. Some of them from Russia, Ukraine, Belarus, Lithuania, Czech Republic, Hungary, Turkey or Poland presented lectures or posters; others actively participated in long lasting discussions with their more experienced colleagues from Western Europe and the USA.

In this book we present papers based on the lectures and selected posters, which constituted the backbone of the Course. Robert Furchgott (USA), with his typical modesty, gave us an account on his "accidental" finding that had led him to the discovery of Endothelium-Derived Relaxing Factor (EDRF), and EDRF was soon identified as NO. Needless to say that since the times of Louis Pasteur everybody has known that "*Dans les champs l'observation le hazard ne favorise que les esprits préparé*". The second Nobel Prize Laureate, Ferid Murad (USA) spent many years studying the physiological role of cyclic gunosine monophosphate (c-GMP) before discovering the historical link between NO, soluble guanylate cyclase and c-GMP; the link that enables signal transduction from NO to the biochemical machinery of cells. Dennis Stuher (USA) presented a complex system of NO synthesis from L-arginine and activated oxygen. He also pointed out to essential factors required for activity of NO synthases (NOS). Bernd Mayer (Austria) in his two lectures developed the above subject and added his recent data on the mode of formation of a deleterious species of peroxynitrite ($ONOO^-$) from $NO^•$ and superoxide anion (O_2^-). Ted Malinski (USA) presented a deep insight into kinetics of generation of NO, O_2^- and $ONOO^-$ in ischaemic brain followed by the reperfusion period. Patrick Valance (UK) introduced the audience to the clinical importance of endogenous inhibitor of NOS, namely asymmetric dimethylarginine (ADMA) and routes of its degradation by corresponding genetic variants of its hydrolase (DDAH). Ryszard Gryglewski (Poland) presented the evidence for pleiotropic NO-mediated pharmacological action of cardiovascular drugs such as statins, inhibitors of angiotensin converting enzyme and some β adrenoceptor antagonists. Also of clinical importance are the findings of Ágnes Végh (Hungary) and James Parratt (UK), who revealed the significance of NO in ischaemic preconditioning of the heart, especially in respect to its antiarrhythmic action. Bohuslav Ošt'ádal (Czech Republic) gave an excellent review on the role of NO in the adult myocardium and on perspectives for studying the physiological function of NO in developing heart. Galen Pieper (USA) presented experimental and clinical problems concerning the role of NO (especially that made by inducible NOS) in cardiac transplant rejection. This was an outstanding example of clinical exploitation of recent findings in basic research on NO. Pietro Minuz (Italy) addressed an important question of interaction between oxidized low density lipoproteins (ox-LDL) and isoprostanes with endothelial NO synthesizing system in respect to interaction with platelets, while Luciano Cominacini (Italy) studied effects of ox–LDL on production of nitrotyrosine in endothelial cells. Christoph Thiemermann (UK) and Salvatore Cuzzocrea (Italy) in two separate lectures not only reviewed the role of NO in circulatory shock but also presented their recent original data obtained with the use of molecular techniques. Finally, Stefan Chłopicki (Poland) clearly proved in blood perfused endotoxic lungs the protective role of NO against destructive action of endogenous toxic lipid mediators on pulmonary endothelium.

Next to the general discussion out of 32 posters 21 summaries were chosen to be published. These include data of M. Bucci *et al.* (Italy); S. Ignatov *et al.* (Russia); C. Paolucci *et al.* (Italy); K. Hansen and O.A. Nedergaard (Denmark); V. Roman *et al.* (Romania); S.A Golubev *et al.* (Belarus); A. Smith *et al.* (USA); A. P. Sistetsky *et al.* (Ukaraine); O. Bulayev and K. Bulayev (Russia); V. Kecskeméti *et al.* (Hungary); B.S. Uydeş-Doğan *et al.* (Turkey); M. Łomnicka *et al.* (Poland); F. Fabi *et al.* (Italy); S. Nedeianu and T. Páli (Hungary); D. Uracz *et al.* (Poland); O. Konopacka *et al.* (UK); H.V Evsyukova (Russia); S. M. Sequeire *et al.* (Portugal); M. Lähde and E. Moilanen (Finland), M. Tunçel *et al.* (Turkey) and G. Benoni and L. Cuzzolin (Italy).

The Directors of the 59th NATO ASI Course
Professor Ryszard Gryglewski, M.D, h.c. D.Sc. (Cracow, Poland)
Pietro Minuz , M.D. (Verona, Italy)

Contents

Nitric Oxide
R.J. Gryglewski and P. Minuz (Eds.)
IOS Press, 2001

Research Leading to Nitric Oxide: the Importance of Accidental Findings

Robert F. Furchgott
Department of Pharmacology, State University of New York Downstate Medical Center, 450 Clarkson Ave., Brooklyn, NY 11203, USA

Abstract. This paper describes a number of accidental findings made in the author's laboratory and discusses how they led to the discovery of the endothelium-derived relaxing factor (EDRF) and its identification as nitric oxide. The accidental finding leading to the discovery of EDRF occurred in 1978 when the author's technician failed to follow directions to washout norepinephrine from two organ chambers containing rings of rabbit thoracic aorta before adding carbamylcholine, and observed relaxation with this muscarinic agonist rather than the expected contraction. The accidental findings next described, which occurred many years earlier, were those of photorelaxation of preparations of rabbit aorta, and the potentiating effect of nitrite on photorelaxation. The last accidental finding described here occurred when a new postdoctoral fellow mistakenly used acidified rather then neutral diluting fluid for making standard solutions of sodium nitrite, and found that such solutions gave greatly potentiated but transient relaxations of aorta as compared to neutral nitrite solutions. How this last accidental finding, along with other findings, led to the proposal that EDRF is NO is briefly discussed.

1. The Accidental Finding of Endothelium-Dependent Relaxation

The accidental finding that led to the discovery of endothelium-dependent relaxation of blood vessels and to the endothelium-derived releasing factor (EDRF) has been described in detail in previous papers [1,2] and therefore will only be summarized here. The finding was made in an *in vitro* experiment on rabbit aorta in 1978 in which my technician failed to follow the sequence of additions of drugs and washouts called for in my written protocol and added carbachol to two organ chambers before washing out the norepinephrine that had been added just previously. The carbachol, like acetylcholine in numerous older experiments, was expected to produce contraction of the aortic preparations, but instead it produced relaxation of the norepinephrine-induced contraction. Acetylcholine was also found to produce relaxation of the aortic preparation pre-contracted with norepinephrine, and atropine was found to block relaxation by both of the muscarinic agonists. The immediate question was why had these muscarinic agonists produced relaxation of pre-contracted aortic preparations in this experiment, whereas they had only produced extra contraction in numerous earlier experiments. The major difference was that the aortic preparation now being used was the transverse ring (recording through a force transducer) rather than the helical strip (recording through an isotonic lever) used in our earlier experiments. After many additional experiments, we found out that endothelial cells had an obligatory role in the relaxation of aortic preparations by acetylcholine, and that these cells were present in the transverse ring preparation but not in the helical strip preparation, from which they had been inadvertently rubbed off during our standard method of preparation. We also found that acetylcholine stimulates the endothelial cells to produce and release a non-prostanoid labile factor that diffuses to the smooth muscle cells and causes them to relax [3]. This factor was soon called the endothelium-derived relaxing factor or EDRF.

2. The Accidental Finding of Photorelaxation and Its Potentiation by the Nitrite Ion

Many years ago I briefly reported on a newly discovered phenomenon - namely, that strips of rabbit aorta undergo reversible relaxation when exposed to light of proper wavelength and intensity [4]. This photorelaxation was an accidental finding that was made during an experiment in which we were using four helical strips of an aorta, two in chambers in one constant temperature bath and two in chambers in a second bath. For some unknown reason these strips, unlike those in practically all previous experiments, spontaneously developed a low level of contractile tone after the first hour or so of equilibration. It was also noted that the contractile tone of the two strips in one bath fluctuated with time, whereas that of two strips in the other bath remained steady. Moreover, the two strips showing fluctuations did so synchronously, and when my technician stood in front of the bath, I noticed that both strips simultaneously developed a small increase in tone. Those two strips, but not the other two, were in a bath near a window through which they were exposed to skylight. Suspecting that the fluctuations in tone were due to fluctuations in light intensity on the strips near the window (it was a cloudy-bright day), I closed the shade on the window and both strips increased in tone. I opened the shade and both decreased in tone. From that point on, we never allowed our strips to be exposed to direct skylight. (The usual overhead fluorescent lights do not produce photorelaxation.)

At a later date we were able to obtain an accurate action spectrum for photorelaxation of rabbit aorta, which exhibited a peak close to 310 nm, a shoulder near 350 nm, and a steady falling off into the visible wavelength range [5]. In the course of our studies on photorelaxation, we were running experiments to compare the relaxing potency of near ultraviolet with that of sodium nitrite, and encountered another accidental finding - namely, that photorelaxation is greatly potentiated in the presence of the nitrite ion. Also, nitrite shifts the peak of the action spectrum to about 355 nm [6].

Many years later in the early 1980's, after the discovery of endothelium-derived relaxing factor (EDRF), I again began research on photorelaxation. Although photorelaxation did not depend on the presence of endothelium on the strip or ring of rabbit aorta, we found many similarities between it and endothelium-dependent relaxation (as produced by acetylcholine or A23187). Not only was photorelaxation, like endothelium-dependent relaxation, causally dependent on the elevation of cyclic GMP as a result of stimulation of guanylate cyclase, but both were inhibited by hemoglobin and by methylene blue [7]. Then, after EDRF was identified in 1986 as nitric oxide, Kazuki Matsunaga (a postdoctoral fellow) and I reinvestigated the potentiation of photorelaxation by sodium nitrite. Using a cleverly designed perfusion-bioassay type apparatus, Matsunaga clearly demonstrated that potentiation was due to the photoactivated release of NO from nitrite [8]

It is tempting to hypothesize that light (in the absence of added nitrite) produces relaxation of vascular smooth muscle by photoactivating release of NO from some endogenous compound stored in the muscle - perhaps a compound formed in a reaction between NO of endothelial origin and some muscle constituent.

3. The Accidental Finding that Acidification of Nitrite Enhances Its Relaxing Effect

Around 1970, I assigned to a new postdoctoral fellow an experiment to familiarize him with procedures for obtaining cumulative dose-response curves with strips of rabbit aorta. Since sodium nitrite was a well-studied relaxing agent in our laboratory, I instructed him to make the desired dilutions of that salt for cumulative additions to the organ chambers, and told him that he could expect fifty percent relaxation of a phenylephrine-induced contraction at a concentration of about 0.1 mM sodium nitrite. After he had carried out the experiment, he came to me with his physiograph records and said that I had greatly underestimated the potency of sodium nitrite as a relaxing agent. I looked at his records, and, indeed, he had obtained about fifty percent relaxation at only 3 μM nitrite and almost complete relaxation at 0.1 mM. However, relaxation on each addition was transient, reaching a peak and then fading over several minutes back to a level close to what we had observed in previous experiments at the same nitrite concentration. I knew that the fellow must have made some mistake. I asked him what diluting fluid he had used for making up his standard nitrite solutions. He pointed to a bottle on the laboratory shelf labeled "acidified diluting fluid." That diluting fluid, containing 0.01 M HCl, was there for making stabile solutions of acetylcholine and of catecholamines. His use of the acidified diluting fluid rather than neutral diluting fluid (in a bottle on the same laboratory shelf) had somehow given rise to strong but transient relaxations in response to relatively low concentrations of nitrite. Looking into the literature on the chemistry of inorganic nitrite, I became aware that the HNO_2 (pK_a of 3.2) produced by acidification of $NaNO_2$ solutions immediately generates low concentrations of NO as a result of a reversible dismutation.

$$NO_2^- + H^+ \longleftrightarrow HNO_2$$

$$2HNO_2 \longleftrightarrow NO + NO_2 + H_2O$$

However, when a small volume of a solution of acidified nitrite is added to the oxygenated, well buffered physiological salt solution used in our organ chambers, any NO present will be removed very rapidly - thus accounting for the transiency of the relaxation of the aortic rings receiving the acidified nitrite.

4. How the Accidental Findings Contributed to the Identification of EDRF as NO

Between 1983 and 1985, with Billy Martin working with us, a number of key observations were made in my laboratory. Among them were that hemoglobin (but not methemoglobin) and methylene blue were potent inhibitors of both the relaxation and the increases in cyclic GMP produced by endothelium-dependent relaxing agents [8]. From Martin I also learned first hand about the work he and others in John Gillespie's laboratory in Glasgow had done in trying to identify the non-adrenergic, non-cholinergic (NANC) neurotransmitter that elicited relaxation of the bovine retractor penis muscle and certain other smooth muscle preparations. Relaxation produced by NANC nerve stimulation and relaxation produced by a partly purified, acid-activated extract made from the bovine retractor penis were both inhibited by hemoglobin. As discussed in a paper which I presented at a symposium on mechanisms of vasodilatation held in July 1986, I was struck by the similarities of the transient relaxations produced by Martin's acid-activated bovine retractor penis inhibitory factor (BRPIF) and the transient relaxations of rabbit aorta produced by acidified solutions of sodium nitrite [9]. Also, earlier in 1986 it had been reported by Rubanyi and Vanhoutte and by Gryglewski, Palmer and Moncada that superoxide anion (O_2^-) rapidly inactivates EDRF, and that superoxide dismutase (SOD) protects against this inactivation. Realizing that acidification of $NaNO_2$ solutions produces HNO_2 which then generates low concentrations of NO and NO_2 as a result of a reversible dismutation (see above), and assuming that the free radical NO might rapidly react with the free radical O_2^-, I began to speculate on the possibility that the acid-activated BRPIF is inorganic nitrite and that EDRF is nitric oxide. It was already well known from much earlier work in the laboratories of Murad and Louis Ignarro that nitrovasodilators (first considered by Murad to be NO generators) and NO itself produced relaxation of smooth muscle by activating the synthesis of cyclic GMP.

In the few months before the symposium in July of 1986, my associates (D. Jothianandan and M.T. Khan) and I carried out a large number of experiments on rabbit aorta comparing the characteristics of relaxation by EDRF (released by acetylcholine) and by NO (released by acidified nitrite). The characteristics (inhibition by hemoglobin and by methylene blue, inhibition by superoxide generators, protection by SOD, association with increases in cyclic GMP) were so similar that I had no hesitancy in proposing at the symposium that EDRF is NO. I also proposed that the acid-activatible BRPIF of Martin and Gillespie was inorganic nitrite, and noted that our findings along with the findings of others on some of the properties of the relaxing factor released on stimulation of NANC nerves made it tempting to consider "the possibility that the smooth muscle inhibitory factor released on nerve stimulation is NO" [9].

At the 1986 symposium at which I proposed that EDRF is NO, Louis Ignarro, on the basis of his studies on bovine pulmonary arteries, independently made the same proposal.

Acknowledgement: Most of the research reported in this paper was carried out with the support of grants from the National Heart, Lung and Blood Institute of the U.S.P.H.S.

References

[1] Furchgott, R.F., The discovery of endothelium-dependent relaxation. Circulation **87**: Suppl. V, V3-V8 (1993).

[2] Furchgott, R.F., The discovery of endothelium-derived relaxing factor and its importance in the identification of nitric oxide. J. Amer. Med. Assn. **276**, 1186-1188 (1996).

[3] Furchgott, R.F., The pharmacology of vascular smooth muscle. Pharmacol. Rev. **7**: 183-265 (1955).

[4] Furchgott, R.F., Ehrreich, S.J. and Greenblatt, E., The photoactivated relaxation of smooth muscle of rabbit aorta. J. Gen Physiol., **44**: 499-519 (1961).

[5] Furchgott, R.F., Effect of various agents on photorelaxation of rabbit aortic strips. In Physiology and Pharmacology of Vascular Neuroeffector systems, ed. by J.A. Bevan, R.F. Furchgott, R.A. Maxwell and A.P. Somlyo, pp. 247-261 (Karger, Basel, 1971).

[6] Furchgott, R.F., Martin, W., Cherry, P.D., Jothianandan, D. and Villani, G.M., Endothelium-dependent relaxation, photorelaxation and cyclic GMP. In 5th International Symposium, Vascular Neuroeffector Mechanisms, ed. by J. Bevan, T. Godfraind, R. Maxwell, J.C. Stoclet and M. Worcel, pp. 105-114 (Elsevier, Amsterdam) (1985).

[7] Matsunaga, K. and Furchgott, R.F., Interactions of light and sodium nitrite in producing relaxation of rabbit aorta. J. Pharmacol. Exp.Ther. **248**: 687-695 (1989).

[8] Martin, W., Villani, G.M., Jothianandan, D. and Furchgott, R.F., Selective blockade of endothelium-dependent and glyceryl trinitrate-induced relaxation by hemoglobin and by methylene blue in the rabbit aorta. J. Pharmacol. Exp. Ther. **232,** 708-716 (1985).

[9] Furchgott, R.F., Studies on relaxation of rabbit aorta by sodium nitrite: the basis for the proposal that the acid-activatable inhibitory factor from retractor penis is inorganic nitrite and the endothelium-derived relaxing factor is nitric oxide. In Vasodilatation: Vascular Smooth Muscle, Peptides, and Endothelium. ed. by P.M. Vanhoutte pp. 401-414 (Raven Press, New York, 1988).

Nitric Oxide
R.J. Gryglewski and P. Minuz (Eds.)
IOS Press, 2001

Signal Transduction with Nitric Oxide, Guanylyl Cyclase and Cyclic Guanosine Monophosphate

Aurora R. Seminara, Joshua S. Krumenacker, Ferid Murad

Department of Integrative Biology and Pharmacology and Institute of Molecular Medicine, University of Texas-Houston Medical School, 6431 Fannin St., Houston, TX 77030, USA

Abstract

Studies surrounding the formation and biological actions of nitric oxide (NO) have grown immensely over the past twenty years. Following the understanding of the biological effects of NO and nitrovasodilators on the regulation of cyclic guanosine monophosphate (cGMP), numerous studies have exposed the complexity of this cell-signaling cascade in various systems. NO itself has been demonstrated to act as an intracellular second messenger, a local autocrine or paracrine factor, a neurotransmitter, and possibly a hormone acting at distant targets. The biological effects of NO can be paradoxically beneficial or detrimental, depending on the effector, its concentration and the surrounding microenvironment. The diverse actions of this signaling cascade make it a good target for drug discovery and research interest.

1. Introduction and Background

While working with cAMP as a cell signaling second messenger molecule in the late 1960s to early 1970s, the emergence of cGMP as another potential second messenger occurred. Our laboratory subsequently became interested in cell signaling through cGMP, since at the time, cGMP was being characterized with respect to its measurement in body fluids, its synthesis by guanylyl cyclase, and its hydrolysis by phosphodiesterases. We began by focusing our efforts on understanding how various hormones, neurotransmitters, and other ligands lead to the stimulation of cGMP synthesis by soluble guanylyl cyclase and describing what functions are physiologically regulated by cGMP. We have now spent approximately three decades answering the above and many additional important biological questions surrounding this unique and diverse cell signaling cascade.

Our initial experiments surrounding cGMP were what we referred to as "dumping" experiments, in which we would add various agents and hormones to cell homogenates, cell cultures or tissue preparations in attempt to correlate the accumulation of cGMP with some biological response of the preparation. Unfortunately, while us and others were able to produce multiple manuscripts documenting the results, these initial experiments were not very helpful to identify the actual function of cGMP accumulation in physiology. We decided to focus our studies on the enzyme that converts GTP to cGMP, guanylyl cyclase. We hypothesized that if we could gain a better understanding of guanylyl cyclase, it would lead us in the right direction to elucidate its biological function and how it may be regulated by hormonal stimulation.

We learned quickly that most tissue preparations contained catalytic activity to produce cGMP, and this catalytic activity existed in both high-speed soluble and particulate fractions (1-5). In addition, these fractions exhibited different enzyme kinetics and properties. Detergents or mild protease treatments were able to solubilize and enhance the activity of the particulate fraction. We suspected there may be multiple isoforms of the enzyme, and subsequent work led to the elucidation of several different isoforms and independent gene products. We know today that several isoforms of both the soluble and particulate guanylyl cyclases exist.

2. Effects of NO donors on the synthesis of cGMP

We suspected we were working with at least two isoforms of guanylyl cyclase, and continued to work with the enzyme in cell-free systems. It was difficult to prove, however, that some of our results were not due to contaminating factors in the crude preparations, such as phosphatases, nucleotidases, and cyclic nucleotide phosphodiesterases. The approach we took was to add general inhibitors of these contaminants in our preparations, such as azide, hydroxylamine, phosphodiesterase inhibitors, fluoride, etc. We were surprised to find that azide, hydroxylamine and nitrite activated most, but not all, of our crude preparations (6). This was an exciting period in the laboratory since we recognized that the addition of many hormones and other ligands would fail to increase cGMP synthesis in our crude preparations, and only had their effects in an intact cell system. We thought the intact cell systems most likely maintained a complex working signaling cascade that regulated cGMP that was disrupted during the crude preparations. An understanding of how azide activated cGMP synthesis, we determined, would allow us to also understand how hormonal regulation of the enzyme occurred (7, 8). Over the next several years, we began to fully understand the signaling cascade (3,9) (Figure 1).

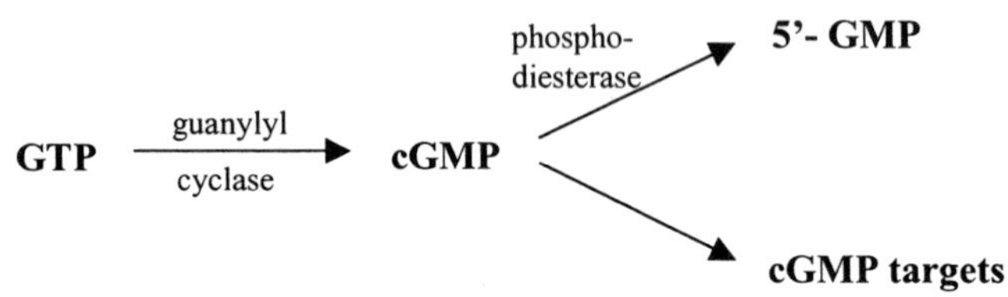

Figure 1. cGMP synthesis, metabolism and function

The activation of guanylyl cyclase by azide occurred following a brief time lag, which suggested azide may be modified or converted to another compound. Also, azide did not stimulate the enzyme in all our preparations. Furthermore, altering the levels of thiols, oxygen, and various other oxidizing and reducing agents (6,10) changed the activation. Because of the differential regulation of guanylyl cyclase by synthetic agents, we also mixed crude supernatant fractions and then assayed them for activity. We found some very interesting results: If we mixed preparations of liver with cerebral cortex, we observed an increased activation of the enzyme. The opposite effect was observed when we mixed the liver preparation with a heart preparation; the enzyme activation by azide was lost (Table 1). We hypothesized that there was a requirement for a factor other than azide to activate the enzyme in the liver extract that also allowed activation of the cerebral cortex extract. Also, another factor either directly blocked guanylyl cyclase or the enzyme was inhibited in another way in the heart extract, since mixing the heart and liver extract attenuated the activity of the liver extract (6,11). We later purified these activating and inhibiting factors, and determined that they were catalase and hemoglobin, respectively (10-17). It became apparent that heme-containing proteins could regulate the enzyme soluble guanylyl cyclase by stimulating or inhibiting its catalytic activity.

Tissue	cGMP (pmol/min./mg protein)	
	(-) azide	(+) azide
liver	38.8	595.4
heart	23	23.2
cerebral cortex	46	42
liver + heart	27.3	23.1
liver + cerebral cortex	23	899

Table 1. Effects of azide on soluble guanylyl cyclase activity

3. Azide and nitrovasodilator effects in tracheal smooth muscle

Earlier, we had developed a fairly homogenous bovine tracheal smooth muscle preparation in which we could measure the biological activity of motility and cGMP accumulation in an organ bath following stimulation by hormones. We expected that the cGMP accumulation preceded the biological activity during hormonal stimulation, and suspected that the activity was a result of the increased cGMP levels (18-21). Because azide, hydroxylamine and nitrite had already been shown by us to increase cGMP accumulation in various tissue preparations, we decided to stimulate the bovine tracheal smooth muscle preparation with these guanylyl cyclase activators and measure motility and cGMP levels (10-21). We found that not only did these activators stimulate cGMP accumulation, but also led to the relaxation of the pre-contracted smooth muscle preparation. We found similar results in a gastrointestinal smooth muscle cell preparation as well (19-21). After determining that these agents led to smooth muscle relaxation, and that cGMP levels correlated beautifully with the relaxation, we began to analyze other smooth muscle relaxants such as nitroglycerin, nitroprusside, and hydrazines among others (15-21). All these agents led to elevated cGMP levels in smooth muscle as well as other tissues, activated guanylyl cyclase, and caused smooth muscle relaxation in tissue preparations. We termed these new guanylyl cyclase activators "nitrovasodilators" because of their biological effects. The effects of the nitrovasodilators did not require catalase to activate soluble guanylyl cyclase, but could be inhibited by hemoglobin and myoglobin.

4. Biological effects of NO

The growing list of functional nitrovasodilators and the effects of additional heme-containing factors that altered the activity of guanylyl cyclase led us to believe that the direct activator of the enzyme and active intermediate compound could be NO (16-22). Indeed, our attempts to directly ventilate NO gas to our tissue preparations were a success. Not only did it increase cGMP levels and caused relaxation in the bovine tracheal smooth muscle system, but also increased cGMP levels in almost all tissue preparations examined (18-22). In an effort to better define the effects of NO on guanylyl cyclase activation, we purified the guanylyl cyclase isoforms to homogeneity and further examined the effects.

Even after purification of soluble guanylyl cyclase, we observed the activation of the enzyme by NO in the preparations (23). Essentially, the more contaminants that were removed from the preparations that trapped or complexed the NO, the less NO was required for its activation. We determined the EC_{50} for activation of soluble guanylyl cyclase by NO to be between 1 and 10 nM, depending on the conditions of the assay (23). We became aware of the magnitude of this discovery, as we found the first activation of an enzyme by a free radical molecule.

5. Could NO be an endogenous activator of guanylyl cyclase?

We proposed in 1978 that the effects of various hormones, autocrine and paracrine factors and neurotransmitters to elevate cGMP levels could be explained through the endogenous production of NO or a NO precursor in intact systems (15, 17). And, the appropriate hormonal stimulation could influence the production, release, or metabolism of NO. Furthermore, if NO was an endogenous activator of the enzyme, we thought that the endogenous levels would be around 1 to 10 nM, near the EC_{50} for activation of guanylyl cyclase. This made sense since higher levels of NO would suggest that the enzyme would always be activated, and make the physiological relevance more questionable. However, the colorometric assays to measure NO and its oxidation products, nitrate and nitrite, were not sensitive enough to measure the concentration needed (nM vs. μM to mM). Therefore, several years passed before we could reliably test this hypothesis.

6. L-arginine Activation of Soluble Guanylyl Cyclase

Deguchi and colleagues, who were examining guanylyl cyclase regulation, found that a factor from brain or neuroblastoma extracts could activate crude or partially purified preparations of soluble guanylyl cyclase (24). This endogenous substance was later identified as L-arginine. L-arginine mimicked nitrovasodilators in its activation of sGC. We confirmed this observation, but found that L-arginine failed to activate purified sGC, suggesting that other components were required for its activation (unpublished results). Hibbs et al found that the cytotoxic effects of macrophages on tumor cells correlated with the accumulation of nitrite and nitrate in the conditioned media (25). L-arginine increased the cytotoxic effects and NO metabolite accumulation whereas L-arginine analogs such as L-N-methyl arginine (LNMA) blocked these effects.

L-arginine was found to be the limiting factor in cGMP accumulation. Cells exposed to oxytocin, a reagent that increases calcium influx, demonstrated calcium-dependent NO production by L-arginine (26). The effect was maintained with 1 mM L-arginine whereas without it, cGMP levels diminished. In addition, administration of D-arginine, the inactive analog of L-arginine, caused a decrease in cGMP levels. It was evident that NO participates in a second messenger system whereby it forms NO via L-arginine to account for cGMP accumulation. The experiment also proved that L-arginine could regulate calcium influx for signaling mechanisms. Within several years, many laboratories including our own began to characterize, purify, and delineate this novel pathway that converted L-arginine to NO and citrulline (figure 2).

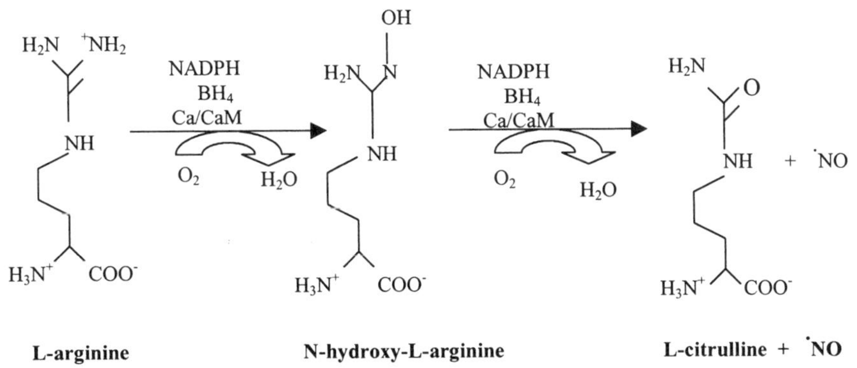

Figure 2. L-arginine forms L-citrulline + ·NO

7. The Discovery of the Nitric Oxide Synthases

One of the projects in our laboratory in the early and mid 1980s was to define the biochemical mechanism of nitroglycerin tolerance. We found that rat lung fibroblast cultures (RFL-6 cells), as well as other cell types, could be used as model systems of tolerance (1,3,4,5). The RFL-6 cells possessed both soluble and particulate isoforms of guanylyl cyclase that could be activated to produce high levels of cGMP in response to nitrovasodilators and atriopeptins, respectively. We realized that these cells could be used as an assay for measuring NO production in various cell lines (27). This permitted us greater sensitivity for the quantification of NO in different tissue extracts and other samples. Using this method and other equally sensitive methods that were developed around the same time, we along with our colleagues learned that numerous cells, tissues, and cell free extracts could produce NO (28). This surge of activity resulted in the characterization and purification of NOS from a number of cells and tissue (1-5). This work elucidated at least three isoforms of the enzyme.

The first NOS isoform to be purified was the neuronal or brain NOS (nNOS, Type I NOS, or NOS-1). This was followed by inducible NOS (iNOS) or NOS-2 (Type II NOS), and then endothelial NOS (eNOS) or NOS-3 (Type III NOS). Some properties of the NOS isoforms are summarized in Table 2. Both NOS-1 and NOS-3 have been called constitutive or housekeeping enzymes, but their expression and synthesis can be regulated under various conditions. NOS-2 is probably absent in most cells until it is induced by endotoxin and/or various cytokines. Exposure to endotoxin (LPS), Interferon-γ, Interleukin-1, Tumor Necrosis Factor-α, and other pro-inflammatory cytokines causes a rise in NOS-2 mRNA levels, protein expression, and catalytic activity within 1 hour and reaches maximal levels in 6-18 hours (29). Various anti-inflammatory cytokines and glucocorticoids can decrease the induction of NOS-2. It is thought that some of the pharmacological and biological effects of these factors may be explained by increased or decreased NO formation. For example, the hypotension observed in septic shock as well as the ensuing multiple organ failure might be due to excessive NO formation.

Nitric Oxide Synthase Isoforms			
Type	**Alternative Names**	**Tissue/Cell Distribution**	**Activation**
NOS –1 (155kD)	Neuronal NOS Brain NOS Type I NOS	Central and peripheral neurons, NANC neurons, islets, endometrium, skeletal muscle	Constitutive, Calcium/calmodulin dependent
NOS-2 (125kD)	Inducible NOS Type II-NOS	Macrophages, liver, smooth muscle, endothelium, heart	Inducible by LPS, cytokines, and glucocorticoids Calcium/calmodulin independent
NOS-3 (133kD)	Endothelial NOS Type III-NOS	Endothelium, brain, heart	Constitutive Calcium/calmodulin dependent Modified by acylation and phosphorylation

Table 2. Properties of Nitric Oxide Synthase Isoforms

The NOS isoforms have considerable sequence homology, about 50-60%. In addition, they are about 90% conserved among species (1-5, references therein). The carboxy terminal domain has considerable homology between the isoforms and is homologous to cytochrome P450. The amino terminal domain has less homology. The NOS enzyme is quite complex in that it uses heme as a prosthetic group and flavin mononucleotides, flavine adenine dinucleotide phosphate as cofactors or cosubstrates. Calcium and/or calmodulin can activate NOS-1 and NOS-3, while NOS-2 has no apparent calcium dependence since calmodulin is tightly associated with this isoform. The catalytically active isoforms exist as homodimers, with tetrahydrobiopterin and heme serving to facilitate dimer formation. The homologous reductase and oxidase domains of the proteins with electron transfer between the substrate L-arginine and the redox cofactors result in the oxidation of the guanidine nitrogen of arginine to form L-hydroxyarginine, and then subsequently forms NO and citrulline.

Transgenic animals deficient for one or more NOS isoforms are viable and have no life threatening conditions associated with the deficiency. NOS 1 knockouts experience hypertrophy of the intestinal pylorus and have a smooth muscle defect (30). They may also have some memory impairments. NOS 2 knockouts appear to be developmentally normal, but have susceptibility to bacterial infections (31,32). NOS 3 knockouts are hypertensive and also have a smooth muscle defect, which impairs blood vessel relaxation (33). The NOS 1 and 3 double knockout (33) has some memory defects.

8. NOS Inhibitors

Recently, the development of selective NOS inhibitors has been a hot topic for research. Both arginine and non-arginine based antagonists have been reported as inhibitors with variable selectivity (1-5, references therein). Most inhibitors to date are competitive antagonists and may be partly selective for one or another isoform. The most commonly used NOS inhibitors are the nonspecific N (G)-nitro-L-arginine methyl ester (L-NAME) and N (G)-monomethyl-L-arginine (L-NMMA). Other somewhat specific inhibitors often used are the neuronal NOS inhibitor 7-nitroindazole (7-NI) and the inducible NOS inhibitors aminoguanidine and L-N- (6)-(1-iminoethyl) lysine (L-NIL). There are now several other NOS inhibitors commercially available, and many more that will soon be available. Recently, very specific inhibitors have been developed for at least two of the NOS isoforms, some of which are currently being used in clinical trials for the treatment of septic shock and certain cancers. This is a new and exciting field emerging in which pharmacological targeting of the NO pathway may soon be utilized in the clinic.

9. Targets of NO and soluble guanylyl cyclase

Over the next several years and up to the present, we along with other laboratories have demonstrated that numerous cells, tissues, and cell-free extracts could produce NO. Multiple biological and potential biological targets and processes that are regulated by NO or cGMP have emerged as a result (Figure 3). The most thoroughly studied process regulated by this signaling pathway is smooth muscle relaxation. However, the list is growing with not only NO and cGMP targeting enzymes and macromolecules, but also playing a critical role in several biological phenomena. NO is now known to be involved in many integral processes such as intestinal secretion and ion transport, renal tubular-glomerular feedback, hormone production and secretion, retinal phototransduction, neurotransmission, tissue injury, inflammation, pathogen cytotoxicity, tumor cytotoxicity, and transcriptional regulation of a variety of genes. These and other areas are currently subject to intense research interest.

Enzymes and macromolecules
- Guanylyl cyclase
- Cyclic nucleotide protein kinases
- Cyclic nucleotide phosphodiesterases
- Cyclooxygenase (COX II)
- Heme proteins, iron centers, and thiol groups
- DNA modifications and repair
- Glutamate (*N*-methyl-D-aspartate) receptor regulation*
- Phospholipase C

Processes
- Smooth, cardiac,* and skeletal* muscle relaxation
- Retinal phototransduction
- Intestinal secretion and ion transport
- Renal tubular-glomerular feedback*
- Endothelial permeability*
- Smooth muscle proliferation
- Platelet adhesion and aggregation
- Insulin secretion*
- Hormone production and secretion*
- Neurotransmission
- Long-term potentiation and memory*
- Transcriptional regulation*
- Tissue injury and inflammation
- Pathogen cytotoxicity
- Tumor cytotoxicity
- Calcium transport and redistribution

*In some cases, the effects of nitric oxide and/or cyclic GMP are well substantiated. However, some of the effects can be considered interesting but preliminary and are designated with an asterisk.

Figure 3. Targets of NO, soluble guanylyl cyclase, and cGMP

10. Conclusions

The NO field, as we know it today, is a culmination of three decades of diligent research, beginning with the discovery of various substances that can increase cGMP levels by activation of guanylyl cyclase to the elucidation of the biochemical mechanisms of NO synthesis. As more sensitive methods of NO detection became available, NO was found in a variety of tissues and cell types. Additional components of the NO signaling pathway were revealed, creating numerous targets for pharmacological manipulation. An escalation of interest in the field came with the delineation of this pathway. Presently, there is a myriad of agents that activate or inhibit one or more components of the NO cascade. These drugs are being studied for their effects on a variety of disease states such as hypertension, septic shock, inflammatory conditions, and cancer. We now know that NO is involved in countless fundamental biological phenomena, and the list continues to flourish. NO continues to be an intense field of study, and will likely remain so throughout the next decade.

References

1. Murad F. Cyclic GMP synthesis, metabolism and function. Adv Pharmacol. 1994;26:1-335.
2. Ignarro L, Murad F. Nitric oxide: biochemistry, molecular biology, and therapeutic implications. Adv Pharmacol. 1995;34:1-516.
3. Murad F. The nitric oxide-cyclic GMP signal transduction system for intracellular and intercellular communication. Recent Prog Horm Res. 1994;49:239-248.
4. Murad F. Nitric oxide signaling: would you believe that a simple free radical could be a second messenger, autacoid, paracrine substance, neurotransmitter, and hormone? Recent Prog Horm Res. 1998;53:43-59.

5. Murad F. The 1996 Albert Lasker Medical Research Awards. Signal transduction using nitric oxide and cyclic guanosine monophosphate. JAMA. 1996 Oct 9;276(14):1189-92.
6. Kimura H, Mittal CK, Murad F. Activation of guanylate cyclase from rat liver and other tissues with sodium azide. J Biol Chem. 1975;250:8016-8022.
7. Murad F. Cyclic guanosine monophosphate as a mediator of vasodilation. J Clin Invest. 1986;78:1-5.
8. Waldman SA, Murad F. Cyclic GMP synthesis and function. Pharm Rev. 1987;39:163-196.
9. Murad F, Forstermann U, Nakane M, et al. The nitric oxide-cyclic GMP signal transduction system for intracellular and intercellular communication. Adv Second Messenger Phosphoprotein Res. 1993;28:101-109.
10. Katsuki S, Arnold W, Mittal CK, Murad F. Stimulation of guanylate cyclase by sodium nitroprusside, nitroglycerin and nitric oxide in various tissue preparations and comparison to the effects of sodium azide and hydroxylamine. J Cyclic Nucl Res. 1977;3:23-35.
11. Murad F, Mittal CK, Arnold WP, Katsuki S, Kimura H. Guanylate cyclase: activation by azide, nitro compounds, nitric oxide, and hydroxyl radical and inhibition by hemoglobin and myoglobin. Adv Cyclic Nucl Res. 1978;9:145-158.
12. Kimura H, Mittal CK, Murad F. Increases in cyclic GMP levels in brain and liver with sodium azide, an activator of guanylate cyclase. Nature. 1975;257:700-702.

13. Katsuki S, Murad F. Regulation of cyclic 3', 5'-adenosine monophosphate and cyclic 3', 5'-guanosine monophosphate levels and contractility in bovine tracheal smooth muscle. Mol Pharmacol. 1977;13:330-341.
14. Katsuki S, Arnold WP, Murad F. Effect of sodium nitroprusside, nitroglycerin and sodium azide on levels of cyclic nucleotides and mechanical activity of various tissues. J Cyclic Nucl Res. 1977;3:239-247.
15. Murad F, Mittal CK, Arnold WP, Braughler JM. Effect of nitro-compound smooth muscle relaxants and other materials on cyclic GMP metabolism: proceedings of the 7th International Congress of Pharmacology, July, 1978, Paris, France. In: Stocklet JC, ed. Advances in Pharmacology and Therapeutics, Volume 3: Ions, Cyclic Nucleotides, Cholinergy. Elmsford, NY: Pergamon Press Inc; 1978:123-132.
16. Arnold WP, Mittal CK, Katsuki S, Murad F. Nitric oxide activates guanylate cyclase and increases guanosine 3', 5'-monophosphate levels in various tissue preparations. Proc Natl Acad Sci U S A. 1977;74:3203-3207.
17. Diamond J, Holmes T. Effects of potassium chloride and smooth muscle relaxants on tension and cyclic nucleotide levels in rat myometrium. Can J Physiol Pharmacol. 1975;53:1099-1107.
18. Craven P, DeRubertis F. Restoration of the responsiveness of purified guanylate cyclase to nitrosoguanidine, nitric oxide and related activators by heme and heme proteins. J Biol Chem. 1978;253:8433-8443.
19. Katsuki S, Arnold W, Mittal CK & Murad F. Stimulation of guanylate cyclase by sodium nitroprusside, nitroglycerin and nitric oxide in various tissue preparations and

comparison to the effects of sodium azide and hydroxylamine. J Cyclic Nucl Res. 1977;3: 23-35.

20. Katsuki S, Arnold WP, Mittal CK & Murad F. Stimulation of formation and accumulation of cyclic GMP by smooth muscle relaxing agents. Proceedings of the 2nd Japanese Cyclic Nucleotide Conference, July 7-9, Osaka, Japan, 1977; 44-50.
21. Katsuki S & Murad F. Regulation of cyclic 3',5'-adenosine monophosphate and cyclic 3',5'-guanosine monophosphate levels and contractility in bovine tracheal smooth muscle. Mol Pharmacol., 1977;13: 330-341.
22. Katsuki S, Arnold WP & Murad F. Effect of sodium nitroprusside, nitroglycerin and sodium azide on levels of cyclic nucleotides and mechanical activity of various tissues. J Cyclic Nucl Res. 1977;3: 239-247.
23. Arnold WP, Mittal CK, Katsuki S & Murad F. Nitric oxide activates guanylate cyclase and increases guanosine 3',5'-monophosphate levels in various tissue preparations. Proc Natl Acad Sci U S A. 1977;74: 3203-3207.
24. Deguchi T, Yoshiako M. L-Arginine identified as an endogenous activator for soluble guanylate cyclase from neuroblastoma cells. J Biol Chem. 1982;257:10147-10157.
25. Hibbs J, Traintor R, Varrin Z. Macrophage cytoxicity: Role for L-arginine deaminase and imino nitrogen oxidation to nitrite. Science. 1987;235:473-476.
26. Leitman DC, Agnost VL, Catalano RM, Schroder H, Waldman SA, Bennett BM, Tuan JJ, Murad F. Atrial natriuretic peptide, oxytocin, and vasopressin increase guanosine 3',5'-monophosphate in LLC-PK1 kidney epithelial cells. Endocrinology. 1988;122:1478-85.

27. Ishii K, Sheng H, Warner T, Forstermann U, Murad F. A simple and sensitive bioassay method for detection of EDRF with RFL6 rat lung fibroblasts. Am J Physiol. 1991;261:H598-H603.

28. Ishii K, Gorsky L, Forstermann U, Murad F. Endothelium-derived relaxing factor (EDRF): the endogenous activator of soluble guanylate cyclase in various types of cells. J Appl Cardiol. 1989;4:505-512.

29. Schmidt HHHW, Warner TD, Forstermann U, Murad F. Regulation and subcellular location of nitric oxide synthases in RAW 264.7 macrophages. Mol Pharmacol. 1992;41:615-664.

30. Huang PL, Dawson TM, Bredt DS, Snyder SH, Fishman MC. Targeted disruption of the neuronal nitric oxide synthase gene. Cell. 1993 Dec 31;75(7):1273-86.

31. MacMicking JD, Nathan C, Hom G, Chartrain N, Fletcher DS, Trumbauer M, Stevens K, Xie QW, Sokol K, Hutchinson N, et al. Altered responses to bacterial infection and endotoxic shock in mice lacking inducible nitric oxide synthase. Cell. 1995 May 19;81(4):641-50.

32. Wei XQ, Charles IG, Smith A, Ure J, Feng GJ, Huang FP, Xu D, Muller W, Moncada S, Liew FY. Altered immune responses in mice lacking inducible nitric oxide synthase. Nature. 1995 Jun 1;375(6530):408-11.

33. Son H, Hawkins RD, Martin K, Kiebler M, Huang PL, Fishman MC, Kandel ER. Long-term potentiation is reduced in mice that are doubly mutant in endothelial and neuronal nitric oxide synthase. Cell. 1996 Dec 13;87(6):1015-23.

Nitric Oxide
R.J. Gryglewski and P. Minuz (Eds.)
IOS Press, 2001

Mechanism of Oxygen Activation in the NO Synthases, and a Kinetic Model for Catalysis

D. J. STUEHR, S. ADAK, J. SANTOLINI, C.-C. WEI, and Z. WANG
Immunology, Lerner Research Institute, Cleveland Clinic, Cleveland, Ohio, 44195, USA

Abstract. The mechanism of NO synthesis, and how the enzymes regulate catalysis, are topics of current interest. Drawing on recent works we present a model describing how the essential cofactor tetrahydrobiopterin and enzyme heme groups cooperate to activate O_2 during NO synthesis. In addition, we describe a general kinetic model that can explain how NO synthase activity is regulated.

1. Introduction

Nitric oxide (NO) is generated from L-arginine (Arg) by the NO synthases (NOSs). Three main isoforms are expressed in animals and differ regarding their primary sequence, tissue location, and means of regulation. All NOS catalyze an identical two-step reaction to make NO. In the first step Arg undergoes mixed function hydroxylation to form water and N^{ω}-hydroxy-L-Arg (NOHA) as an enzyme-bound intermediate. NOHA then undergoes mixed function oxidation in the second step to generate water, NO, and citrulline. Both reactions take place within the oxygenase domain of NOS, which contains iron protoporphyrin IX (heme), the cofactor 6R-tetrahydrobiopterin (H_4B), and the Arg binding site. Electrons derived from NADPH are provided to the oxygenase domain by an attached reductase domain that binds FMN, FAD, and NADPH. A ~20 amino acid consensus site for calmodulin (CaM) binding is located between the reductase and oxygenase domains of each NOS. CaM binding activates NO synthesis by triggering electron transfer between the reductase and oxygenase domains of NOS.

The mechanism of NO synthesis, and how the enzymes regulate their catalysis, are currently topics of interest. Drawing on recent works, we present a model showing how H4B and heme cooperate to activate O_2 during NO synthesis. In addition, we describe a general kinetic model that explains how NOS enzyme activity is regulated.

2. Role for H_4B in Oxygen Activation

Understanding how heme and H_4B participate in NO synthesis is important because it represents novel cooperation between these two groups in biologic oxidation. The NOS heme is ligated to a cysteine thiolate and probably activates oxygen as in the cytochrome P450 monooxygenases. In fact, both steps of NO synthesis are envisioned to involve heme-based oxygen activation. However, NOS also requires H_4B, and given that H_4B is not required for heme reduction, it has been puzzling why this cofactor is essential for NO synthesis. Aromatic amino acid hydroxylases use H_4B as an electron donor for oxygen activation, and a similar role for H_4B in NOS has long been considered. Recent crystal structures of NOS oxygenase domains show that H_4B binds near the heme edge and forms a hydrogen bond between N3 of its pterin ring and a heme propionate, consistent with a possible role in heme reduction. Indeed, H_4B can convert to a radical during reaction of ferrous NOS oxygenase domain with O_2.

We recently investigated Arg and NOHA oxidation by H_4B-free neuronal NOS (nNOS) [1]. We saw that nNOS can catalyze oxidation of both Arg and NOHA in the absence of H_4B, but the reaction is uncoupled with regard to NADPH oxidation. We also recently completed a stopped-flow kinetic analysis of H_4B radical formation during Arg hydroxylation [2]. This work established kinetic parameters for H_4B electron transfer during a step in oxygen activation by the heme that allows NOS to hydroxylate Arg. Together, these studies lead to a unified kinetic mechanism for H4B and heme cooperation during oxygen activation in both steps of NO synthesis (Fig. 1).

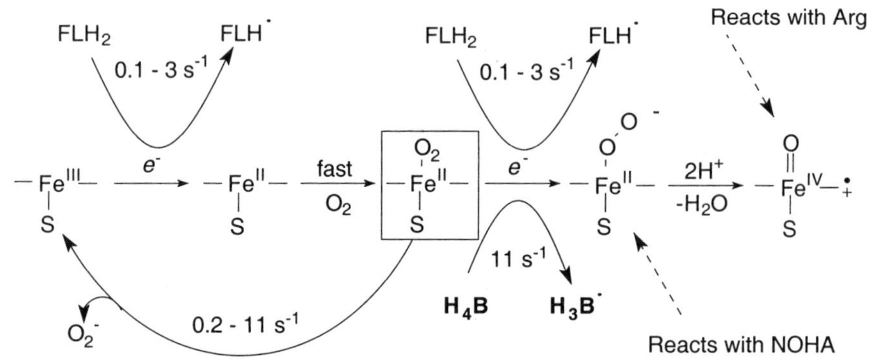

Figure 1. Role for H_4B in supporting oxygen activation by the NOS heme. The reaction starts with ferric heme on the left. The ferric heme first receives an electron from the reductase domain (FLH_2). This enables rapid O_2 binding and buildup of the $Fe^{II}O_2$ intermediate. Either H_4B or the flavins can then reduce the $Fe^{II}O_2$ intermediate, but reduction by H_4B is faster. This allows formation of the ultimate oxidants that react with NOHA or Arg. Rates of electron transfer from the flavins and from H_4B were determined at 10 °C. See text for details.

It is clear that the reductase domain gives the first electron to the heme irrespective of bound H_4B, but the second electron could come either from the reductase domain or from H_4B (Fig. 1). Our current work shows that the reductase domain can provide the second electron when H_4B is missing, but this leads to uncoupling between NADPH oxidation and product formation. Because the reductase domain can transfer both electrons to the heme that are required for oxygen activation and Arg oxidation in nNOS (1), it is puzzling why NOS would also use H_4B as an electron donor. The answer lies in their different rates of electron transfer: H_4B simply provides an electron to the $Fe^{II}O_2$ intermediate more quickly than can the reductase domain (Fig. 1). This is important because the NOS $Fe^{II}O_2$ intermediate is unreactive and will decay to generate superoxide if a second electron is not provided in a timely manner. In the absence of H_4B, decay of the $Fe^{II}O_2$ complex at 10 °C varies from 0.1 s^{-1} in the nNOS oxygenase domain to 11 s^{-1} in full-length nNOS (Fig. 1). NOS reductase domains transfer an electron to the ferric heme at rates ranging from 0.1 to 3 s^{-1} at 10 °C, and this rate does not increase in the absence of H_4B. Because heme reduction by the reductase domain is slower than either H_4B radical formation or $Fe^{II}O_2$ decay, the NOS seem to utilize H_4B as a kinetically-preferred source of the second electron. H4B thus circumvents a kinetic problem in stepwise heme reduction during oxygen activation, and enables NOS to tightly couple NADPH oxidation to Arg hydroxylation.

3. Kinetic model for NOS catalysis

We developed a kinetic model [3] to understand the unique behavior of nNOS among all isoforms: A fast build-up of ferrous-NO complex to include about 80% of the steady-state population, with an associated decrease between initial and steady-state activities. Our model (Fig. 2) integrates the latest results in the literature, particularly the concept of fast recombination between ferric heme and NO newly formed in the active site [4]. Simulations of our kinetic model closely mimic all catalytic features of wild-type or mutant nNOS, including their initial and steady-state behaviors, populations of heme species, and apparent K_mO_2 values. Together, this strongly argues that the fundamental features of our current model are correct.

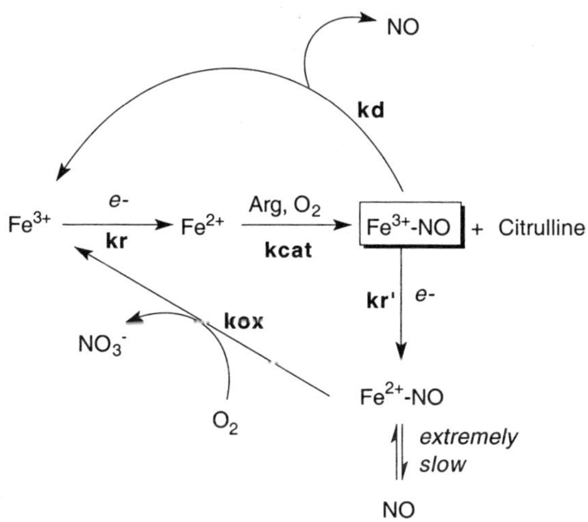

Figure 2. Kinetic model for NOS catalysis. See text for details.

In our model catalytic turnover is regulated by two distinct features: (A) partitioning of an immediate FeIII heme-NO product between a futile and productive regenerating pathway, and (B) the relative rate of the futile regenerating pathway. These two features share no common influences and thus operate completely independent of one another. Partitioning between the futile and productive pathway simply reflects the relative rates of NO dissociation (kd) versus reduction (kr') of the immediate FeIII heme-NO product (Fig. 2). Because these are fundamental characteristics of the enzyme, they are immune to the external environment when the enzyme operates under V_{max} conditions (*i.e.*, sufficient NADPH, Arg, *etc.*). However, because they oppose each other, changes in these two rates can have a relatively large effect on enzyme partitioning. The relative rate of the futile pathway is determined primarily by the oxidation rate of the FeII heme-NO species (kox).

One imagines that intrinsic parameters of nNOS have been set for its proper physiologic function. This may be illustrated by its measured rate of heme reduction corresponding to an activity optimum predicted in our model [3]. Intrinsic parameters can also govern the relative impact of extrinsic factors such as O_2 concentration. This is illustrated by the high apparent K_mO_2 that is unique to nNOS: Our model reveals that a relatively fast heme reduction, coupled with a relatively slow cycling through the futile pathway, create this special regulation in nNOS [3]. Functionally, this allows the enzyme to minimize its catalytic sensitivity to O_2 concentration changes, while generating NO in a near-linear manner as a function of O_2 concentration. Why this evolved is still unclear, but it must be physiologically important, because the enzyme sacrifices catalytic efficiency in order to create this response.

In contrast to nNOS, the FeII heme-NO species is not observed during steady state catalysis in eNOS or iNOS. At this point, our model can provide an explanation for this difference. eNOS is distinguished by a low steady-state activity, a slow heme reduction, and an undetectable level of FeII heme-NO species. Its low catalytic activity is therefore not linked to eNOS cycling through the futile regenerating pathway. In fact, according to our simulations, the slow heme reduction rate in eNOS is sufficient to explain its low proportion of FeII heme-NO species during the steady state and its overall lower activity. This suggests that eNOS follows the same kinetic model as nNOS, but its slower heme reduction minimizes partitioning into the futile regenerating cycle and diminishes the global speed of catalysis. In iNOS, one observes a similar catalytic activity and a slightly slower heme reduction rate compared to nNOS, but little or no FeII heme-NO complex formation in the steady state. This phenotype is similar to what is seen for the W409 nNOS mutants. Indeed, our preliminary data suggest that iNOS exhibits accelerated FeII heme-NO oxidation (Santolini and Stuehr, manuscript in preparation), as occurs in the W409 nNOS mutants, and this can account for such results. Thus, our kinetic model seems capable of simulating behaviors of all three NOS isoforms.

References

[1] S. Adak *et al.*, Arginine conversion to nitroxide by tetrahydrobiopterin-free neuronal NO synthase. Implications for mechanism. *J. Biol. Chem.* **275** (2000) 33554-33561.

[2] C.-C. Wei, *et al.*, Rapid kinetic studies link tetrahydrobiopterin oxidation, heme-dioxy reduction, and arginine hydroxylation in inducible NO synthase. *J. Biol. Chem.* (2001), in press.

[3] J. Santolini *et al.*, A kinetic simulation model that describes catalysis and regulation in NO synthase. *J. Biol. Chem.* (2001), in press.

[4] S. Boggs *et al.*, Formation and reactions of the heme-dioxygen intermediate in the first and second steps of NO synthesis as studied by stopped-flow spectroscopy under single turnover conditions *Biochemistry* **39** (2000) 2332-2339.

Nitric Oxide
R.J. Gryglewski and P. Minuz (Eds.)
IOS Press, 2001

The Role of Tetrahydrobiopterin in Nitric Oxide Synthesis

Antonius C.F. Gorren and Bernd Mayer
Institut für Pharmakologie und Toxikologie, Karl-Franzens-Universität, Universitätsplatz 2, 8010 Graz, Austria

Abstract. The role of tetrahydrobiopterin (BH4) in nitric oxide synthase (NOS) has been one of the most contentious issues with regard to the mechanism of NO synthesis. In this review we will give a brief account of how recent research has led to surprising new insights into this topic. For many years BH4 has been known to exert several allosteric and structural effects, such as increasing substrate affinity and stabilizing the essential dimeric enzyme structure. However, these actions fail to explain the absolute dependence of NO synthesis on BH4. A series of low-temperature absorbance spectroscopic and electron paramagnetic studies has now revealed a novel and hitherto unanticipated function of BH4. These studies showed that BH4 serves as a 1-electron donor to the oxyferrous ($Fe(II)O_2$) heme complex of NOS during catalysis. Thus, NO synthesis involves redox transitions between the BH4 and BH3• radical pterin species, which constitutes a new paradigm for the biological actions of pteridines.

1. Nitric Oxide Synthase Structure and Function

The last couple of decades of the 20th century witnessed the remarkable metamorphosis of nitric oxide (NO) from a noxious gas into an important biomolecule. Biogenesis of NO is

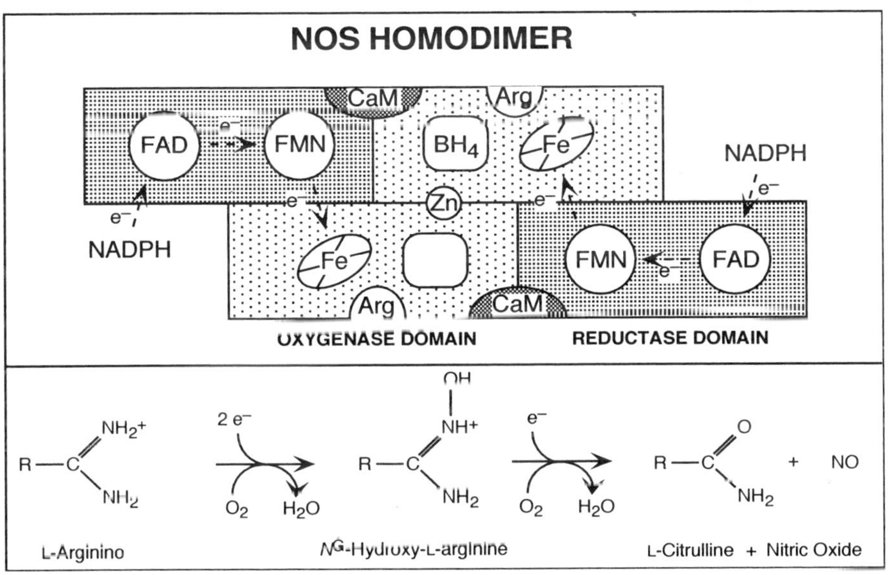

Figure 1. The upper panel provides a schematic representation of the NOS dimer. The lower panel shows the two-step conversion of L-arginine into L-citrulline and NO.

catalyzed by nitric oxide synthase (NOS), which transforms L-arginine into L-citrulline and nitric oxide. NOS exists in distinct neuronal, endothelial, and inducible isoforms (nNOS, eNOS, and iNOS), which differ with respect to function, localization, and regulation (for reviews covering most of the topics discussed in this section see Refs. 1-6). The activity of the constitutively expressed nNOS and eNOS is controlled by Ca^{2+}, which enables these isoforms to translate a transient increase in Ca^{2+} levels into a short burst of NO synthesis. In this capacity NO is involved in smooth muscle (vaso)relaxation and anti-platelet aggregation (eNOS), and in neurotransmission (nNOS). Expression of iNOS requires induction by cytokines; this isoform is not regulated by Ca^{2+}. Consequently, once iNOS is expressed, it will produce sustained high levels of NO, in accordance with its role in the immune response.

NOS is a modular protein that consists of a C-terminal reductase domain and a N-terminal oxygenase domain. The reductase domain, which contains FAD, FMN, and a binding site for NADPH, is a member of the family of cytochrome P450 reductases. Its function is to shuttle electrons from NADPH to the oxygenase domain. The latter contains a P450-type heme and binding sites for substrate and pterin. In between the reductase and oxygenase domains is a calmodulin-binding sequence, which varies between the isoforms, with iNOS having a much higher affinity for calmodulin. Interdomain electron transfer only occurs with calmodulin bound to the enzyme, which confers Ca^{2+} sensitivity on the constitutive isoforms.

NOS is only active as a dimer, because electrons are transferred from the reductase domain of one subunit to the oxygenase domain of the other. NOS binds one Zn^{2+} ion per dimer in the interface between the monomers, which appears to play a pivotal role in maintaining dimeric structure. Figure 1 shows a schematic model of the NOS dimer.

2. Mechanism of NO Synthesis

The transformation of L-arginine into L-citrulline and NO takes place at the heme in two discrete steps, with intermediate formation of *N*-hydroxy-L-arginine (NHA). Both steps require O_2 and NADPH-derived electrons (Fig. 1). The first reaction consumes one molecule

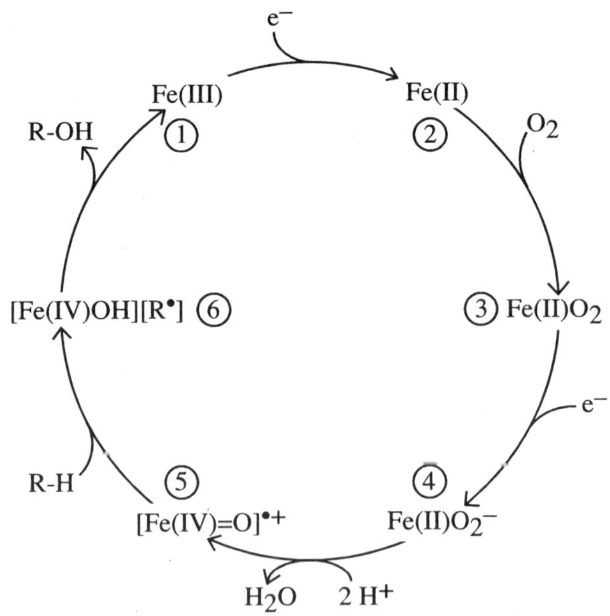

Figure 2. Postulated catalytic cycle of cytochrome P450.

of O_2 and two electrons, the second reaction only one electron and one molecule of O_2. The presence of BH4 is required for both steps.

Although NOS cannot synthesize NO in the absence of substrate or pterin, it may still oxidize NADPH [7]. In this so-called uncoupled reaction O_2 is reduced to O_2^- [8], while H_2O_2 can be detected as the final product [9]. Whether all observed H_2O_2 derives from disproportionation of O_2^-, or if part of it is formed by NOS directly, is not clear yet [10].

The conversion of Arg into NHA is believed [1, 5, 6, 11] to resemble the reaction cycle proposed for cytochrome P450 (Fig. 2). Accordingly, catalysis starts by heme reduction and O_2 binding, resulting in formation of an oxyferrous (superoxy ferric) complex (**3**). Reductive activation of this complex is thought to yield a superoxy ferrous/peroxy ferric state (**4**) which is transformed into a compound I-type structure with an oxyferryl heme and a porphyrin-based radical (**5**). This species is believed to abstract hydrogen from substrate, already bound at the active site. The resulting unstable species (**6**) rapidly regenerates ferric heme (**1**) and NHA.

The conversion of NHA into L-citrulline and NO is thought to proceed differently. Since this reaction requires only one electron, it is assumed that no further reduction occurs after formation of the oxyferrous complex (**2**), and no oxyferryl heme is formed. Instead, it was proposed that NHA itself activates the heme [1, 3, 5, 6, 11].

3. Tetrahydrobiopterin

NOS is unique among cytochrome P450-type enzymes in utilizing tetrahydrobiopterin (BH4, Fig. 3) as an additional cofactor. At a time when the main functions of the other prosthetic groups and cofactors (heme, FAD, FMN, Ca^{2+}/calmodulin, and Zn^{2+}) had been elucidated, the role of BH4 remained enigmatic [4-6, 12]. BH4 is a dissociable cofactor of the aromatic amino acid hydroxylases [13, 14]. In these enzymes BH4 plays a prominent part in binding and reduction of O_2. In the process 4a-hydroxydihydrobiopterin is produced, which dissociates from the enzyme and is reconverted into BH4 in two enzymatic reactions with intermediate formation of quinonoid dihydrobiopterin (q-BH2, Fig. 3). In this way the pteridine cofactor exhibits a 2-electron redox cycle between BH4 and q-BH2. A similar role was anticipated for BH4 in NOS, but it soon became clear that BH4 behaves differently in NOS. As in cytochrome P450, binding and activation of oxygen are catalyzed by a thiolate-ligated heme, and the electrons are furnished by NADPH via two flavin moieties. Indeed, NOS catalyzes the reduction of O_2 to O_2^- in the absence of BH4. Furthermore, BH4 remains tightly bound to NOS during catalysis, and there is no need for enzymatic regeneration of BH4.

4. Structural and Allosteric Effects of BH4

When it became clear that BH4 is not a dissociable, enzymatically redox-cycling cofactor during NO synthesis, the focus of attention shifted more and more to regulatory effects. BH4 increases the affinity of NOS for its substrate [15, 16]. It also affects the heme by shifting the spin-state equilibrium from inactive low spin to active high spin [16-18].

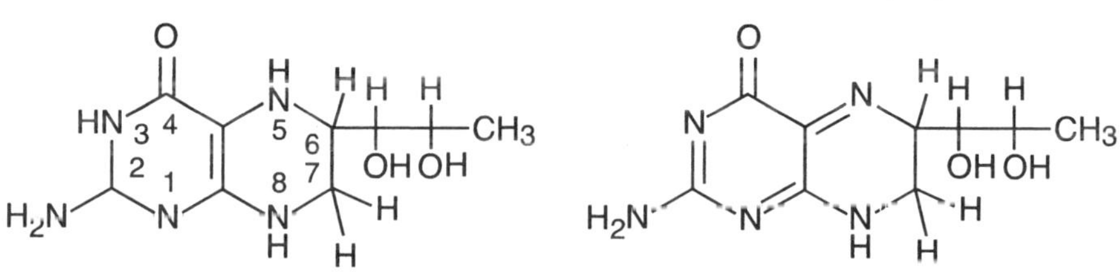

tetrahydrobiopterin (BH4) quinonoid dihydrobiopterin (q-BH2)

Figure 3. Structures of tetrahydrobiopterin (BH4) and quinonoid dihydrobiopterin (q-BH2).

Most significantly, it was reported that enzyme dimerization depends critically on the presence of BH4 [19, 20]. Since only the NOS dimer is able to synthesize NO, this would provide an explanation for the dependence of NO synthesis on BH4. Early X-ray crystallographic studies supported this hypothesis, for although BH4 was found to bind in the vicinity of the heme, it was not bound in the distal, substrate-binding pocket, but at a site which seemed to preclude direct participation in electron transfer [21]. Rather, it bound at the dimer interface, and exhibited interactions with amino acids of both monomers, which provided a rationale for the reported structural effects and supported a predominantly structural role.

However, although the structural and allosteric effects discussed above may have important regulatory implications, they fail to explain the absolute dependence of NO synthesis on BH4. Substrate binding will occur in the absence of BH4, stabilization of the heme high-spin state can be achieved by substrate (Arg or NHA) alone, and although BH4 dramatically stimulates NOS dimerization, its presence is not a prerequisite for dimer formation [18, 22, 23]. Nevertheless, substrate-saturated, high-spin heme-containing, dimeric NOS is incapable of NO synthesis in the absence of BH4.

The enzyme as isolated usually contains only about one BH4 molecule per dimer [19, 24, 25], and BH4 binding to the dimer is strongly anticooperative [16, 26]. For a discussion of the important implications that this property of BH4 may have for NOS catalysis, the reader is referred to Refs. 2 and 4.

5. Alternative Pterin Cofactors and Inhibitors

Important clues to the function of BH4 have been deduced from the use of alternative pteridines [10, 15, 27-33]. It was found that a number of pteridines, including 6*S*-BH4, tetrahydroneopterin, and 6-methyl-tetrahydropterin, could substitute for BH4 as a cofactor [15, 27, 30]. A second group of pteridines, including 7,8-BH2, 4-amino-BH4, and sepiapterin, acted as BH4-competitive inhibitors, suggesting that these compounds bind to the same site as BH4, but are unable to sustain catalysis [15, 27, 30, 31]. Significantly, several of these inactive pteridines mimicked the allosteric and structural effects of BH4 [28-30, 32]. Binding of some of these compounds to the same site as BH4 has been verified by X-ray crystallography [34]. This study showed that exchanging BH4 for 7,8-BH2 or 4-amino-BH4 hardly affected protein structure, in line with the previous observation that the presence of BH4 did not alter NOS structure [35].

All of these observations argue against a structural basis for the indispensability of BH4 during NO synthesis. All of the catalysis-supporting BH4 analogues are tetrahydropteridines, whereas most of the pterin-based inhibitors are 7,8-dihydropteridines. One crucial distinction between these classes of compounds lies in their electrochemical properties, since 7,8-dihydropteridines are far more resistant to redox cycling than tetrahydropteridines [36]. Therefore, the results obtained with BH4 analogues suggest a redox-active role for BH4, despite the lack of evidence for redox cycling.

Recent studies with 5-methyl-BH4, the N5-methylated derivative of BH4, yielded important additional information [33]. Because of the absence of a dissociable proton at the N5-position, this compound cannot be reversibly oxidized to a quinonoid dihydro species. Surprisingly, it was found that 5-methyl-BH4 supported NO synthesis, which ruled out 2-electron redox cycling between BH4 and q-BH2 as an essential feature of NOS catalysis. It should be noted, however, that methylation at N-5 does not preclude reversible 1-electron oxidation of the pteridine.

6. BH4 Functions as a 1-Electron Donor during Reductive Activation of the Oxy-Ferrous Complex

A breakthrough in the elucidation of the role of BH4 in NOS catalysis was achieved by low-temperature absorbance spectroscopic studies of the reaction between ferrous NOS and O_2 [37]. Since these studies were performed with the isolated oxygenase domain, or with full-length NOS in the absence of calmodulin, only the electron originating from ferrous

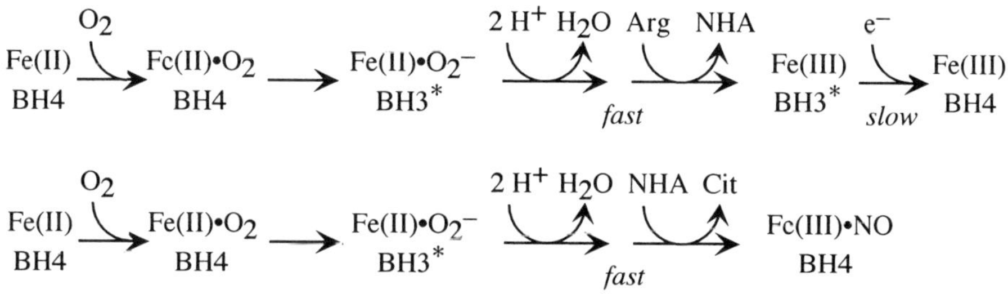

Figure 4. Reactions of ferrous NOS with O_2 in the presence of BH4 and Arg (top) or NHA (bottom).

heme would be expected to be available for catalysis. Without substrate the reaction resulted in rapid formation of a spectrum that was ascribed to oxy-ferrous ($Fe(II){\bullet}O_2$) heme [37-41]. The same intermediate was detected with Arg in the absence of BH4 [37]. However, when Arg and BH4 were both present, a different intermediate spectrum was observed, which was tentatively ascribed to a species further along the catalytic cycle. This suggested that continuation of the cycle beyond the stage of the oxy-ferrous complex required Arg and BH4. In line with this interpretation NHA was formed in these experiments in the presence of both BH4 and Arg, but not when either compound was absent [37].

In the absence of BH4, NADPH oxidation is uncoupled from NO synthesis, and O_2 reduction results in O_2^- formation [7-10]. The low-temperature optical studies provided an explanation for this phenomenon: BH4 is required for reductive activation of the oxy-ferrous complex. By enabling the reaction with O_2 to continue beyond the $Fe(II){\bullet}O_2$ complex, BH4 prevents decomposition of that complex into Fe(III) and O_2^-. Similar conclusions were arrived at on the basis of stopped-flow studies [38, 42]. However, the specific conditions of the low-temperature studies allowed a second, more surprising, conclusion. Since formation of NHA from Arg requires two electrons, whereas only one electron appeared to be available, it was postulated that BH4 serves as the 1-electron donor for the reduction of the $Fe(II){\bullet}O_2$ complex (Fig. 4, top) [37].

Initial response to this novel concept was predominantly sceptical [5, 31, 39, 42-44], and the key observation of NHA formation from 1-electron reduced NOS could not be confirmed at first [42, 45]. Indirect support for the validity of the hypothesis came from crystallographic studies, which suggested that the pterin-binding site might stabilize $BH4{\bullet}H^+$ and $BH3^{\bullet}{\bullet}H^+$ cations [35]. Direct evidence for the postulated mechanism has since then come from a study that demonstrated the formation of the $BH3^{\bullet}$ radical species by electron paramagnetic resonance (EPR), and that also confirmed the formation of stoichiometric amounts of NHA from Arg [11]. These observations have now been corroborated independently by two research groups [46, 47], and simulation of the EPR signals has identified the radical as the N-5 protonated cationic species (Fig. 5) [48].

Since BH4 is also absolutely required for the second catalytic cycle, in which NHA is transformed into L-citrulline and NO, one is tempted to propose a similar function for BH4 in

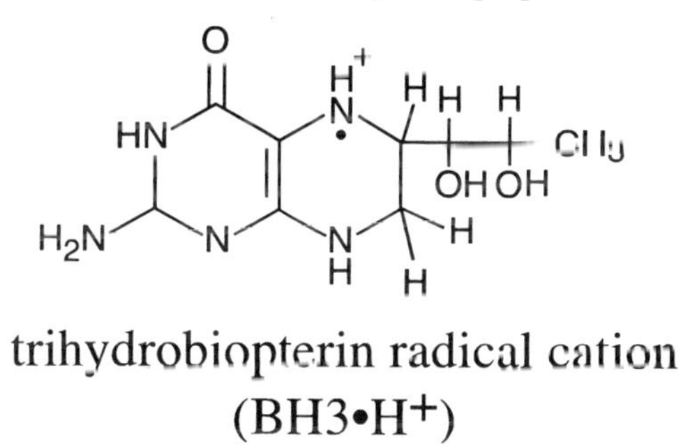

Figure 5. Structure of the cationic radical trihydrobiopterin intermediate.

this reaction as well. Indeed, similar to what was found with Arg, low-temperature optical studies with NHA showed that the reaction was halted at the level of the oxy-ferrous complex, unless BH4 was present [49]. In the latter case the cycle could be completed, as was illustrated by the rapid formation of a complex of Fe(III) with the reaction product NO. Virtually identical results were obtained in stopped-flow studies [42]. As in the first cycle, stoichiometric formation of product, in this case L-citrulline, in single-turnover experiments requires the presence of BH4 [11, 45]. However, for the second cycle no radical EPR signal was found [11, 46, 48]. Although the observations by the various research groups were quite similar, interpretations varied considerably. According to one school of thought [11, 42, 45] it was argued that since only one electron is required for the second cycle, BH4 does not function as an electron donor in this reaction, in line with the absence of an EPR-detectable radical signal. This hypothesis fails to provide a satisfactory explanation for the strict dependence of the second cycle on BH4. Moreover, in our opinion it is unlikely that the enzyme would utilize different electron donors in the two cycles at this stage of the reaction. We therefore propose that in both cycles BH4 donates the electron that reduces the oxyferrous complex [46, 49]. In the first cycle this electron is consumed to form NHA, which results in formation of a EPR-detectable $BH3^{\bullet}$ radical. In the second cycle the electron is not consumed during NHA conversion, and therefore is available for rapid regeneration of BH4 (Fig. 4, bottom). However, direct validation of the proposed mechanism for the second cycle is lacking presently and will require further studies.

7. Outlook

Although the unprecedented function of BH4 as a 1-electron donor to the oxyferrous complex during catalysis seems secured by the observations discussed above, there are numerous remaining questions.

For instance, the present model explains how BH4 prevents uncoupling by reducing the oxyferrous species, but fails to explain why substrate is needed for that as well. We postulate that reduction of the oxyferrous complex by BH4 is an uphill reaction, with the $Fe(II)O_2\cdots BH4 \longleftrightarrow Fe(II)O_2^{-}\cdots BH3^{\bullet}$ equilibrium favouring the tetrahydropterin state. Accordingly, BH4 cannot prevent accumulation of the oxyferrous species unless substrate is present to allow rapid continuation of the reaction cycle (Fig. 6). This model has the additional advantage of rationalizing the otherwise puzzling observation that BH4 only seems to donate the second electron in the NOS cycle. However, alternative explanation cannot be excluded and additional studies will be necessary to resolve this isuue.

It is also conceivable that the role of BH4 extends beyond formation of the $Fe(II)\bullet O_2^-$ complex. Since BH4 may be deprotonated upon oxidation, it might serve as a proton donor in the subsequent reaction step (see Fig. 2), as was recently suggested on the basis of crystallographic studies [34]. The highly reactive $[Fe(IV)O]^-$ that is formed in this reaction step is stabilized by transfer of an electron to the heme from its surrounding, yielding a Compound I type structure ($Fe(IV)=O$ -- $X^{\bullet+}$). In cytochrome P450 the electron is thought to

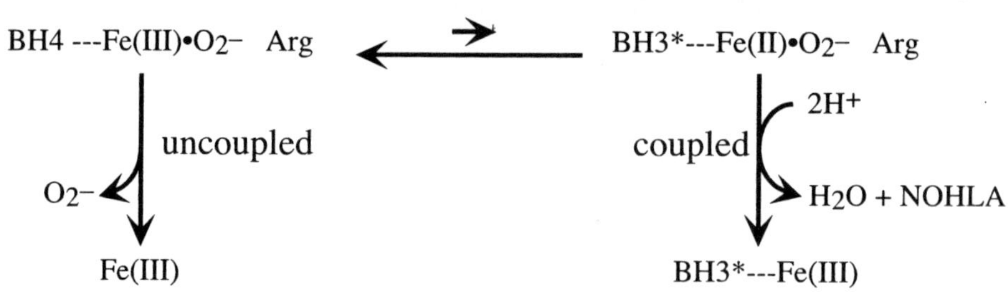

Figure 6. Illustration how pterin and substrate might cooperate to prevent uncoupling. BH4 is required for reduction of the oxyferrous complex. This reaction is probably very fast because of the vicinity of the heme and the pterin, but may be thermodynamically unfavourable, because of the high oxidation potential of the $BH4/BH3^{\bullet}$ couple. In that case BH4 will not prevent uncoupling unless the reaction is pulled forward by rapid subsequent steps, which only occur in the presence of substrate (Arg or NHA).

be furnished by the heme heterocycle, resulting in the formation of a porphyrin π cation [50]. However, in other proteins tryptophyl and tyrosyl radical cations have been implicated [51, 52]. The realization that relatively stable pterin-centered radicals are formed during NOS catalysis has opened up the intriguing possibility that a BH3• radical is also involved in stabilization of the putative oxy-ferryl complex. These additional functions of BH4 are as yet hypothetical, but the recent developments suggest that further research into the involvement of BH4 in NOS catalysis may have some more surprises in store for us.

Acknowledgements The experimental work in the authors' laboratory was supported by the Fonds zur Förderung der Wissenschaftlichen Forschung in Österreich (grant numbers 13013-MED and 13586-MED).

References

[1] O. W. Griffith and D. J. Stuehr, *Annu. Rev. Physiol.* **57** (1995) 707-736.
[2] B. Mayer and B. Hemmens, *Trends Biochem. Sci.* **22** (1997) 477-481.
[3] M. A. Marletta, A. R. Hurshman and K. M. Rusche, *Curr. Opin. Chem. Biol.* **2** (1998) 656-663.
[4] A. C. F. Gorren and B. Mayer, *Biochemistry (Moscow)* **63** (1998) 870-880.
[5] D. J. Stuehr, *Biochim. Biophys. Acta* **1411** (1999) 217-230.
[6] S. Pfeiffer, B. Mayer and B. Hemmens, *Angew. Chem. Int. Ed.* **38** (1999) 1714-1731.
[7] B. Mayer, M. John, B. Heinzel, E. R. Werner, H. Wachter, G. Schultz and E. Böhme, *FEBS Lett.* **288** (1991) 187-191.
[8] S. Pou, W. S. Pou, D. S. Bredt, S. H. Snyder and G. M. Rosen, *J. Biol. Chem.* **267** (1992) 24173-24176.
[9] B. Heinzel, M. John, P. Klatt, E. Böhme and B. Mayer, *Biochem. J.* **281** (1992) 627-630.
[10] J. Vásquez-Vivar, N. Hogg, P. Martásek, H. Karoui, K. A. Pritchard Jr. and B. Kalyanaraman, *J. Biol. Chem.* **274** (1999) 26736-26742.
[11] A. R. Hurshman, C. Krebs, D. E. Edmondson, B. H. Huynh and M. A. Marletta, *Biochemistry* **38** (1999) 15689-15696.
[12] B. Mayer and E. R. Werner, *Naunyn-Schmiedeberg´s Arch. Pharmacol.* **351** (1995) 453-463.
[13] T. J. Kappock and J. P. Caradonna, *Chem. Rev.* **96** (1996) 2659-2756.
[14] P. F. Fitzpatrick, *Annu. Rev. Biochem.* **68** (1999) 355-381.
[15] P. Klatt, M. Schmid, E. Leopold, K. Schmidt, E. R. Werner and B. Mayer, *J. Biol. Chem.* **269** (1994) 13861-13866.
[16] A. C. F. Gorren, B. M. List, A. Schrammel, E. Pitters, B. Hemmens, E. R. Werner, K. Schmidt and B. Mayer, *Biochemistry* **35** (1996) 16735-16745.
[17] K. McMillan and B. S. S. Masters, *Biochemistry* **34** (1995) 3686-3693.
[18] I. Rodríguez-Crespo, N. C. Gerber and P. R. Ortiz de Montellano, *J. Biol. Chem.* **271** (1996) 11462-11467.
[19] K. J. Baek, B. A. Thiel, S. Lucas and D. J. Stuehr, *J. Biol. Chem.* **268** (1993) 21120-21129.
[20] H. J. Cho, E. Martin, Q. Xie, S. Sassa and C. Nathan, *Proc. Natl. Acad. Sci U.S.A.* **92** (1995) 11514-11518.
[21] B. R. Crane, A. S. Arvai, D. K. Ghosh, C. Wu, E. D. Getzoff, D. J. Stuehr and J. A. Tainer, *Science* **279** (1998) 2121-2126.
[22] P. Klatt, S. Pfeiffer, B. M. List, D. Lehner, O. Glatter, H. P. Bächinger, E. R. Werner, K. Schmidt and B. Mayer, *J. Biol. Chem.* **271** (1996) 7336-7342.
[23] R. C. Venema, H. Ju, R. Zou, J. W. Ryan and V. J. Venema, *J. Biol. Chem.* **272** (1997) 1276-1282.
[24] B. M. List, P. Klatt, E. R. Werner, K. Schmidt and B. Mayer, *Biochem. J.* **315** (1996) 57-63.
[25] C. F. B. Witteveen, J. Giovanelli, M. B. Yim, R. Gachhui, D. J. Stuehr and S. Kaufman, *Biochem. Biophys. Res. Commun.* **250** (1998) 36-42.

[26] W. K. Alderton, A. Boyhan and P. N. Lowe, *Biochem. J.* **332** (1998) 195-201.

[27] E. R. Werner, E. Pitters, K. Schmidt, H. Wachter, G. Werner-Felmayer and B. Mayer, *Biochem. J.* **320** (1996) 193-196.

[28] S. Pfeiffer, A. C. F. Gorren, E. Pitters, K. Schmidt, E. R. Werner and B. Mayer, *Biochem. J.* **328** (1997) 349-352.

[29] B. Mayer, C. Wu, A. C. F. Gorren, S. Pfeiffer, K. Schmidt, P. Clark, D. J. Stuehr and E. R. Werner, *Biochemistry* **36** (1997) 8422-8427.

[30] A. Presta, U. Siddhanta, C. Wu, N. Sennequier, L. Huang, H. M. Abu-Soud, S. Erzurum and D. J. Stuehr, *Biochemistry* **37** (1998) 298-310.

[31] H. M. Bömmel, A. Reif, L. G. Fröhlich, A. Frey, H. Hofmann, D. M. Marecak, V. Groehn, P. Kotsonis, M. La, S. Köster, M. Meinecke, M. Bernhardt, M. Weeger, S. Ghisla, G. D. Prestwich, W. Pfleiderer and H. H. H. W. Schmidt, *J. Biol. Chem.* **273** (1998) 33142-33149.

[32] A. Reif, L. G. Fröhlich, P. Kotsonis, A. Frey, H. M. Bömmel, D. A. Wink, W. Pfleiderer and H. H. H. W. Schmidt, *J. Biol. Chem.* **274** (1999) 24921-24929.

[33] C. Riethmüller, A. C. F. Gorren, E. Pitters, B. Hemmens, H.-J. Habisch, S. J. R. Heales, K. Schmidt, E. R. Werner and B. Mayer, *J. Biol. Chem.* **274** (1999) 16047-16051.

[34] B. R. Crane, A. S. Arvai, S. Ghosh, E. D. Getzoff, D. J. Stuehr and J. A. Tainer, *Biochemistry* **39** (2000) 4608-4621.

[35] C. S. Raman, H. Li, P. Martásek, V. Král, B. S. S. Masters and T. L. Poulos, *Cell* **95** (1998) 939-950.

[36] A. C. F. Gorren, A. J. Kungl, K. Schmidt, E. R. Werner and B. Mayer, *submitted for publication.*

[37] N. Bec, A. C. F. Gorren, C. Völker, B. Mayer and R. Lange, *J. Biol. Chem.* **273** (1998) 13502-13508.

[38] H. M. Abu-Soud, R. Gachhui, F. M. Raushel and D. J. Stuehr, *J. Biol. Chem.* **272** (1997) 17349-17353.

[39] H. Sato, I. Sagami, S. Daff and T. Shimizu, *Biochem. Biophys. Res. Commun.* **253** (1998) 845-849.

[40] A. P. Ledbetter, K. McMillan, L. J. Roman, B. S. S. Masters, J. H. Dawson and M. Sono, *Biochemistry* **38** (1999) 8014-8021.

[41] M. Couture, D. J. Stuehr and D. L. Rousseau, *J. Biol. Chem.* **275** (2000) 3201-3205.

[42] S. Boggs, L. Huang and D. J. Stuehr, *Biochemistry* **39** (2000) 2332-2339.

[43] S. Ghosh, D. Wolan, S. Adak, B. R. Crane, N. S. Kwon, J. A. Tainer, E. D. Getzoff and D. J. Stuehr, *J. Biol. Chem.* **274** (1999) 24100-24112.

[44] P. Kotsonis, L. G. Fröhlich, Z. V. Shutenko, R. Horejsi, W. Pfleiderer and H. H. H. W. Schmidt, *Biochem. J.* **346** (2000) 767-776.

[45] H. M. Abu-Soud, A. Presta, B. Mayer and D. J. Stuehr, *Biochemistry* **36** (1997) 10811-10816.

[46] N. Bec, A. C. F. Gorren, B. Mayer, P. P. Schmidt, K. K. Andersson and R. Lange, *J. Inorg. Biochem.* **81** (2000) 207-211.

[47] C.-C. Wei, Z. Wang, Q. Wang, A. L. Meade, C. Hemann, R. Hille and D. J. Stuehr, *J. Biol. Chem.* **in press** (2000)

[48] P. P. Schmidt, R. Lange, A. C. F. Gorren, E. R. Werner, B. Mayer and K. K. Andersson, *J. Biol. Inorg. Chem.* **in press** (2000)

[49] A. C. F. Gorren, N. Bec, A. Schrammel, E. R. Werner, R. Lange and B. Mayer, *Biochemistry* **39** (2000) 11763-11770.

[50] C. Larroque, R. Lange, L. Maurin, A. Bienvenue and J. E. van Lier, *Arch. Biochem. Biophys.* **282** (1990) 198-201.

[51] R. Karthein, R. Dietz, W. Nastainczyk and H. H. Ruf, *Eur. J. Biochem.* **171** (1988) 313-320.

[52] L. A. Fishel, M. F. Farnum, J. M. Mauro, M. A. Miller, J. Kraut, Y. J. Liu, X. L. Tan and C. P. Scholes, *Biochemistry* **30** (1991) 1986-1996.

Nitric Oxide
R.J. Gryglewski and P. Minuz (Eds.)
IOS Press, 2001

The Biological Chemistry of Nitric Oxide and Peroxynitrite

Silvia Pfeiffer and Bernd Mayer

Department of Pharmacology and Toxicology, Karl-Franzens-Universität Graz, Universitätsplatz 2, A-8010 Graz, Austria

Abstract. The free radical nitric oxide (•NO) functions as a specific signaling molecule in biological systems as diverse as blood vessels, the brain, and the immune system. The biological activity of •NO is almost exclusively mediated by direct chemical interactions with target molecules such as oxygen, superoxide anion radicals ($O_2^{\bullet-}$) or biomolecules, including structural and functional proteins, lipids, and DNA. Many of the cytotoxic effects of •NO are thought to be caused by peroxynitrite ($ONOO^-$), the product of the very fast reaction between •NO and $O_2^{\bullet-}$. This article summarizes the chemical properties of •NO and peroxynitrite with a special focus on the mechanisms underlying both the beneficial and harmful effects of these molecules.

1. Introduction

Unlike most other biological mediators, the free radical nitric oxide (•NO) exerts its biological activity through chemical reactions with target molecules rather than through interaction with specific cellular receptor proteins [1]. Therefore, understanding the chemical reactivity of •NO appears to be essential to understand the diverse biological effects of this wide-spread signaling molecule. High affinity binding to the prosthetic heme group of soluble guanylyl cyclase (sGC) resulting in intracellular accumulation of cyclic guanosine 3',5'-monophosphate (cGMP) is certainly the most relevant reaction of •NO with respect to physiology but there are several other targets of •NO that need to be considered to fully account for its various physiological and pathophysiological effects.

The fast, nearly diffusion-controlled reaction of •NO with superoxide ($O_2^{\bullet-}$) to yield peroxynitrite (PN) has received special attention in the past few years, because $O_2^{\bullet-}$ formation is increased in a number of pathological situations, especially cardiovascular and neuronal diseases associated with oxidative stress, resulting in reduced •NO bioavailability and PN-triggered tissue injury. The potential targets of PN are virtually all classes of biomolecules, including proteins, DNA, and lipids. One of the most remarkable reactions of PN is nitration of free and protein-bound tyrosine to 3-nitrotyrosine. Nitrated proteins have been detected by immunostaining of tissue slices in numerous models of human disease, and nitration of specific tyrosine residues was shown to impair the function and/or regulation of several target proteins. However, despite the nitrating potential of PN and the unambigous identification of tyrosine nitration in tissues, the causal link between PN formation and tyrosine nitration has not been clearly established so far. In this article we briefly describe the chemical properties of •NO and peroxynitrite pertinent to the physiology and pathophysiology of the L-arginine/•NO pathway in mammalian tissues.

2. Chemistry of NO in aqueous solutions

•NO is freely diffusible in both aqueous and lipid environment. This makes •NO an ideal messenger molecule crossing cell membranes without significant diffusion barrier: the tissue volume that becomes affected by a point source of •NO is limited by inactivation reactions rather than hindrance of diffusion. Depending on the location and rate of •NO synthesis, molecular oxygen, metal centers and $O_2^{\bullet-}$ are the main physiologically occurring molecules rapidly reacting with free •NO. Some of these reactions mediate important physiological effects, others result in tissue injury or serve as efficient sinks reducing NO bioavailability.

2.1. Autoxidation

Oxygen is the most abundant cellular target of •NO. In both gas phase and aqueous solution the rate limiting reaction yields $\bullet NO_2$. The reaction is second order with respect to •NO and first order with respect to O_2 with a rate constant of 6.3×10^6 M^{-2} s^{-1} [2]. As a consequence of this third order rate law, the half-life of •NO increases with decreasing •NO concentrations. In air, for instance, the time required for half of a given concentration of •NO to react with O_2 is 24 s for 10,000 ppm of •NO, but 7 h for 10 ppm. Authentic $\bullet NO_2$ dimerizes to N_2O_4, giving stoichiometric amounts of nitrite (NO_2^-) and nitrate (NO_3^-). The observation that NO_2^- is the sole stable end product of •NO autoxidation [3] indicates that $\bullet NO_2$ does not accumulate but rapidly reacts with another molecule of •NO to give N_2O_3, hydrolysis of which yields NO_2^-. The sum equation ($4\bullet NO + O_2 + 2H_2O \rightarrow 4H^+ + 4NO_2^-$) agrees well with the overall stoichiometry of NO autoxidation.

2.2. S-Nitrosation

Various tissues, including human plasma, airways, and white blood cells, have been reported to contain endogenous S-nitrosothiols (thionitrites; RS-NO), which may function as relatively stable cellular storage or transport forms of •NO [4]. In the absence of O_2, •NO does not react with thiols at significant rates, but S-nitrosation does occur in O_2 containing solutions [5]. Most likely, N_2O_3, which is formed as intermediate of •NO autoxidation, reacts with thiols to yield the corresponding S-nitroso compounds ($2RSH + N_2O_3 \rightarrow 2RS\text{-}NO + H_2O$). Although •NO autoxidation is probably too slow to account for S-nitrosation at physiological •NO concentrations, accumulation of •NO in lipid compartments, e.g. cell membranes, may give rise to sufficient formation of N_2O_3 to sustain this type of nitrosation reaction [6]. Alternatively, a radical mechanism based on the reaction of thiyl radicals (RS•) with •NO might contribute to nitrosation *in vivo*. RS• could be formed when thiols are used as cellular antioxidants or through thiol oxidation by $\bullet NO/O_2^{\bullet-}$ [7].

S-nitrosothiols can transfer nitrosonium (NO^+) equivalents to other thiols ($R_1S\text{-}NO + R_2SH \rightarrow R_1SH + R_2S\text{-}NO$) [8]. These transnitrosation reactions enable the rapid exchange of nitroso groups between low and high molecular mass thiols. The reaction doesn't require metal ions and appears to be driven exclusively by thiol abundance, whereas homolysis of S-nitrosothiols to give •NO and RS• is catalyzed by reduced metal ions, in particular Cu(I). Thus, nitrosothiols can be stabilized by chelating trace metal ions and rapidly decompose to release free •NO in the combined presence of copper or iron salts and reducing agents such as GSH or ascorbate [9,10]. In addition to this non-enzymatic reaction, decomposition of S-nitrosothiols is catalyzed by enzymes such as GSH peroxidase [11], γ-glutamyltranspeptidase [12] or thioredoxin reductase [13]. Further studies are needed to address the potential involvement of S-nitrosothiols as •NO carriers in cellular signaling.

2.3. Reaction with metal centers

•NO reacts with high affinity with a variety of metal centers in proteins. The most prominent metal proteins interacting with •NO are hemoglobin (Hb), sGC and cytochrome c oxidase. The reaction with oxyHb follows second order kinetics with a rate constant of ~3 x 10^7 M^{-1} s^{-1} and yields metHb and NO_3^- (Hb-Fe^{II}(O_2) + •NO → Hb-Fe^{III} + NO_3^-) [14]. The reaction with desoxyHb, which occurs at low O_2 tension, results in formation of a ferrous nitrosyl-heme complex (Hb-Fe^{II}-NO). Dissociation of this complex is extremely slow but in the presence of O_2 it decays rapidly to metHb and NO_3^-. The high affinity binding of NO to both oxy- and desoxyHb renders Hb in circulating erythrocytes the most efficient sink for •NO in the body [15,16].

Many of the biological effects of •NO as a physiological signaling molecule result from activation of sGC by several hundred-fold above basal activity. sGC catalyzes the conversion of Mg^{2+}-GTP to the intracellular messenger cGMP which connects the •NO signal to the cellular response by activating cGMP-dependent protein kinases and regulating the activity of phosphodiesterases and ion channels [17]. sGC is a heterodimer of α and ß subunits with a prosthetic heme group bound to histidine-105 of the ß subunit [18]. Although both the heme binding sequence motif and the spectral proerties of the enzyme indicate that its heme pocket resembles that of hemoglobins and myoglobins, sGC binds O_2 very poorly, allowing for formation of the corresponding penta-coordinated ferrous nitrosyl-heme complex (sGC-Fe^{II}-NO) even in the presence of O_2 [19]. Also, the dissociation of the nitrosyl-heme complex is unusually fast (k_{obs} at 20 °C = ~7 x 10^{-4} s^{-1} and 0.04 s^{-1} in the absence and presence of Mg^{2+}-GTP, respectively) [20]. Extrapolated to 37 °C, this would give a half-life of about 5 s in the presence of the substrate [21]. In the presence of effective physiological •NO sinks, this fast rate of •NO dissociation may be crucial for the rapid deactivation of sGC in •NO/cGMP signaling, but recent evidence suggests that additional mechanisms may facilitate inactivation of the •NO-stimulated enzyme in cells. A possible mechanism would be oxidation of the heme iron: this is the mechanism of action of the sGC inhibitor 1*H*-[1,2,4]oxadiazolo[4,3-*a*]quinoxalin-1-one (ODQ) [22]. Although there is no evidence yet for a controlled heme oxidation of sGC under physiological conditions, a recent study suggests that vascular tissue contains a NADH-dependent flavoprotein that catalyzes reduction of ferric sGC back to the •NO binding ferrous form [23,24]

•NO competes with O_2 for binding to cytochrome c oxidase, resulting in inhibition of mitochondrial respiration [25,26]. This could be a significant toxic effect of •NO under inflammatory conditions associated with increased NO formation or in hypoxic states with reduced cellular O_2 tension. However, the discovery of an •NO synthase in mitochondria [27,28] raises the possibility that the balance of •NO versus O_2 binding to cytochrome c oxidase may also be used to regulate respiration under normal physiological conditions.

2.4. Reaction with superoxide anion

An important reaction of •NO in biological media is the bimolecular reaction with $O_2^{\bullet-}$, yielding peroxynitrite ($ONOO^-$, PN, oxoperoxonitrate(1-)) [29]. Formation of PN from NO and $O_2^{\bullet-}$ occurs at a nearly diffusion-controlled rate of ~2 x 10^{10} M^{-1} s^{-1}. This rate constant is about ten times higher than dismutation of $O_2^{\bullet-}$ by superoxide dismutase (SOD), indicating that •NO outcompetes the SOD reaction at physiological steady-state concentrations. PN is a potent oxidant and has been shown to react with virtually all classes of biomolecules *in vitro*. In the following section we briefly describe the biological chemistry of PN and its potential contribution to tissue injury in pathophysiological conditions.

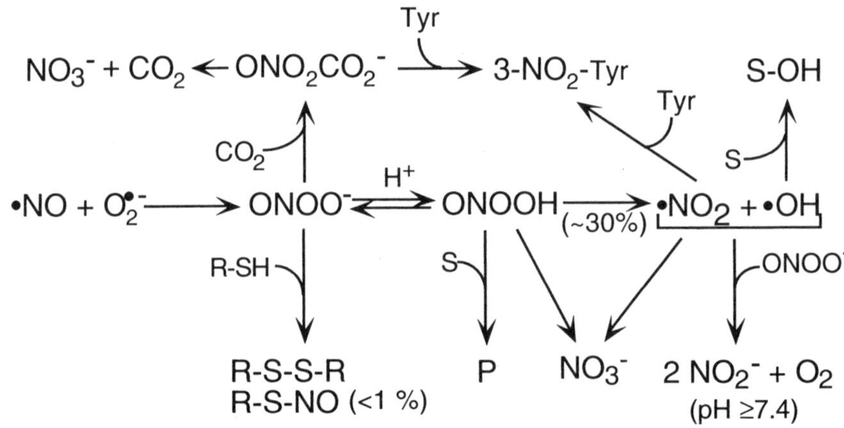

Fig. 1: Chemical reactions of peroxynitrite ($ONOO^-$) and peroxynitrous acid (ONOOH). S, substrates for 1- and 2-electron oxidations; P, products; Tyr, tyrosine.

3. Chemistry of peroxynitrite

The chemical reactions of PN relevant to its biological effects are summarized in Fig. 1.

3.1. Isomerization/decomposition to nitrate, nitrite and oxygen

Peroxynitrite is stable in alkaline solution and can be stored for weeks at pH ≥13.0 and -70 °C. Protonation of PN (pKa of 6.8 at 37 °C) results in formation of peroxynitrous acid (ONOOH) which is unstable and undergoes two different pathways of decay: isomerization to NO_3^- and decomposition to NO_2^- and O_2 [30]. The partitioning between the two pathways is determined by the pH but not the PN concentration. Isomerization to NO_3^- is predominant at low pH (<7.0) and involves homolysis of ONOOH to yield caged $\bullet NO_2$ and •OH radicals. About 30 % of the radicals can escape the cage before isomerization and may be responsible for the observed radical properties of PN [31]. At pH values ≥7.0, the decomposition pathway becomes significant. It may involve reaction of •OH with $ONOO^-$ giving •NO, O_2 and OH^-, followed by reaction of •NO with $\bullet NO_2$ to yield NO_2^- via formation of N_2O_3 as an intermediate. The overall stoichiometry of this hypothetical pathway ($ONOOH + ONOO^- + H_2O \rightarrow 2NO_2^- + O_2 + 2H^+ + OH^-$) agrees well with the 2:1 ratio of NO_2^- and O_2 formation observed in PN decomposition at alkaline pH [30]. In the presence of bicarbonate, the decomposition pathway is prevented by the reaction of $ONOO^-$ with CO_2, rendering NO_3^- the exclusive stable metabolite of PN at all pH values [32].

Homolysis of ONOOH to yield $\bullet NO_2$ and •OH was questioned because of both calculations rendering this reaction "thermodynamically impossible" and data showing that neither NO_2^- (a product of $\bullet NO_2$ hydrolysis) nor H_2O_2 (the product of •OH dimerization) are formed when PN decomposes at acidic pH [33]. However, the thermodynamics of ONOOH homolysis has recently been revised and the lack of NO_2^- and H_2O_2 formation has been explained by a rapid electron transfer reaction occurring between NO_2^- and •OH that gives OH^- and $\bullet NO_2$ [34]. Thus, initially formed NO_2^- and •OH are rapidly consumed to recycle $\bullet NO_2$, such that NO_3^- is finally formed as exclusive product of the overall reaction without concomitant production of H_2O_2. It is now widely accepted that homolysis of ONOOH does occur and yields about 30 % of free $\bullet NO_2$ and •OH radicals [35].

3.2. Reaction with CO_2

PN anion reacts with CO_2 to yield nitroso-peroxycarbonate, which decomposes to NO_3^- and CO_2 in the absence of other reactive molecules ($ONOO^- + CO_2 \rightarrow ONO_2CO_2^- \rightarrow CO_2 + NO_3^-$) [36]. The rate constant of this reaction ($3 \times 10^4\ M^{-1}\ s^{-1}$ at pH 7.4 and 37 °C) is large enough that it could be the predominant pathway of PN decay in biological fluids, where the total carbonate concentration is between 1 and 25 mM. The adduct, $ONO_2CO_2^-$, exhibits different reactivity than PN: in the presence of CO_2 nitration yields are increased by 2- to 4-fold [37], whereas thiol oxidation is reduced [38]. This may be explained by homolysis of $ONO_2CO_2^-$ to give $\bullet CO_3^-$ and $\bullet NO_2$ [39]. In the absence of other reactive molecules, these radicals combine to form NO_3^- ($k = 10^9\ M^{-1}\ s^{-1}$), but one-electron oxidation of tyrosine by $\bullet CO_3^-$ in concert with a termination reaction of the tyrosyl radical with $\bullet NO_2$ would yield the nitration product 3-nitrotyrosine [32]. On the other hand, decomposition of $ONO_2CO_2^-$ does not give $\bullet OH$ radicals, presumably explaining the significant reduction of PN-triggered oxidation reactions by CO_2 [38].

3.3. Reaction with physiological scavengers

Many physiologically occurring antioxidants, metallocompounds and proteins were shown to react with PN, suggesting that some of these compounds may act as PN scavengers *in vivo*. The biologically most relevant reactants are low and high molecular mass thiols, ascorbate, uric acid, Se compounds, as well as oxyHb and other metalloproteins.

The reaction of PN with SH-groups results mainly in oxidation to disulfides (>90 %), but small amounts of other products (sulfenic and sulfinic acid, S-nitrosothiols) are also formed [40]. The apparent second order rate constants for the reactions of PN ($ONOO^-$ + ONOOH) with L-cysteine and GSH at pH 7.4 and 37 °C are $4.5 \times 10^3\ M^{-1}\ s^{-1}$ and $1.35 \times 10^3\ M^{-1}\ s^{-1}$, respectively, with the PN anion presumably reacting with the non-dissociated form of the thiols [40,41]. Despite the moderate reaction rates, thiol oxidation may be an important route for PN detoxification because of the high concentration of GSH in tissues (1 - 10 mM). In blood, serum albumin is the most abundant protein that becomes oxidized at one single cysteine (Cys-34 in human serum albumin) and six methionine residues [42].

Ascorbate, one of the most abundant antioxidants, was believed to play only a minor role as PN scavenger because the cellular ascorbate concentration is at least one order of magnitude lower than that of GSH and the reaction with ascorbate ($k = 235\ M^{-1}\ s^{-1}$) is about 6-fold slower than that with GSH [43,44]. However, recent evidence suggests that ascorbate may be an important scavenger of PN in the presence of CO_2 [45,46]. This effect may be due to the nearly diffusion-controlled reaction of ascorbate with $\bullet CO_3^-$, the product of $ONO_2CO_2^-$ homolysis, resulting in a dramatic reduction of the oxidative potential of PN. An additional protective effect of ascorbate may result from its so-called repair function, i.e. the reduction of oxidized substrate (cation) radicals [47,48]. Both mechanisms may contribute significantly to PN scavenging by ascorbate *in vivo*.

Uric acid is present in plasma in relatively high concentrations of up to 0.3 mM [49]. With an apparent second order rate constant of $155\ M^{-1}\ s^{-1}$ at pH 7.4 and 37 °C [50], the reaction appears to be too slow to be of biological significance, but recent studies showed that urate inhibits PN-triggered oxidation and nitration reactions in the presence of CO_2 [51,52] and exhibits beneficial effects in inflammation [53-55]. Thus, urate may interfere with PN cytotoxicity *in vivo* through a mechanism similar to that described for ascorbate.

It was shown that selenocysteine containing GSH peroxidase can act as a PN reductase and thereby prevents both oxidation and nitration reactions [56]. The enzyme utilizes reducing equivalents from GSH for the two-electron reduction of PN to nitrite. The relatively fast reaction of PN with GSH peroxidase ($k_{app} = 8.0 \times 10^6\ M^{-1}\ s^{-1}$) [57] renders this pathway another efficient way of PN detoxification *in vivo*. Similarly, selenoprotein P [58] and synthetic Se-compounds such as ebselen were shown to act as efficient PN decomposition catalysts *in vitro* and in various animal models of inflammatory disease [59,60].

PN oxidizes oxyHb to metHb with a second order rate constant of $1.0 \times 10^4\ M^{-1}\ s^{-1}$ at pH 7.4 and 25 °C [61,62]. The pH dependence of the reaction suggests that the active species is peroxynitrous acid (ONOOH), but the mechanism appears to be quite complex [63]. Both the high concentration of Hb in blood and the rapid uptake of PN in erythrocytes [62] may render vascular oxyHb an important target of PN *in vivo*.

3.4. Tyrosine modification

The reaction of PN with tyrosine yields three products: 3-hydroxytyrosine (L-dopa), 3-nitrotyrosine, and dityrosine [64]. The initial event is the one-electron oxidation of tyrosine to yield tyrosyl radicals (•tyr) which either dimerize to give dityrosine, react with $•NO_2$ to give 3-nitrotyrosine, or combine with •OH radicals to give L-dopa. Tyrosine oxidation is caused by peroxynitrous acid or more efficiently (in the presence of CO_2) by $•CO_3^-$ radicals formed in the course of $ONO_2CO_2^-$ homolysis (tyr + $•CO_3^- \rightarrow$ •tyr + CO_2) [37]. As expected from the lack of •OH formation from $ONO_2CO_2^-$, hydroxylation of tyrosine does not occur in the presence of CO_2 at physiological pH [65].

3-Nitrotyrosine has been detected in injured tissues for a wide variety of disease states associated with increased oxidative stress, including neurodegenerative diseases, atherosclerosis, infections with pathogens, and chronic inflammation [66]. These diseases are associated with increased expression of inducible •NO synthase in several cell types, causing a high output of •NO over long periods of time. Together with $O_2^{\bullet-}$ formed in the respiratory burst of white blood cells infiltrating injured tissues, this sustained •NO formation is thought to result in PN-triggered tyrosine nitration. Since PN causes nitration and consequent loss of function of numerous proteins, this pathway was proposed to be causally involved in the pathogenesis of certain diseases. For instance, nitration of neurofilament L has been shown to interfere with filament polymerization in amyotrophic lateral sclerosis [67], and nitration of cardiac actin may impair contractile function of the heart in myocarditis [68]. In addition to these structural proteins, tyrosine nitration may affect the function of enzymes such as Mn-SOD [69] and glutamine synthetase [70] and disrupt signaling cascades that depend on reversible tyrosine phosphosphorylation [71,72].

The assumption that 3-nitrotyrosine is a footprint of PN formation *in vivo* is based on the fact that free and protein-bound tyrosine becomes nitrated with a bolus of authentic PN added to biological samples, such as isolated tissues, cultured cells, or protein preparations. Since this approach is a poor approximation to *in vivo* situations, in which PN is formed by continuous reaction of •NO and $O_2^{\bullet-}$ at physiological pH, we developed a method that should simulate these conditions more closely by combining •NO-generating and $O_2^{\bullet-}$-generating systems. Surprisingly, tyrosine nitration by •NO and $O_2^{\bullet-}$ generated simultaneously at equal rates was found to be 10- to 100-fold less efficient than the reaction triggered by authentic PN, even though the potential to trigger oxidation reactions was similar in both cases [73]. Virtually identical observations were made with several different •NO/$O_2^{\bullet-}$ generating systems including pulse radiolysis [52,74]. While the initial reports on the poor nitrating efficiency of •NO/$O_2^{\bullet-}$ were limited to the *in vitro* nitration of free tyrosine, we recently extended these findings to cultured cells and obtained compelling evidence that protein tyrosine nitration in cytokine-activated macrophages is triggered by peroxidase/ H_2O_2-catalyzed oxidation of NO_2^- rather than PN formed from •NO and $O_2^{\bullet-}$ (Pfeiffer, S., Lass, A., Schmidt, K. & Mayer, B., manuscript in preparation). Myeloperoxidase in neutrophils [75] and eosinophil peroxidase [76] have been identified as catalysts of cellular tyrosine nitration in the presence of NO_2^- and H_2O_2, suggesting that oxidation of nitrite may be an efficient alternative pathway of protein nitration *in vivo*.

Kinetic simulations suggest that the inefficient tyrosine nitration by continuous fluxes of •NO and $O_2^{\bar{\bullet}}$ may be a consequence of the resulting low (subnanomolar) PN steady-state concentrations [52]. As pointed out in the beginning of this section, nitration requires oxidation of tyrosine to tyrosyl radical (•tyr). At low PN concentrations, the model predicts, dimerization of •tyr to give dityrosine outcompetes the reaction of •tyr with •NO_2, yielding 3-nitrotyrosine. This model agrees with experimental findings, showing that dityrosine is the major product of tyrosine reacting with PN at low steady-state concentrations, whereas nitration becomes significant only at high PN concentrations that are unlikely to occur *in vivo* [52]. Thus, PN formed from continuous fluxes of •NO and $O_2^{\bar{\bullet}}$ may cause cellular injury through oxidative modification of biomolecules, but tyrosine nitration appears to result from alternative routes, such as peroxidase-catalyzed oxidation of nitrite or other as yet unknown enzymatic reactions.

Acknowledgements The experimental work in the author's laboratory was supported by the "Fonds zur Förderung der Wissenscahftlichen Forschung in Österreich". Silvia Pfeiffer is a recipient of an Austrian Academy of Sciences APART fellowship (APART 7/98).

References

[1] Pfeiffer, S., Mayer, B. and Hemmens, B. (1999) Angew. Chem. Int. Edit. 38, 1715-1731.

[2] Ford, P.C., Wink, D.A. and Stanbury, D.M. (1993) FEBS Lett. 326, 1-3.

[3] Ignarro, L.J., Fukuto, J.M., Griscavage, J.M., Rogers, N.E. and Byrns, R.E. (1993) Proc. Natl. Acad. Sci. U.S.A. 90, 8103-8107.

[4] Butler, A.R. and Rhodes, P. (1997) Anal. Biochem. 249, 1-9.

[5] Wink, D.A., Nims, R.W., Darbyshire, J.F., Christodoulou, D., Hanbauer, I., Cox, G.W., Laval, F., Laval, J., Cook, J.A., Krishna, M.C., Degraff, W.G. and Mitchell, J.B. (1994) Chem. Res. Toxicol. 7, 519-525.

[6] Liu, X.P., Miller, M.J.S., Joshi, M.S., Thomas, D.D. and Lancaster, J.R. (1998) Proc. Natl. Acad. Sci. U.S.A. 95, 2175-2179.

[7] Mayer, B., Pfeiffer, S., Schrammel, A., Schmidt, K., Koesling, D. and Brunner, F. (1998) J. Biol. Chem. 273, 3264-3270.

[8] Stamler, J.S., Toone, E.J., Lipton, S.A. and Sucher, N.J. (1997) Neuron 18, 691-696.

[9] Gorren, A.C.F., Schrammel, A., Schmidt, K. and Mayer, B. (1996) Arch. Biochem. Biophys. 330, 219-228.

[10] Singh, R.J., Hogg, N., Joseph, J. and Kalyanaraman, B. (1996) J. Biol. Chem. 271, 18596-18603.

[11] Freedman, J.E., Frei, B., Welch, G.N. and Loscalzo, J. (1995) J. Clin. Invest. 96, 394-400.

[12] Hogg, N., Singh, R.J., Konorev, E., Joseph, J. and Kalyanaraman, B. (1997) Biochem. J. 323, 477-481.

[13] Nikitovic, D. and Holmgren, A. (1996) J. Biol. Chem. 271, 19180-19185.

[14] Doyle, M.P. and Hoeksta, J.W. (1981) J. Inorg. Biochem. 14, 351-358.

[15] Liu, X.P., Miller, M.J.S., Joshi, M.S., SadowskaKrowicka, H., Clark, D.A. and Lancaster, J.R. (1998) J. Biol. Chem. 273, 18709-18713.

[16] Vaughn, M.W., Huang, K.T., Kuo, L. and Liao, J.C. (2000) J. Biol. Chem. 275, 2342-2348.

[17] Mayer, B., Koesling, D. and Böhme, E. (1993) Adv. Second Messenger Phosphoprotein Res. 28, 111-119.
[18] Wedel, B., Humbert, P., Harteneck, C., Foerster, J., Malkewitz, J., Böhme, E., Schultz, G. and Koesling, D. (1994) Proc. Natl. Acad. Sci. U.S.A. 91, 2592-2596.
[19] Stone, J.R. and Marletta, M.A. (1994) Biochemistry 33, 5636-5640.
[20] Kharitonov, V.G., Sharma, V.S., Magde, D. and Koesling, D. (1997) Biochemistry 36, 6814-6818.
[21] Kharitonov, V.G., Russwurm, M., Magde, D., Sharma, V.S. and Koesling, D. (1997) Biochem. Biophys. Res. Commun. 239, 284-286.
[22] Schrammel, A., Behrends, S., Schmidt, K., Koesling, D. and Mayer, B. (1996) Mol. Pharmacol. 50, 1-5.
[23] Iesaki, T., Gupte, S.A., Kaminski, P.M. and Wolin, M.S. (1999) Am. J. Physiol. 46, H978-H985.
[24] Gupte, S.A., Rupawalla, T., Phillibert, D. and Wolin, M.S. (1999) Am. J. Physiol. 277, L1124-L1132.
[25] Cleeter, M.W.J., Cooper, J.M., Darley-Usmar, V.M., Moncada, S. and Schapira, A.H.V. (1994) FEBS Lett. 345, 50-54.
[26] Brown, G.C. and Cooper, C.E. (1994) FEBS Lett. 356, 295-298.
[27] Ghafourifar, P. and Richter, C. (1997) FEBS Lett. 418, 291-296.
[28] Tatoyan, A. and Giulivi, C. (1998) J. Biol. Chem. 273, 11044-11048.
[29] Beckman, J.S. and Koppenol, W.H. (1996) Am. J. Physiol. 40, C1424-C1437.
[30] Pfeiffer, S., Gorren, A.C.F., Schmidt, K., Werner, E.R., Hansert, B., Bohle, D.S. and Mayer, B. (1997) J. Biol. Chem. 272, 3465-3470.
[31] Merenyi, G., Lind, J., Goldstein, S. and Czapski, G. (1998) Chem. Res. Toxicol. 11, 712-713.
[32] Lymar, S.V. and Hurst, J.K. (1998) Inorg. Chem. 37, 294-301.
[33] Koppenol, W.H. (1998) Free Rad. Biol. Med. 25, 385-391.
[34] Coddington, J.W., Hurst, J.K. and Lymar, S.V. (1999) J. Am. Chem. Soc. 121, 2438-2443.
[35] Gerasimov, O.V. and Lymar, S.V. (1999) Inorg. Chem. 38, 4317-4321.
[36] Lymar, S.V. and Hurst, J.K. (1995) J. Am. Chem. Soc. 117, 8867-8868.
[37] Lymar, S.V., Jiang, Q. and Hurst, J.K. (1996) Biochemistry 35, 7855-7861.
[38] Denicola, A., Freeman, B.A., Trujillo, M. and Radi, R. (1996) Arch. Biochem. Biophys. 333, 49-58.
[39] Bonini, M.G., Radi, R., FerrerSueta, G., Ferreira, A.M.D. and Augusto, O. (1999) J. Biol. Chem. 274, 10802-10806.
[40] Radi, R., Beckman, J.S., Bush, K.M. and Freeman, B.A. (1991) J. Biol. Chem. 266, 4244-4250.
[41] Quijano, C., Alvarez, B., Gatti, R.M., Augusto, O. and Radi, R. (1997) Biochem. J. 322, 167-173.
[42] Alvarez, B., FerrerSueta, G., Freeman, B.A. and Radi, R. (1999) J. Biol. Chem. 274, 842-848.
[43] Bartlett, D., Church, D.F., Bounds, P.L. and Koppenol, W.H. (1995) Free Rad. Biol. Med. 18, 85-92.
[44] Squadrito, G.L., Jin, X. and Pryor, W.A. (1995) Arch. Biochem. Biophys. 322, 53-59.
[45] Scorza, G. and Minetti, M. (1998) Biochem. J. 329, 405-413.
[46] Kirsch, M. and deGroot, H. (2000) J. Biol. Chem. 275, 16702-16708.
[47] Hunter, E.P., Desrosiers, M.F. and Simic, M.G. (1989) Free Rad. Biol. Med. 6, 581-585.

[48] Sturgeon, B.E., Sipe, H.J.J., Barr, D.P., Corbett, J.T., Martinez, J.G. and Mason, R.P. (1998) J. Biol. Chem. 273, 30116-30121.
[49] Halliwell, B. and Gutteridge, J.M. (1989) Clarendon Press, Oxford.
[50] Squadrito, G.L., Cueto, R., Splenser, A.E., Valavanidis, A., Zhang, H.W., Uppu, R.M. and Pryor, W.A. (2000) Arch. Biochem. Biophys. 376, 333-337.
[51] Gow, A., Duran, D., Thom, S.R. and Ischiropoulos, H. (1996) Arch. Biochem. Biophys. 333, 42-48.
[52] Pfeiffer, S., Schmidt, K. and Mayer, B. (2000) J. Biol. Chem. 275, 6346-6352.
[53] Hooper, D.C., Spitsin, S., Kean, R.B., Champion, J.M., Dickson, G.M., Chaudhry, I. and Koprowski, H. (1998) Proc. Natl. Acad. Sci. USA 95, 675-680.
[54] Hooper, D.C., Scott, G.S., Zborek, A., Mikheeva, T., Kean, R.B., Koprowski, H. and Spitsin, S.V. (2000) FASEB J. 14, 691-698.
[55] Spitsin, S.V., Scott, G.S., Kean, R.B., Mikheeva, T. and Hooper, D.C. (2000) Neurosci. Lett. 292, 137-141.
[56] Sies, H., Sharov, V.S., Klotz, L.O. and Briviba, K. (1997) J. Biol. Chem. 272, 27812-27817.
[57] Briviba, K., Kissner, R., Koppenol, W.H. and Sies, H. (1998) Chem. Res. Toxicol. 11, 1398-13401.
[58] Arteel, G.E., Mostert, V., Oubrahim, H., Briviba, K., Abel, J. and Sies, H. (1998) Biol. Chem. Hoppe-Seyler 379, 1201-1205.
[59] Masumoto, K., Kissner, R., Koppenol, W.H. and Sies, H. (1996) FEBS Lett. 398, 179-182.
[60] Sies, H. and Arteel, G.E. (2000) Free Rad. Biol. Med. 28, 1451-1455.
[61] Schmidt, K., Klatt, P. and Mayer, B. (1994) Biochem. J. 301, 645-647.
[62] Denicola, A., Souza, J.M. and Radi, R. (1998) Proc. Natl. Acad. Sci. U.S.A. 95, 3566-3571.
[63] Exner, M. and Herold, S. (2000) Chem. Res. Toxicol. 13, 287-293.
[64] van der Vliet, A., Eiserich, J.P., O'Neill, C.A., Halliwell, B. and Cross, C.E. (1995) Arch. Biochem. Biophys. 319, 341-349.
[65] Santos, C.X.C., Bonini, M.G. and Augusto, O. (2000) Arch. Biochem. Biophys. 377, 146-152.
[66] Ischiropoulos, H. (1998) Arch. Biochem. Biophys. 356, 1-11
[67] Estevez, A.G., Spear, N., Manuel, S.M., Radi, R., Henderson, C.E., Barbeito, L. and Beckman, J.S. (1998) J. Neurosci. 18, 923-931.
[68] Ungureanu-Longrois, D., Balligand, J.-L., Kelly, R.A. and Smith, T.W. (1995) J. Mol. Cell. Cardiol. 27, 155-167.
[69] MacMillan-Crow, L.A. and Thompson, J.A. (1999) Arch. Biochem. Biophys. 366, 82-88.
[70] Berlett, B.S., Levine, R.L. and Stadtman, E.R. (1998) Proc. Natl. Acad. Sci. U.S.A. 95, 2784-2789.
[71] Kong, S.-K., Yim, M.B., Stadtman, E.R. and Chock, P.B. (1996) Proc. Natl. Acad. Sci. USA 93, 3377-3382.
[72] Gow, A.J., Duran, D., Malcolm, S. and Ischiropoulos, H. (1996) FEBS Lett. 385, 63-66.
[73] Pfeiffer, S. and Mayer, B. (1998) J. Biol. Chem. 273, 27280-27285.
[74] Goldstein, S., Czapski, G., Lind, J. and Merenyi, G. (2000) J. Biol. Chem. 275, 3031-3036.
[75] Eiserich, J.P., Hristova, M., Cross, C.E., Jones, A.D., Freeman, B.A., Halliwell, B. and vanderVliet, A. (1998) Nature 391, 393-397.
[76] Wu, W.J., Chen, Y.H. and Hazen, S.L. (1999) J. Biol. Chem. 274, 25933-25944.

Nitric Oxide
R.J. Gryglewski and P. Minuz (Eds.)
IOS Press, 2001

Kinetics of Nitric Oxide and Superoxide Release in the Brain During Ischemia/Reperfusion

L.W. Dobrucki, I. Dobrucki and T. Malinski

Department of Chemistry and Biochemistry
Konneker Research Laboratories, Ohio University
Athens, OH 45701-2979, USA
Tel./Fax. (740) 593-9641
E-mail: malinski@ohio.edu

A role of nitric oxide in ischemia/reperfusion (I/R) injury of brain in normotensive Wistar-Kyoto (WKY rats), stroke-prone spontaneously hypertensive (SHR-SP rats), control Sprague-Dawley (SD rats) and L-NNA-treated rats was studied. Cerebral ischemia was produced in rats by occlusion of the middle cerebral artery (MCA). NO and O_2^- releases in the brain in response to MCA occlusion followed by reperfusion were simultaneously monitored (3h) using electrochemical microsensors. The size of infarct was evaluated in the course of I/R from images of brain slices stained with 2,3,5-triphenyltetrazolium chloride. The concentration of NO was significantly lower during I/R in SHR than in WKY rats (the maximal NO concentration was 2.61±0.22 μmol/L for WKY and 1.51±0.16 μmol/L for SHR rats; $P<0.01$). In contrast, the concentration of O_2^- release during cerebral ischemia was higher in SHR than in WKY rats (the maximal increase was 122±24 nmol/L for SD and 220±44 nmol/L for SHR-SP rats; $P<0.01$). The concentration of NO release in L-NNA-treated rats was lower than in SD control rats (the maximal NO concentration was 1.49±0.13 μmol/L for SD treated rats and 2.23±0.18 μmol/L for SD control rats; $P<0.01$).
The concentration of O_2^- release during cerebral ischemia was significantly higher in SHR than in WKY rats (the maximal increase was 122±24 nmol/L for WKY and 220±44 nmol/L for SHR rats; $P<0.01$) but similar for untreated (116±20 nmol/L) and treated (88±19 nmol/L) SD rats ($P<0.01$). The infarct sizes 180 minutes after onset of ischemia were larger in SHR than WKY rats (9.2±1.1% vs. 5.8±0.6%) while in L-NNA-treated SD rats the infarct size was 23% smaller (4.1±1.0%) comparing with untreated SD rats (5.3±0.7%).
The increase of cerebral I/R injury correlates positively with the increase of superoxide production.

Introduction

In the central nervous system nitric oxide (NO) is regarded as a factor involved in neurotransmitting processes [1] and regulation of cerebral blood flow [2]. NO has also been implicated in the pathophysiology of cerebral ischemia on the basis of its actions as a mediator of tissue injury [3]. During brain ischemia, a reduction in the supply of oxygen and energy substrates results in a cascade of events in brain tissues, including alterations in the pool of oxygen free radicals and other reactive oxygen species, which may lead to development of infarct. It is also known that reperfusion of ischemic tissue during revascularizing procedures paradoxically causes tissue injury beyond the damage already developed. However, there is conflicting evidence related to the possible contribution of NO in pathogenesis of ischemia/reperfusion in brain injury [3]. It has been indicated in several studies that NO may have neurotoxic effects [4, 5] if present in abnormally high concentrations which takes place during cerebral ischemia [6, 7]. However, a postulated mechanism of ischemia/reperfusion damage is based on high production of superoxide. The excess of NO generated during ischemia/reperfusion can react with superoxide radical (O_2^-) to produce peroxynitrite [6]. An isomerization of peroxynitrite can lead to production of cytotoxic oxygen species like hydroxyl ($OH^•$) or $NO_2^•$ radicals. Such a cytotoxic mechanism initiated by overproduction of peroxynitrite has been recently indicated in the development of ischemia/reperfusion injury in skeletal muscles [8]. It has been suggested that the agents that limit NO and/or O_2^- production may mollify ischemia/reperfusion injury [4, 5, 9].
The aim of the present studies was to examine the role of kinetics of NO and O_2^- release on the ischemic injury of the brain. In these *in vivo* experiments, NO release in the brain was measured with a porphyrinic microsensor and O_2^- concentration was monitored simultaneously with a superoxide-sensitive electrochemical microsensor.

Materials and Methods

Animals

Male, Sprague-Dawley (SD) rats were treated with the nitric oxide synthase inhibitor L-NNA (N^G-Nitro-L-Arginine) in drinking water at a concentration of 120 mg/L, giving a dose of approximately 20 mg/kg/d for 14 days. Control rats were given only tap water.
In addition to SD control and L-NNA-treated rats, male spontaneously hypertensive stroke-prone (SHR-SP) and normotensive Wistar-Kyoto (WKY) rats were used in this study. All animals weighted 250-300g, had access to food and water *ad libitum* and were housed in individual cages.
All experimental procedures were performed in accordance with the National Institutes of Health *Guide for the Care and Use of Laboratory Animals*.

Production of Infarcts

The rats were given anesthesia (ketamine hydrochloride, 150mg/kg) and maintained every hour by subcutaneous injection. The animal was placed on a stereotaxic instrument to immobilize the head since head movements could influence the results by generating piezoelectric noise signals in electrochemical NO and O_2^- sensors. The right middle cerebral artery was occluded according to procedures described previously [10]. Briefly, the artery was exposed using microsurgical techniques and clamped with a microvascular clip. The position of the occlusion was localized near the origin of the striate branch of the MCA. After 120 minutes of brain ischemia, the artery clamp was released to restore the blood flow for about 60 minutes.

Nitric Oxide and Superoxide Sensors

The NO sensor was prepared according to the procedure previously described [6, 8, 11]. Briefly, polymeric film was deposited electrochemically on the carbon fiber electrode's tip from a solution of monomeric Ni(II) tetrakis (3-methoxy-4-hydroxyphenyl) porphyrin. The sensor surface was then coated with a Nafion film (5% w/w in alcohol) for 15-20 s. The microsensor was calibrated by adding known volumes of NO standard (1.76mmol/L at 0°C) prepared by saturating a 25ml degassed phosphate buffer (pH=7.4) with NO gas.

Similarly, a superoxide microsensor was fabricated following modified procedures described elsewhere [12]. The sensor was calibrated before each measurement by adding known concentrations of aqueous solutions of xantine and xantine oxidase to generate O_2^- radicals.

A detection limit was 1nmol/L for both NO and O_2^- sensors and the response time of sensors was 0.1ms.

In Vivo Measurements of Nitric Oxide and Superoxide

NO and O_2^- were continuously measured for 4h (1h of preischemia, 2h of ischemia, 1h of reperfusion) using amperometry. The microsensors were stereotaxically positioned with a micromanipulator 1mm distal along the MCA from the point of occlusion. Using a standard 28-gauge needle, microsensors were implanted 1mm deep in the parietal lobe of the brain.

In amperometric measurements, potential was kept at a constant level (650 mV for NO sensor and –230 mV for O_2^- sensor). A three-electrode system was used for *in vivo* measurements of NO and O_2^- before (baseline recording), during and after brain ischemia. The system consisted of the working electrode (NO or O_2^- sensor), and a standard silver-silver chloride (SSCE) reference electrode combined into one unit with a platinum counter electrode.

Measurements were carried out with an EG&G potentiostat (Perkin-Elmer, USA) coupled with an IBM compatible computer via data acquisition board. The data were continuously collected by a LabView software and recorded for further processing.

Quantification of Infarct Size

The animals were decapitated at distinct time points (30, 100 and 180 minutes after MCA occlusion). Brains were immediately removed, washed with ice-cold saline solution and placed on dry ice. The frozen brains were cut into several 2mm sagittal slices then stained with the mitochondrial activity indicator – 2,3,5-triphenyl-tetrazolium chloride (TTC, 2% solution in saline at 37°C) for 30 minutes as described in details elsewhere [13]. The images of stained sections were captured by a DC210 Kodak Digital Camera and transferred to the computer for further processing. The brain infarct size was calculated by numeric integration of data from individual slices and presented as percentage of the total brain weight.

Statistical Analysis

Results were presented as mean±SD. Statistical analysis was done by an unpaired Student's *t* test. All analyses were performed using a statistical and graphing package Microcal Origin. Means were considered significantly different at $P<0.01$.

Results

Measurements of NO Release

NO release in the region of the infarct was recorded continuously before, during and after MCA occlusion. Figures 1 and 2 show representative tracings of these recordings from SHR and WKY rat and also from an L-NNA-treated and a control SD rat. After clamping the MCA, a rapid increase in NO concentration was observed in

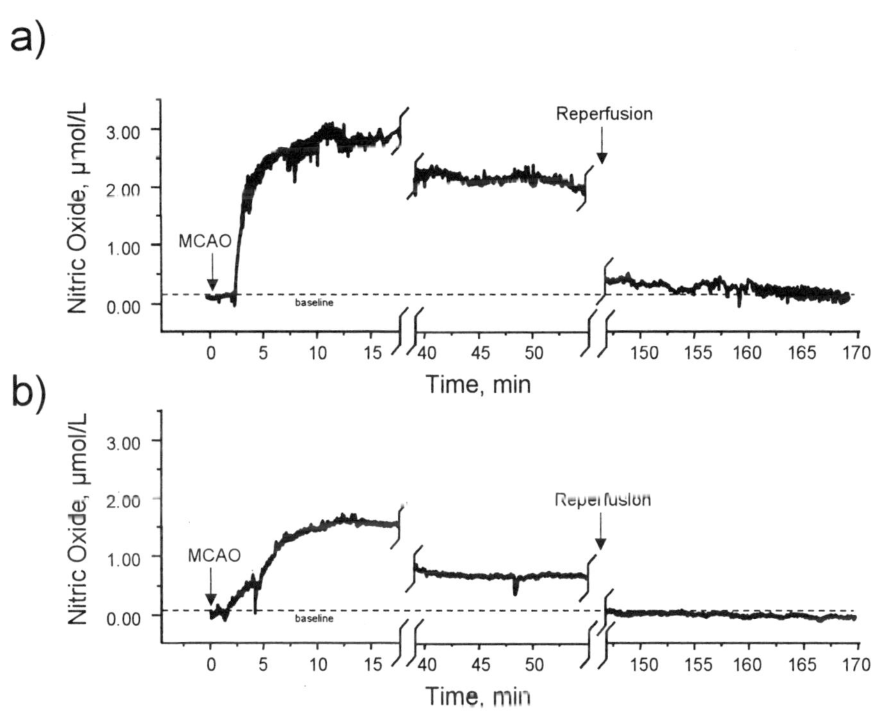

Figure 1. Typical amperograms of nitric oxide release recorded during cerebral ischemia/reperfusion in the striatum of (a) normotensive Wistar-Kyoto and (b) spontaneously hypertensive stroke-prone (SHR) rat.

all rats. The increase in NO concentration following MCAO in SHR and WKY rats was 1.51±0.16 µmol/L and 2.61±0.22 µmol/L ($P<0.01$) respectively.
During the reperfusion NO concentration decreased below the basal level for both WKY and SHR rats. However, only in the brain of SHR the NO concentration after 160 min dropped below the level of 10^{-9} mol/L and was not detectable by the sensor.

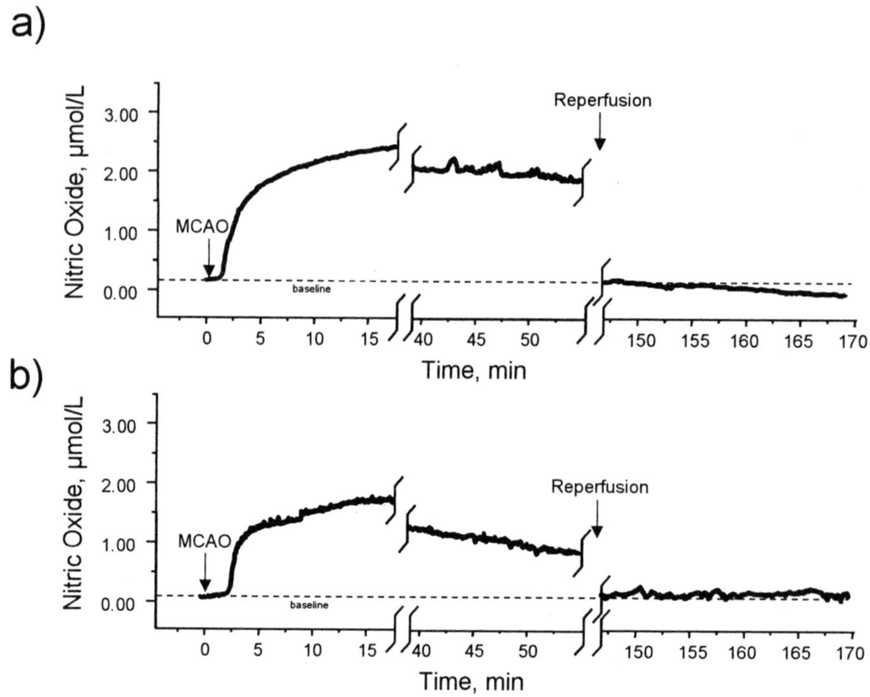

Figure 2. Typical amperograms of nitric oxide release recorded during cerebral ischemia/reperfusion in the striatum of (a) Sprague-Dawley control rat and (b) Sprague-Dawley rat subjected to NO synthase blockade (L-NNA, 20 mg/kg/day).

The maximum NO concentration was observed after 12±3 min for SD control rats and after 14±2 min of the occlusion for L-NNA-treated rats. The mean increment of NO concentration following occlusion in L-NNA-treated rats was 1.49±0.13 µmol/L. This increment in NO concentration was significantly less than that in control SD rats in which the increment was 2.23±0.18 µmol/L ($P<0.01$). The plateau observed during NO release in both treated and untreated SD rats persisted for several minutes and than NO concentration started to decay. A reperfusion (160 min after MCAO) caused a drop in NO concentration below the basal level which was more significant for control SD rats.

Measurements of O_2^- Release

The measurements of O_2^- concentration were conducted in a similar manner to NO measurements. Before MCA occlusion a baseline was recorded followed by measurements of O_2^- during ischemic conditions.

Figure 3 shows O_2^- concentrations recorded as a series of single measurements at distinct time points (0, 20, 40, 100, 120 and 180 minutes from the onset of ischemia). The kinetics of O_2^- release differs from the kinetics of NO. The superoxide concentration reached a semiplateau about 40 minutes after onset of brain ischemia, peaking at 122±24 nmol/L for WKY rats, 220±44 nmol/L for SHR rats, 116±20 nmol/L for SD control rats and 88±19 nmol/L for L-NNA-treated rats.

During reperfusion a significant initial increase in O_2^- concentration was observed reaching 180±36 nmol/L for WKY rats. Significantly higher concentration was observed in SHR rats (297±59 nmol/L). In SD control rats the maximal O_2^- concentration was 191±33 nmol/L (120 min, after reperfusion) which was 71% higher than in L-NNA-treated rats.

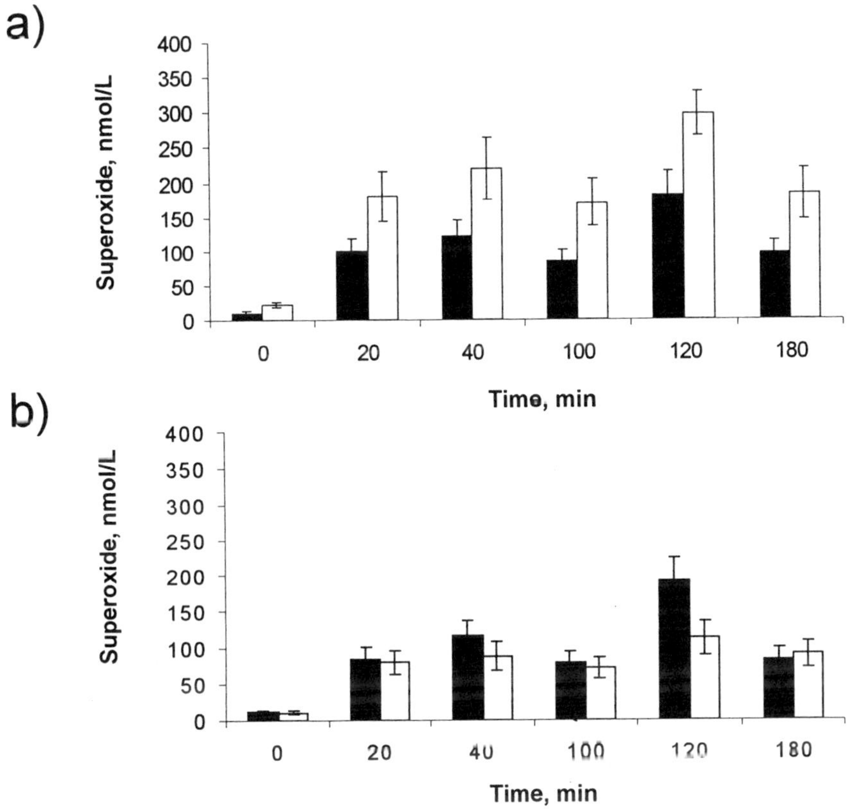

Figure 3. Superoxide concentrations recorded sequentially during ischemia/reperfusion in the striatum of (a) Wistar-Kyoto (solid) and SHR (open) and (b) SD control (solid) and LNNA-treated (open) rats.

Infarct Size Determination

The infarct sizes determined in WKY, SHR, SD control and L-NNA treated rats after brain ischemia are depicted in Figure 4. Infarct sizes were expressed as a percentage of total brain weight. About 30 minutes after MCA occlusion TTC staining showed very small brain damage in WKY (1.1±0.4%), SHR (1.8±0.4%), SD (1.3±0.5%) and L-NNA-treated (1.2±0.5%) rats. After 100 minutes of cerebral ischemia, small but significant increase in brain damage was noticed; 2.2±0.6% for WKY, 2.8±0.8% for SHR, 2.1±0.5% for SD and 1.9±0.5% for L-NNA treated rats. The difference between the infarct size of WKY, SHR and SD control and treated rats was statistically insignificant. The largest infarct size was observed at 180 minute reaching 5.8±0.6%, 9.2±1.1%, 5.3±0.7% and 4.1±1.0 in WKY, SHR, SD control and treated rats, respectively ($P<0.01$).

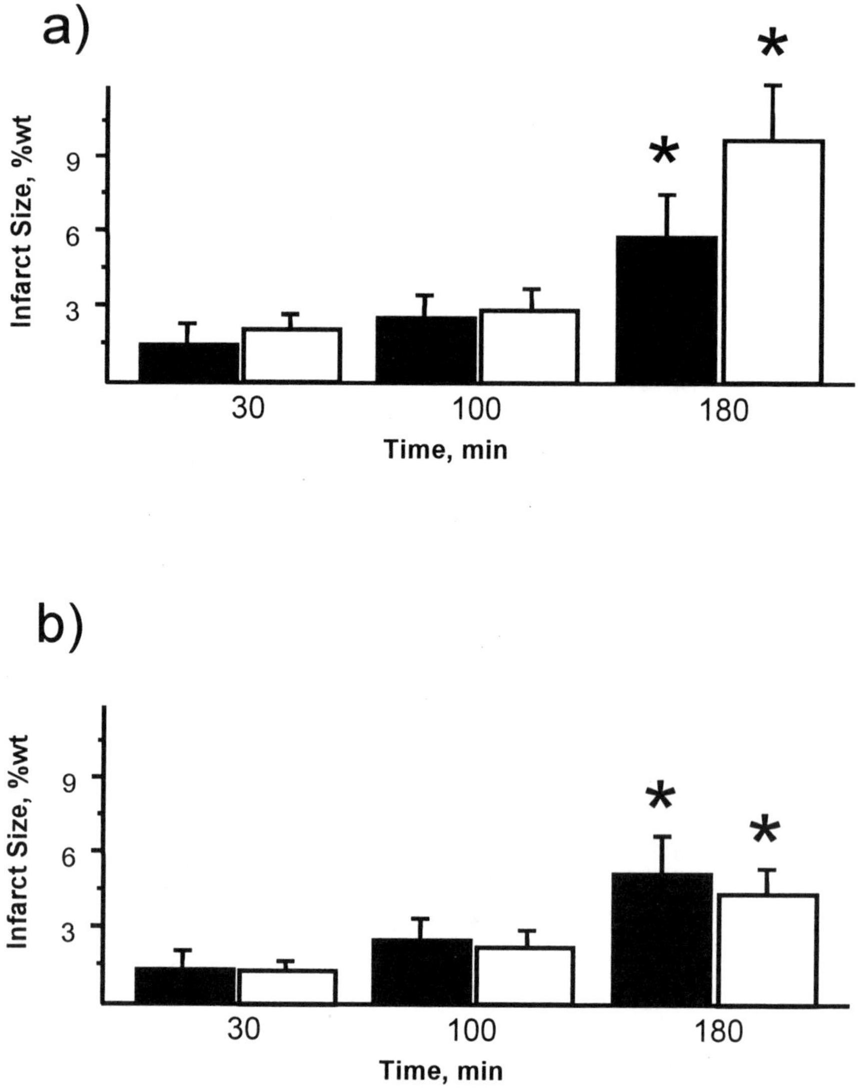

Figure 4. An infarct size (percent of whole brain weight) during cerebral ischemia/reperfusion of (a) Wistar-Kyoto (solid) and SHR (open) and (b) SD control (solid) and LNNA-treated (open) rats.

Discussion

The present study provides the correlation between NO and O_2^- production, and infarct size during transient focal ischemia (2h) and reperfusion (1h).

The rationale underlying our studies is that a deficit in NO release during I/R of the brain is due to significant generation of O_2^- followed by the production of peroxynitrite. This hypothesis is supported by two observations described previously [14]: 1) the measured increase in NO concentration in the brain in response to MCA occlusion in SHR was almost identical to the increase in NO concentration observed in NO-deficient (L-NNA-treated) rats (60% less than the increase in NO concentration observed in control SD and WKY rats). 2) The size of the infarct resulting from MCA occlusion was 23% smaller in NO-deficient (L-NNA-treated) rats than in control (untreated) rats. NO in the brain is released rapidly in response to ischemia. The function of NO release is to decrease the cerebrovascular resistance and to restore blood flow to the ischemic region. An analysis of the kinetics of NO and O_2^- release during I/R indicates a rapid increase in NO concentration during initial period of ischemia. However, prolonged ischemia leads to the decrease of NO release due to generation of superoxide. Parallel to the increase of O_2^- concentration, which reached maximum at 40 min from the onset of ischemia, the rapid decrease in concentration of NO was observed (Figures 1-3). During the reperfusion, the concentration of O_2^- reached maximum causing a significant drop in NO concentration. Nitric oxide and superoxide when produced in close proximity may react together forming the stable peroxynitrite ($ONOO^-$). Peroxynitrite when produced at low concentration usually undergoes protonation and isomerization to form hydrogen cation and nitrate anion. However, under conditions of high $ONOO^-$ production, the protonated peroxynitrite ion can diffuse for a significant distance of several cells and undergo both homolytic and heterolytic cleavage resulting in formation of hydroxyl ($OH^\bullet$) and nitrogen dioxide ($NO_2^\bullet$) radicals and a nitronium cation (NO^+_2). These molecules are among the most reactive and damaging species in biological systems.

During the first minutes of reperfusion, NO release depends not only on availability of substrates for cNOS (L-arginine and O_2) but also on the amount of accumulated O_2^-. Present study shows that the O_2^- generation can be inhibited by certain L-arginine derivatives (like L-NNA) what suggests that like NO, production of O_2^- during pathological conditions is calcium-dependent [8] and is related to the function/dysfunction of NOS. There are also several calcium-independent sources of O_2^-, which may contribute to the basal level of O_2^- such as prostaglandin metabolism, cytochrome P-450, and processes stimulated by protein kinase C or xantine oxidase. These observations are in good agreement with results obtained from infarct size measurements. During the initial phase of cerebral ischemia (at 30 min) the observed infarct size was very small and similar for WKY, SD control, L-NNA-treated and hypertensive SHR rats. 100 minutes after MCA occlusion, shortly before the reperfusion, the infarct size in all groups of rats increased but was still relatively small. However, after 180 minutes of I/R, infarct size measurement revealed significant changes between brain damages of WKY, SHR, SD and L-NNA-treated rats. These results can be clearly correlated with O_2^- concentration.

In comparison to normotensive animals (WKY and SD rats) in SHR rats a deficit in NO concentration was observed probably due to dysfunction of NOS. The dysfunctional NOS can produce high O_2^- concentration. Although in L-NNA-treated SD rats a deficit in NO release was also observed, the O_2^- release was about 50% less than in SHR rats.

The damage of the brain directly correlates with the increase of O_2^- production during ischemia/reperfusion and does not correlate with initial generation of NO.

The production of O_2^- depends on the degree of dysfunction of NOS. The dysfunction of NOS can be due to extensive NO release during prolonged period of ischemia and/or pathological changes of endothelium (SHR). NOS inhibitors change the kinetics of NO release and limit NO production which leads to the decrease of the dysfunction of NOS, decrease of O_2^- production and decrease of ischemic damage of the brain.

Acknowledgements

This study has been supported by a grant from the Public Health Service (HL 60900).

References

[1] Dawson, T.M., J. Zhang, V.L. Dawson, and S.H. Snyder, Nitric oxide: cellular regulation and neuronal injury. Prog Brain Res, 1994. **103**: p. 365-9.

[2] Faraci, F.M. and D.D. Heistad, Regulation of the cerebral circulation: role of endothelium and potassium channels. Physiol Rev, 1998. **78**(1): p. 53-97.

[3] Lipton, P., Ischemic cell death in brain neurons. Physiol Rev, 1999. **79**(4): p. 1431-568.

[4] Buisson, A., I. Margaill, J. Callebert, M. Plotkine, and R.G. Boulu, Mechanisms involved in the neuroprotective activity of a nitric oxide synthase inhibitor during focal cerebral ischemia. J Neurochem, 1993. **61**(2): p. 690-6.

[5] Balkan, E., S. Balkan, T. Ozben, M. Serteser, S. Gumuslu, and N. Oguz, The effects of nitric oxide synthase inhibitor, L-NAME on NO production during focal cerebral ischemia in rats: could L-NAME be the future treatment of sudden deafness? Int J Neurosci, 1997. **89**(1-2): p. 61-77.

[6] Malinski, T., F. Bailey, Z.G. Zhang, and M. Chopp, Nitric oxide measured by a porphyrinic microsensor in rat brain after transient middle cerebral artery occlusion. J Cereb Blood Flow Metab, 1993. **13**(3): p. 355-8.

[7] Zhang, Z.G., M. Chopp, F. Bailey, and T. Malinski, Nitric oxide changes in the rat brain after transient middle cerebral artery occlusion. J Neurol Sci, 1995. **128**(1): p. 22-7.

[8] Huk, I., J. Nanobashvili, C. Neumayer, A. Punz, M. Mueller, K. Afkhampour, M. Mittlboeck, U. Losert, P. Polterauer, E. Roth, S. Patton, and T. Malinski, L-arginine treatment alters the kinetics of nitric oxide and superoxide release and reduces ischemia/reperfusion injury in skeletal muscle [see comments]. Circulation, 1997. **96**(2): p. 667-75.

[9] O'Neill, M.J., C. Hicks, M. Ward, and J.A. Panetta, Neuroprotective effects of the antioxidant LY231617 and NO synthase inhibitors in global cerebral ischaemia. Brain Res, 1997. **760**(1-2): p. 170-8.

[10] Coyle, P., Middle cerebral artery occlusion in the young rat. Stroke, 1982. **13**(6): p. 855-9.

[11] Malinski, T. and Z. Taha, Nitric oxide release from a single cell measured in situ by a porphyrinic-based microsensor [see comments]. Nature, 1992. **358**(6388): p. 676-8.

[12] Lvovich, V. and A. Scheeline, Amperometric Sensors for Simultaneous Superoxide and Hydrogen Peroxide Detection. Anal Chem, 1997. **69**: p. 454-462.

[13] Goldlust, E.J., R.P. Paczynski, Y.Y. He, C.Y. Hsu, and M.P. Goldberg, Automated measurement of infarct size with scanned images of triphenyltetrazolium chloride-stained rat brains. Stroke, 1996. **27**(9): p. 1657-62.

[14] Kidd, G.A., L.W. Dobrucki, V. Brovkovych, D.F. Bohr, and T. Malinski, Nitric oxide deficiency contributes to large cerebral infarct size. Hypertension, 2000. **35**(5): p. 1111-8.

Nitric Oxide
R.J. Gryglewski and P. Minuz (Eds.)
IOS Press, 2001

Endogenous Inhibition of NOS and the ADMA/DDAH Pathway

Patrick Vallance
Centre for Clinical Pharmacology & Therapeutics,
University College London, WC1E 6JJ, UK

1. Introduction

N^G monomethyl-L-arginine (L-NMMA) was the first widely used inhibitor of nitric oxide synthases (NOS). It inhibits all three isoforms of NOS and its effects can be reversed by provision of excess L-arginine. When administered to humans L-NMMA causes arterial and arteriolar vasoconstriction and increases blood pressure. It also reverses the hypotension associated with severe sepsis [1,2]. In animal models or *in vitro* L-NMMA has been shown to increase platelet and white cell adhesion to endothelium, increase vascular smooth muscle cell growth and enhance the formation of artheroscleritic-like lesions [3]. However, L-NMMA is more than just a useful pharmacological tool, it is a naturally occurring amino acid that circulates in plasma and is excreted in urine [4]. Three endogenous methylarginines have been identified – L-NMMA, asymmetric dimethylarginine (ADMA) and symmetric dimethylarginine (SDMA). In each case it is the guanidino nitrogens of arginine that are methylated (Figure 1). Like L-NMMA, ADMA inhibits all three isoforms of NOS [4]. In contrast, SDMA is not a NOS inhibitor. Thus it has become apparent that it is the asymmetric methylation of arginine that determines whether the analogue is a NOS inhibitor. The IC_{50} of L-NMMA and ADMA for NOS is in the order of 1-10μM depending on the prevailing concentration of substrate (L-arginine) [5].

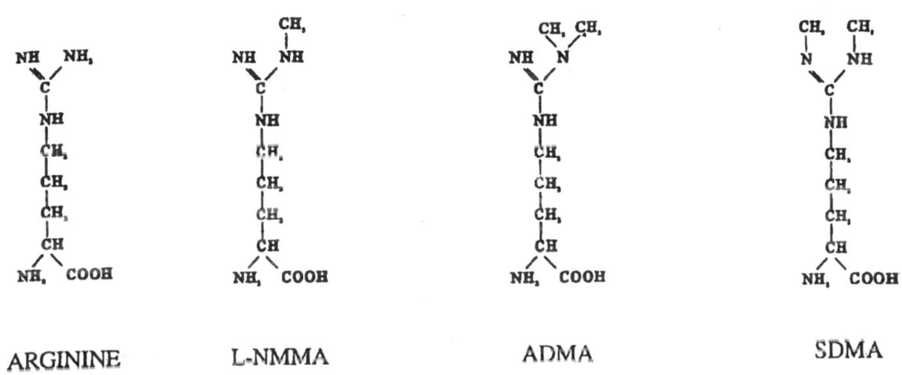

Figure 1. Structure of L-arginine and endogenous methylarginines, N^G monomethyl-L-arginine (L-NMMA), N^G N^G dimethylarginine (ADMA) and N^G N^G dimethylarginine (SDMA). Only L-NMMA and ADMA are nitric oxide synthase inhibitors.

2. Circulating Methylarginines

The concentration of ADMA and SDMA circulating in plasma is in the order of 0.5–2μM. L-NMMA is present in far lower amounts [4]. All of the methylarginines enter cells through the Y^+ amino acid transport system and tend to get concentrated within cells [6]. Although the major route of elimination of methylarginines is by metabolism (see below), free methylargines are also found in urine. The concentration of ADMA in urine may reach 20μM or greater in some situations [4]. Because methylarginines are excreted in urine it was speculated that the plasma concentration would increase in patients with renal failure. This is indeed the case and it is possible that the accumulation of ADMA may be sufficient partially to block NOS and thereby contribute to some of the pathophysiology of this condition [4]. In particular, chronic accumulation of ADMA would be expected to cause vasocosntriction and may enhance atherogenesis. Following this initial observation many groups have detected changes in the levels of ADMA in a wide range of conditions (for review see reference 7). In the cardiovascular field, increased plasma ADMA concentrations have been detected in hypercholesterolaemia, certain types of hypertension, and in individuals with atherosclerotic disease [8]. Indeed, one study has suggested that the ADMA level correlates better with evidence of vascular damage and impaired endothelium-dependent relaxation than the levels of classical risk factors such as cholesterol [9]. However, even when ADMA accumulates the absolute levels remain relatively low, especially in comparison to the amounts of arginine. This led to exploration of what what might be happening inside cells, rather than in the circulating blood.

3. Routes of Synthesis of Methylarginines

Arginine residues in proteins are methylated by the action of protein arginine methyl transefrases (PRMTs). Some PRMTs generate SDMA residues, others asymmetrically methylate arginine to produce L-NMMA and then ADMA [7]. The free methylarginines are liberated upon hydrolysis of the protein. The function of methylation of arginine residues in proteins is not understood but it is possible that methylation affects protein function analogous to the effects of phosphorylation. It is not known whether generation of methylarginines varies greatly over time, but it seem unlikely that variation in rates of synthesis would provide a mechanism to affect local or circulating concentrations of methylarginines. Within the cardiovascular system it has been shown that endothelial cells generate substantial amounts of ADMA , but not SDMA [7].

4. Routes of Degredation of Methylarginines

Whilst SDMA is excreted in urine largely unchanged, the major route of elimination of the asymmetric methylarginines from the body is by metabolism (Figure 2). Two routes of metabolism have been identified of which the major seems to be the enzyme dimethylarginine dimethylaminohydrolase (DDAH) [7]. DDAH metabolises ADMA and L-NMMA to citrulline and dimethylamine or monomethylamine respectively. Thus there is an enzymatic system for metabolising the methylarginines that inhibit NOS, which does not metabolise the "innocent" symmetric methylarginine SDMA. Recently two isoforms of DDAH have been discovered [10]. DDAH-I is expressed predominantly in the brain, whereas DDAH-II has a more cardiovascular pattern of expression. Both isoforms are found in high concentration in the kidney and to a lesser extent the liver. Broadly DDAH-I

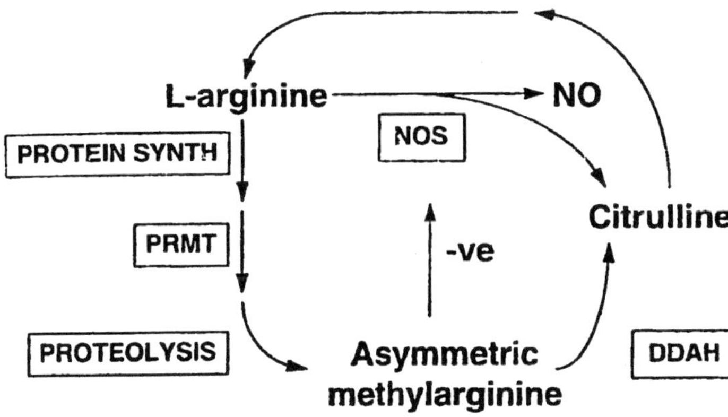

Figure 2. Proposed model for the regulation of intracellular asymmetric methylarginine and NO synthesis. Free arginine is metabolised to nitric oxide (NO) and citrulline by nitric oxide synthase (NOS). Arginine incorporated into proteins during protein synthesis can be methylated by PRMT enzymes to generate asymmetric methylarginines (L-NMMA and ADMA) which are subsequently released into the cytoplasm following proteolysis. Free intracellular L-NMMA and ADMA inhibit the production of NO catalysed by NOS. Asymmetric methylarginine is metabolised by DDAH to citrulline. Citrulline can be converted back to arginine in a two-step reaction catalysed by enzymes of the urea cycle.

expression is more similar to that of neuronal NOS whereas DDAH-II is more similar to that of endothelial NOS [10]. Non-selective inhibition of DDAH I & II in endothleial cells leads to accumulation of ADMA and inhibition of DDAH in intact blood vessels causes an endothelium-dependent contraction, presumably due to indirect inhibition of NOS as the ADMA levels rise. These studies show that an active DDAH system is essential to avoid tonic inhibition of NOS by accumulation of endogenous methylarginines. Interestingly, relatively large amounts of arginine are required to reverse the effects of DDAH inhibition, suggesting that once ADMA or L-NMMA have inhibited NOS rather large amounts of arginine are required to reverse the effect. This is consistent with pharmacological studies using L-NMMA and perhaps might help explain the arginine paradox – the observation that supplying excess arginine boosts nitric oxide generation in certain situations in which the K_m of NOS for arginine would suggest that arginine should not be rate-limiting for nitric oxide generation.

5. How Might Disease Affect DDAH?

Although it is possible that disease states may alter rates of synthesis of methylarginines, it seems that disease effects on DDAH are more plausible (see above). First, genetic variation in DDAH may alter enzyme expression or function and this may contibute to pathophysiology. This is a possibility my group is actively pursuing. Second, the local chemical environment may alter DDAH expression or activity. Already there is some evidence that DDAH-II is transcriptionally regulated and can be induced by retinoids (unpublished observations). In addition, studies have shown that tumour necrosis factor alpha or oxidised LDL can decrease DDAH activity [11]. It remains to be determined whether this effect is mediated by transcriptional regulation or by oxidative stress affecting enzyme function. Again, my group has preliminary data to indicate that DDAH is susceptible to oxidative stress. Perhaps when the crystal structure of DDAH has been fully solved this will shed light on how such an effect may be mediated. The genes for DDAH-I and II map to chromosomes 1 and 6 respectively [11]. The region of chromosome 6 where DDAH-II is found has been linked to diabetes and rheumatoid arthritis [12].

6. Conclusions

The ADMA/DDAH pathway has emerged as a potential regulator of NOS activity, particularly in disease states. However, it is also possible that a selective inhibitor of DDAH-I or II might lead to tissue- or cell-specific decreases in nitric oxide generation that might have therapeutic potential. The importance of this pathway as a regulatory system or as a target for drug treatments will become clearer as kncokout mice become available and specific small molecule inhibitors are developed.

Acknowledgements

Work in my laboratory on DDAH is funded by the British Heart Foundation

References

1. Vallance P, Collier J Moncada S. Effects of endothelium-derived nitric oxide on peripheral arteriolar tone in man. Lancet 1989;8670:997–1000.
2. Petros A, Lamb G, Leone A et al. Effects of a nitric oxide synthase inhibitor in humans with septic shock. Cardiovasc Res 1994;28:34–39.
3. Cayatte AJ, Palacino JJ, Horten K, Cohen RA. Chronic inhibition of nitric oxide production accelerates neointima formation and impairs endothelial function in hypercholesterolemic rabbits. Arterioscler Thromb 1994;14:753–759.
4. Vallance P, Leone A, Calver A et al. Accumulation of an endogenous inhibitor of nitric oxide synthesis in chronic renal failure. Lancet 1992;339:572–575.
5. MacAllister R, Whitley G, Vallance P. Effects of guanidino and uremic compounds on nitric oxide pathways. Kid.Int. 1994;45:737–742.
6. Bogle RG, MacAllister RJ, Whitley G, Vallance P. Induction of N^G-monomethylarginine uptake — a mechanism for differential inhibition of NO synthases. Am J Physiol 1995;269:C750–C765.
7. Leiper J, Vallance P. Biological significance of endogenous methylarginines that inhibit nitric oxide synthases. Cardiovascular Research 1999;43:542–548.
8. Boger RH, Bode-Boger SM, Szuba A et al. Asymmetric dimethylarginine (ADMA): A novel risk factor for endothelial dysfunction. Its role in hypercholesterolaemia. Circulation 1998;98:1842–1847.
9. Miyazaki H, Matsuoka H, Cooke JP et al. Endogenous nitric oxide synthase inhibitor — A novel marker of atherosclerosis. Circulation 1999;99:1141–1146.
10. Leiper J, MacAllister R, Whitley G, Santa Maria J, Chubb A, Charles I, Vallance P. Identification of two human dimethylarginine dimethylaminohydrolases with distinct tissue distributions and homology to microbial arginine deiminases. Biochem. J 1999;343(I):209–214.
11. Ito A, Tsao PS, Adimoolam S, Kimoto M, Ogawa T, Cooke JP. Novel mechanism for endothelial dysfunction: dysregulation of dimethylarginine dimethylaminohydrolase. Circulation 1999;99:3092–3095.

Role of Endothelial Nitric Oxide in Pleiotropic Action of Cardiovascular Drugs: Nebivolol

Ryszard J. Gryglewski, Wojciech Uracz, Ewa Marcinkiewicz, Andrzej Jakubowski, Tadeusz Malinski*
*Chair of Pharmacology, Jagiellonian University Medical College, Grzegórzecka 16, 31-531 Cracow, Poland and *Edison Biotechnology Institute, Konneker Research Laboratories, The Ridge Bldg 25, Ohio University, Athens, Ohio 45701*

Abstract: Some of cardiovascular drugs such as statins (e.g. simvastatin or atorvastatin), "tissue type" inhibitors of angiotensin converting enzyme (quinapril, perindopril), thienopyridines (ticlopidine, clopidogrel) or β-adrenoceptor antagonists (nebivolol, sotalol) are able to exert pleiotropic endothelium-dependent pharmacological actions. Among others, these actions consist in releasing endothelial mediators like nitric oxide (NO) or prostacyclin (PGI2). We proved that in vivo all four classes of drugs would produce pleiotropic thrombolytic effect associated with activation of the vascular endothelium. Here, we focused on nebivolol. In humans nebivolol, apart from its β-adrenoceptor blocking properties, induces endothelium-dependent vasodilatation. In cytosole of cultured bovine aortic endothelial cells (BAEC) nebivolol in a concentration-dependent manner (1–100 μM) increases free cytosolic calcium levels [Ca2+]i. Along with increasing free cytosolic calcium ions nebivolol releases nitric oxide (NO), as detected on the surface of cultured endothelial cells by a porphyrinic electrode microsensor. The release of NO by nebivolol (10 μM) from BAEC is suppressed by pre-treatment with either ethylene glycol-bis(β-aminoethylether)-N,N,N′,N′-tetraacetic acid (EGTA, 2mM), and inhibited by NG-nitro-L-arginine methyl ester (L-NAME, 100 μM). Removal of extracellular calcium ions as a source for a rise in [Ca2+]i affects the nebivolol NO releasing capacity to a lesser degree than that of A 23187 ionophore. Still, the constitutive endothelial NO synthase (NOS-3) seems to be the only isoenzyme involved in the NO release by nebivolol from BAEC, because nebivolol neither activates nor induces NOS-2 in cultured mouse macrophage line J774.2. This selective activation of endothelial NOS-3 by nebivolol may explain its vasodilator effect in the perfused guinea pig heart. Indeed, nebivolol at concentrations of 10 μM and 30 μM significantly increases coronary blood flow in hearts perfused at a constant pressure, and this effect is significantly inhibited by L-NAME (100 μM). *In vivo* in rats nebivolol (1–10 mg kg -1 intravenously) produces short lasting thrombolytic (30 min) and hypotensive (15 min) responses similar to those evoked by NO-donors such as S-nitroso-N-acetyl penicillamine or sodium nitroprusside given intravenously at doses of 30 μg kg-1. Unlike nebivolol, statins (e.g. atorvastatin) and "tissue type" ACE inhibitors (e.g. quinapril) evoke no hypotensive effect, while their thrombolytic action stays much longer (> 3 hrs) and is associated with endothelial PGI2 release. Summing up, both in vitro and in vivo nebivolol shows actions associated with [Ca2+]i - dependent stimulation of endothelial NOS-3.

1. Introduction

At least four classes of cardiovascular drugs produce endothelial pleiotropic effects, which are not inevitably associated with their principal mechanism of action. Statins, such as lovastatin, simvastatin or atorvastatin are hypocholesterolaemic drugs which act as inhibitors of hepatic 3-hydroxy-3-methyl-glutaryl coenzyme A reductase. However, in patients with atherosclerosis statins are claimed to stimulate endothelial NO secretion through "extralipid", peripheral mechanisms [1–4]. These pleiotropic properties of statins might be associated with inhibition of prenylation of enzymic or signalling proteins. "Tissue type" inhibitors of angiotensin I converting enzyme (ACE-I) such as quinapril, perindopril or ramipril are more than just drugs for treatment of arterial hypertension. In high risk patients ramipril reduces death rate from myocardial infarction and ischaemic stroke [5]. In patients with coronary artery atherosclerosis quinapril reverses endothelial dysfunction by improving NO-mediated vasodilator responses to acetylcholine [6]. We reported that "tissue type" ACE-I were exerting *in vivo* thrombolytic action mediated via endogenous bradykinin, endothelial B_2 receptor and prostacyclin [7]. Thienopyridines (ticlopidine and clopidogrel) compete with aspirin as antiplatelet drugs in the treatment of ischaemic stroke, myocardial infarction and peripheral artery disease [8]. The success of thienopyridines in the treatment of patients with atherothrombosis [9;10] can hardly be explained only by a blockade of cysteinyl moiety of $P2t_{AC}$ receptor for ADP on blood platelets [11]. We claim that endothelium-mediated thrombolytic properties of thienopyridines do play an important role in their therapeutic efficacy [12;14]. A number of reports indicate that nebivolol, a lipophylic selective, β_1-adrenoceptor antagonist [15], apart from its β-adrenolytic properties may produce endothelium-dependent vasorelaxation both in canine coronary arteries [16] and in healthy volunteers [17;18], most likely through a NO-dependent mechanism. Here, we study mechanisms by which nebivolol-activated endothelial cells may release NO.

2. Methods

2.1 Cytosolic Free Calcium [Ca2+]i in Cultured Endothelial Cells of Bovine Aorta (BAEC)

Endothelial cells were harvested from bovine thoracic aorta (BAEC) after treatment with 0.05% (w/v) trypsin and 0.02% (w/v) EDTA [19], pooled and seeded in T25 flasks (NUNC, Brand Products, Denmark) precoated with 1% gelatine. After 1 h of incubation at 37°C in CO_2, non-adherent cells were removed. Then BAEC were cultured in OPTI-MEM I, supplemented with foetal bovine serum (4% v/v) and with antibiotics [20], and incubated at 37°C in humidified air containing 5% CO_2. The culture media were replaced three times a week. Upon reaching confluence, cultured endothelial cells were passaged (trypsin/EDTA at a ratio of 0.05%/0.02% w/v for 1 min at 37°C), and reseeded into T80 culture flasks. Primary cultures of endothelial cells usually reached confluence between 4 to 6 days. Cells between 2nd and 4th passage were used for experiments. Their homogeneity was identified by their typical cobblestone morphology under the Axiovert 25 Inverted Microscope (Carl Zeiss Jena GmBH, Germany). Cytosolic calcium ion $[Ca^{2+}]i$ assay was accomplished by the method of Grynkiewicz et al. [21], namely BAEC (0.5×10^6 cells ml^{-1}) were loaded with fura-2 by incubation for 60 min at 25°C with this fluorescent dye in its membrane permeant form, i.e. acetoxymethyl ester (fura-2AM) in presence of bovine serum albumin. Then the extracellular dye was removed by centrifugation and cells were resuspended in HBS containing glucose (5 mM). Fluorescence was measured at 37°C in a spectrofluorimeter with a dual wavelength excitation, and magnetic stirring (LS 50B, Perkin-Elmer, Corporation, Beaconsfield, U.K.); at 500 nm with the excitation wavelength of 340 nm and 380 nm. Calibration was completed using 0.2% Triton X-100 for R_{max} and 5 mM EGTA for R_{min}. Eventually, levels of cytosolic calcium $[Ca^{2+}]i$ were calculated according to the equation of Grynkiewicz et al. [21]. Samples showing basal $[Ca^{2+}]i$ higher than 100 nM were rejected. Net effects of reference substance or nebivolol were calculated by a computer-assisted subtraction of control $[Ca^{2+}]i$. Bradykinin (10 nM) was used as a reference compound, while nebivolol was applied at a range of concentrations of 1 - 30 µM.

2.2 *Endothelial Release of Nitric Oxide as Measured with the Malinski's Electrode*

Nitric oxide (NO) was measured using a porphyrinic microsensor which was prepared by cyclic voltametric scanning that deposited a film of polymeric porphyrin (Ni(II) tetrakis(3-methoxy-4-hydroxyphenyl-porphyrin) on a thermally sharpened carbon fibre electrode, then coated with nafion as described previously [22]. The microsensor working electrode was placed close (20 ± 5 µm) to the surface of confluent monolayer of BAEC along with a platinum wire counterelectrode and a saturated calomel reference electrode. Voltametric analyser (PAR model 264A) with the current-sensitive preamplifier (PAR model 181) were used for amperometric measurements at a potential of 0.68 V. NO measurements were standardised using aqueous NO standard prepared as described by Jia and Furchgott [23]. Nitric oxide was measured over a single BAEC in Hank's balanced solution at 37 °C. The response time was 0.1 ms and the detection limit 10^{-9} M of NO. Surface concentration of NO was expressed in a conventional manner (nM) as measured at the tip of the electrode [24]. Nebivolol was dissolved in dimethylsulfoxide (DMSO) as a stock solution. In the experiments with NO synthase inhibitors, BAEC were pre-treated for 30 minutes with N^G-nitro-L-arginine methyl ester (L-NAME, 100 µM) or N^G-nitro-L-arginine (L-NNA). In some experiments $CaCl_2$ was omitted from the buffer and 2 mM ethylene glycol-bis(β-aminoethylether)-N,N,N′,N′ -tetraacetic acid (EGTA) added.

2.3 *Isolated Perfused Guinea Pig Heart by the Langendorff Method*

Coronary circulation of guinea pig hearts was retrogradely perfused under steady pressure of 60 mm Hg through the aorta. Coronary flow (CF) and myocardial contractility (dt/dp_{max}) were measured by a computerised system (Hugo Sachs Electronics) as previously described [25]. In this system, acetylcholine hydrochloride at a dose of 100 pmoles given as bolus injection was used as a reference substance for the NO-mediated coronary vasodilatation. Nebivolol was infused into coronary circulation during 1 min at concentrations of 10 or 30 µM. To check the endothelium-dependence of coronary responses to Ach or to nebivolol, the hearts were pre-treated with L-NAME at a concentration of 100 µM.

2.4 *Mouse Macrophage Cell Line J774.2*

The mouse macrophage cell line J774.2 was cultured in T75 flasks in DMEM that contained 10% foetal bovine serum supplemented with streptomycin (100 µg ml^{-1}), penicillin (100 U ml^{-1}) and fungizone (0,25 µg ml^{-1}). Flasks were kept at 37°C in atmosphere of humidified air containing 5% CO_2. Then, the cells were seeded in 96-well plates and cultured in 200 µl of culture medium until reaching confluence (approximately

10^5 cells per well). Nitric oxide synthase (NOS-2) in macrophages was induced by lipopolysaccharide from *Escherichia coli* (LPS, 1 μg ml^{-1}). Nitrite accumulation in the J774.2 supernatant was measured 24 hours after LPS administration using the Griess method [26]. Briefly, 100 μl of 1% sulphanilamide in 5% phosphoric acid, followed by 100 μl of 0,1% N-(1-naphtyl)-ethylene diamine in 5% phosphoric acid were added to 100 μl of culture medium. After 10 minutes of incubation at 23°C the absorbance at 550 nm was read. Micromolar concentrations of nitrite were calculated from standard curve constructed with sodium nitrate as a reference compound. Nebivolol (0,1 μM – 30 μM, dissolved in 0,03% DMSO) was given either 15 minutes prior to, or 10 hours after LPS (1 μg ml^{-1}) [27].

2.5 Thrombolysis in Extracorporeal Circulation of Anaesthetised Rats

In vivo cat model for studying of thrombolytic properties of drugs [28] was adapted to rats [13;14;29]. Briefly, male Wistar rats body weight 300 – 350 g were anaesthetised (thiopental 30 mg kg-1 i.p.) and unfractionated heparin at a dose of 800 units kg-1 i.v. was administered. The extracorporeal circulation was established between left carotid artery and left jugular vein, and a collagen strip was superfused with arterial blood at a rate of 1.5 ml min-1. Its weight was continuously monitored by an auxotonic Harvard transducer. Due to deposition of thrombi [28;30] the strip gained in weight by 80–120 mg during the first 20 min of superfusion and remained at that level for the next 3–5 hours of the experiment. Mean arterial blood pressure (BP) was monitored in right carotid artery by a Harvard pressure transducer, and right femoral vein was prepared for drug administration. In the above system dispersion of thrombi occurred after intravenous administration of PGI2 or iloprost (0.1–1.0 mg kg^{-1}), glyceryl trinitrate (30-100 mg kg^{-1}), metacholine hydrochloride (10 mg kg^{-1}) or kallikrein (100 units kg-1) whereas aspirin at doses of 5–50 mg kg^{-1} did not evoke thrombolysis. Streptokinase (3–30 megaunits kg^{-1}) produced biphasic thrombogenic / thrombolytic response [13].

2.6 Statistical Analysis

Arithmetical means are given with s.e. mean. Differences between groups were assessed with paired Student's t test or one-way analysis of variance with Bonferroni's multiple comparison test using GraphPad Statistical Package (Prism 3, USA).

3. Results

3.1 Effects of Nebivolol on [Ca^{2+}]i in Cultured BAEC

In BAEC of the second passage with basal levels of [Ca^{2+}]i varying from 60 nM to 100 nM nebivolol at a range of concentrations from 1 to 30 μM raised [Ca^{2+}]i to the maximum levels ranging from 120 to 260 nM. The maximum effect of nebivolol at the highest concentration used (30 μM) on [Ca^{2+}]i was close to the effect triggered by bradykinin at concentrations of 10 nM. Examples of [Ca^{2+}]i responses of BAEC to bradykinin and nebivolol are shown in Fig. 1.

3.2 Effect of Nebivolol on the Release of NO by Cultured BAEC

Nebivolol in a concentration-dependent manner (0.01 – 100 μM) stimulated NO release from BAEC (Fig. 2). Figure 3 shows a high-resolution amperogram recorded after

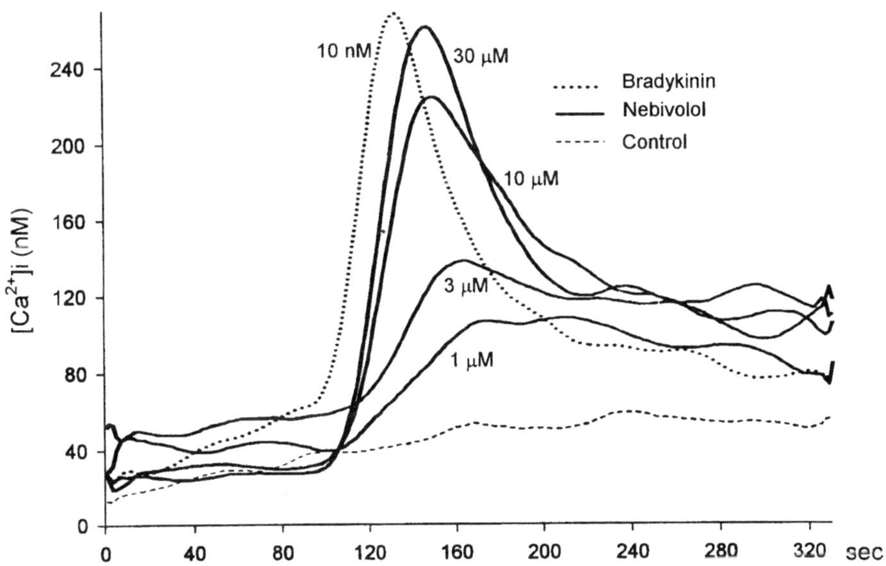

Fig. 1. Increase in intracellular free calcium levels [Ca^{2+}]i in cultured bovine endothelial cells (BAEC, 2nd passage) by nebivolol (1 μM, 3 μM, 10 μM and 30 μM, solid lines), bradykinin (10 nM, dotted line) as compared to control cells (dashed line).

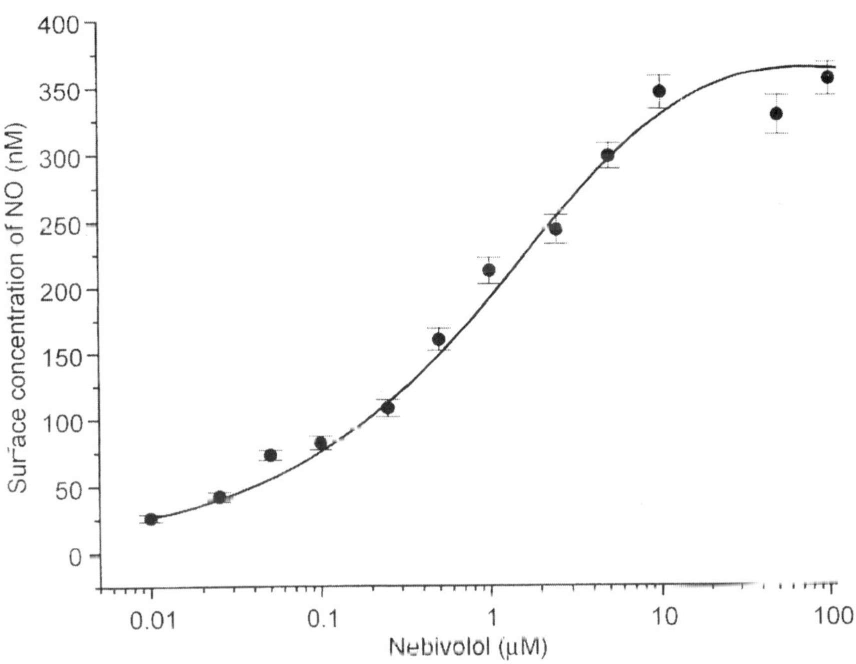

Fig. 2. NO release (nM) from BAEC by various concentrations of nebivolol (μM).

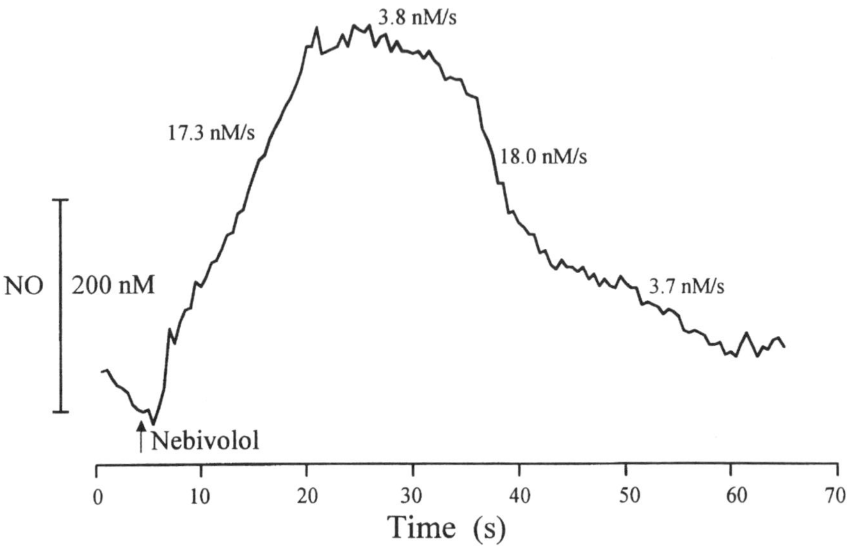

Fig. 3. Typical amperogram showing time-dependent changes in NO concentration (nM) on the surface of a single BAEC after stimulation with nebivolol (10 μM). Figures along the curve represent speed (nM s^{-1}) in increasing or decreasing of NO appearance.

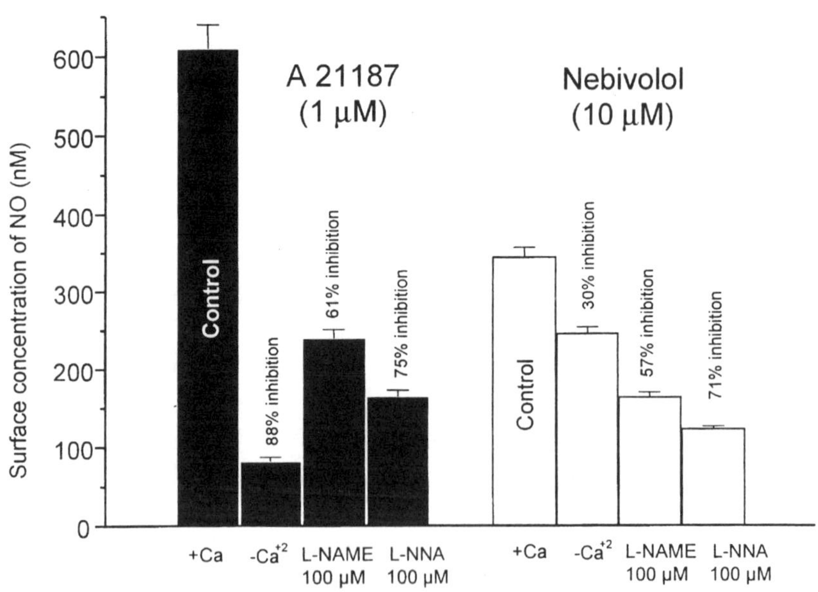

Fig. 4. Comparison of A 23187 (1μM) - and nebivolol (10 μM) – induced NO (nM) release from bovine aortic endothelial cells (BAEC). Effects of extracellular calcium ion removal ($-Ca^{2+}$) or NOS inhibition by L-NAME (100 μM) and L-NNA (100 μM) on the effects of A 23187 and nebivolol. Bars represent SEM. All manipulations resulted in a significant ($p<0.01$) suppression in NO release. Percent of inhibition of NO release vs control values are shown in the figure.

instillation of nebivolol (10 μM) to BAEC. Nitric oxide was released at a peak concentration of 345 ± 12 nM and that represents about 55% of maximum NO concentration being released by calcium ionophore (A23187, 1 μM). The release of NO started one second after instillation of nebivolol. The initial rate of NO release was slow (17.3 nM/s) (above fifteen times slower than recorded for A23187). A semi-plateau was established after 15 s. After about 35 s a slow decrease in NO concentration occurred at a rate of 18.0 nM/s. A linear increase in peak NO concentration was observed along with increasing concentrations of nebivolol (0.01 – 10 μM) (Fig. 2). The peak of released NO was constant for nebivolol at concentrations from 10 to 100 μM, reaching 355 ± 12 nM of NO for nebivolol (100 μM). Pre-treatment with NOS inhibitors decreased NO release both by calcium ionophore A 23187 and by nebivolol to a similar extend: i.e. 61% and 57% for L-NAME and 75% and 71% for L-NNA, whereas the removal of extracellular calcium influences the NO release by A 23187 and by nebivolol unequally (Fig. 4). It suppressed A 23187-induced NO release by 88% and nebivolol-induced NO release by 33% only (Fig. 4) suggesting that calcium influx from extracellular space is only partially involved in the NO-releasing mechanism of nebivolol.

3.3 Effect of Nebivolol on Coronary Flow in Isolated Guinea Pig Heart

Basal coronary flow 6.59 ± 1.13 ml min^{-1} (n = 8) increased immediately during one minute infusions of nebivolol at concentrations of 10 μM or 30 μM. This increase was represented by values of 10.8 ± 1.7 ml min^{-1} and 17.05 ± 2.1, respectively. The vasodilator activity of nebivolol at both concentrations was statistically significant ($p < 0.0005$). Coronary flow also increased following bolus injection of Ach (100 pmoles) to 11.44 ± 1.69 ml min^{-1}. In the presence of L-NAME (100 μM) in the perfusing Krebs' buffer coronary flow decreased to 4.31 ± 0.7 (Fig. 5). Also vasodilator responses to nebivolol at concentrations of 10 μM and 30 μM and Ach (100 pmoles) were reduced by 68%, 65% and by 77%, respectively (Table 1), pointing out to NO as a major mediator of these responses. Basal heart contractility was 745 ± 42 mmHg sec^{-1} (n = 8) and decreased significantly during infusions of nebivolol at concentrations of 10 μM (to 702 ± 34 mmHg sec^{-1}; $p < 0.05$) and 30 μM (to 607 ± 39 mmHg sec^{-1}; $p < 0.01$). Fifteen minutes perfusion with L-NAME decreased dp/dt_{max} to 570 ± 79 mmHg sec^{-1} ($p < 0.01$ vs. basal contractility). Inhibition of nitric oxide synthase by L-NAME caused decrease of inotropic negative effect of nebivolol at both concentrations by about 50% (Table 1).

3.4 Effect of nebivolol on NOS-2 in J774.2 macrophage line

Nebivolol influenced neither activity nor induction of the calcium-independent inducible isoform of nitric oxide synthase (NOS-2) in cultured J774.2 macrophages following their induction by LPS (1 μg/ml). Nebivolol at a range of concentrations from 0,1 μM to 30 μM did not affect nitrite accumulation in cultured macrophages either when given 15 min prior to LPS (Fig. 6, upper panel) or 10 hours after LPS challenge (Fig. 6, lower panel). These results point out to nebivolol as a selective NOS-3 activator (see 3.2 and 3.3) unable to activate or to induce NOS-2.

3.5 Thrombolytic effect of nebivolol in anaesthetised rats

Nebivolol at the doses of 0.1 mg kg^{-1}, 0.3 mg kg^{-1} and 1 mg kg^{-1} caused loss of thrombus weight from 9.7% to 16.2% and 23.3%, respectively (Table 2). The duration of

Table 1. Effects of one minute infusions of nebivolol at concentrations of 10 μM and 30 μM and acetylcholine hydrochloride (100 pmoles, bolus injection) in the absence or presence of L-NAME (100 μM) on the coronary flow (ml min^{-1}) in isolated guinea pig heart. Mean ± SEM from 8 experiments is shown (* $p < 0.001$ as compared to control by paired Student's t test)

Treatment	Coronary Flow (ml min^{-1})
none	**6.59 ± 1.13**
L-NAME	**4.31 ± 0.76**
Nebivolol 10 μM	**10.8 ± 1.70 ***
L-NAME + Nebivolol 10 μM	**5.58 ± 0.81**
Nebivolol 30 μM	**17.05 ± 2.10 ***
L-NAME + Nebivolol 30 μM	**8.01 ± 1.03**
Acetylcholine	**11.44 ± 1.69 ***
L-NAME + Acetylcholine	**5.34 ± 0.95**

Table 2. Effects of intravenous administration of nebivolol at doses of 0.1 mg kg^{-1}, 0.3 mg kg^{-1} and 1 mg kg^{-1} on the maximum loss of thrombus weight (%) and duration of thrombolytic response (min). Mean ± SEM out of 10 experiments are shown. Dose dependency of responses as analyzed by one way ANOVA was significant at a level p<0.001.

	Nebivolol (mg kg^{-1})		
	0.1	0.3	1
Maximum thrombolysis	**9.7** ± 0.5 %	**16.2** ± 1.1 %	**23.3** ± 1.1 %
Duration of thrombolytic response	**67** ± 9 min	**123** ± 12 min	**171** ± 13 min

thrombolytic action of nebivolol was dose-dependent and lasted from 67 to 123 and 171 min, respectively. Fig. 7 shows original tracing of thrombolytic action (THR) of nebivolol (1mg kg^{-1}). Bolus injection of nebivolol caused immediate decrease in blood pressure (BP) that lasted 45–60 min.

4. Discussion

Nebivolol is a new selective β_1-adrenoceptor antagonist combining in its molecule two mechanisms of action. Apart from acting as β_1 receptor blocking agent in myocardium [31], it also shows up as an endothelium-dependent vasodilator [16;17;32–35]. This last

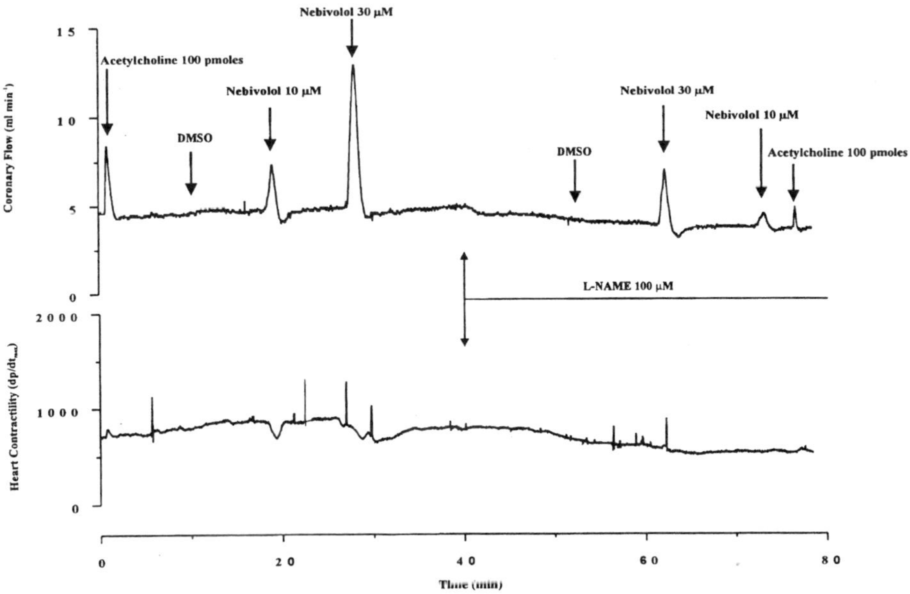

Fig. 5. Original recording of coronary flow (CF) and myocardial contractility (dp/dt) in the guinea pig heart perfused by the Langendorff methods. Effects of acetylcholine hydrochloride (100 pmoles, bolus injection), DMSO (20 μL), nebivolol (10 μM and 30 μM, one minute infusion) in the absence and presence of L-NAME (100 μM) in the buffer.

pharmacological property is likely to be associated with activation of the L-arginine / nitric oxide pathway [18;36;37]. Nebivolol is a d,l-racemic mixture. It seems that d-enantiomer is responsible for its β_1-adrenoceptor antagonistic action while l-enantiomer mediates its endothelium-dependent vasodilator effect [35;38–41]. Interestingly, in the ischaemic reperfused working rabbit heart l-nebivolol improves the cardiodynamic function, revealing dissociation between β-adrenolytic and cardioprotective effects of nebivolol [32]. Nebivolol was also reported to prevent myocardial dysfunction mediated by hydroxyl radicals [42].

Nebivolol at the therapeutical daily doses of 2.5 – 10 mg is well tolerated by patients with arterial hypertension [43;44], coronary insufficiency or congestive heart failure [35]. In contrast with the most of classical β adrenoceptor antagonists, nebivolol at a concentration of 0.1 μM inhibits proliferation of human coronary smooth muscle, perhaps through damping secretion of endothelin [45]. In healthy volunteers the chronic administration of nebivolol decreases systemic oxidative stress as evidenced by a fall in urinary excretion of isoprostanes [46].

Here we report that in cultured endothelial cells (BAEC) d,l-nebivolol causes a rapid, sustained and concentration-dependent (1, 10, 30 and 100 μM) rise in $[Ca^{2+}]i$ (Fig. 1) accompanied by the release of NO on the surface of endothelial cells (Figs. 2,3). In absence of extracellular calcium ions the NO generating potency of nebivolol is suppressed to a much smaller degree than that of A 23817 ionophore (Fig. 4), thus pointing out to a possibility of involvement of intracellular activation of inositol phosphate metabolism by nebivolol as it was freshly proposed [47]. On the other hand, only a partial dependence on extracellular calcium ions for NO-releasing capacity of nebivolol could suggest a partial involvement of inducible NO synthase (NOS-2) along with the major pathway of NOS-3 activation. The presence of transcripts for NOS-2 was reported in HUVEC [7]. However,

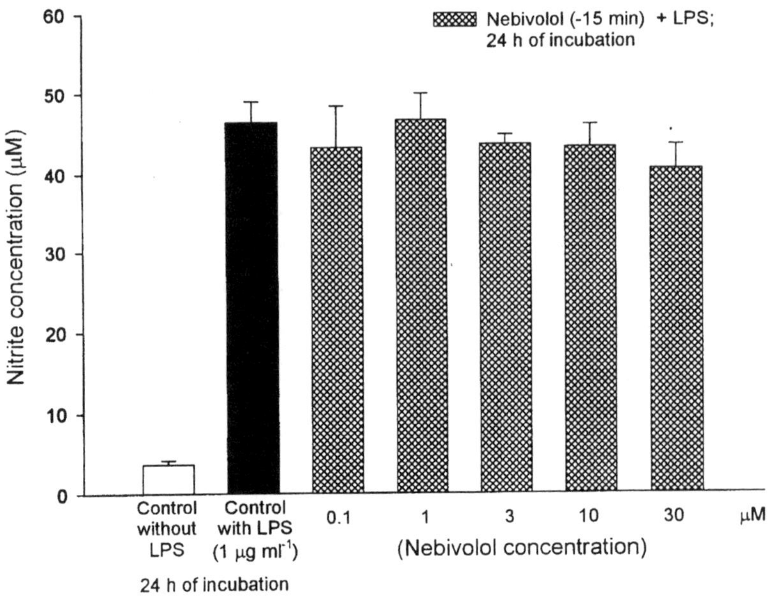

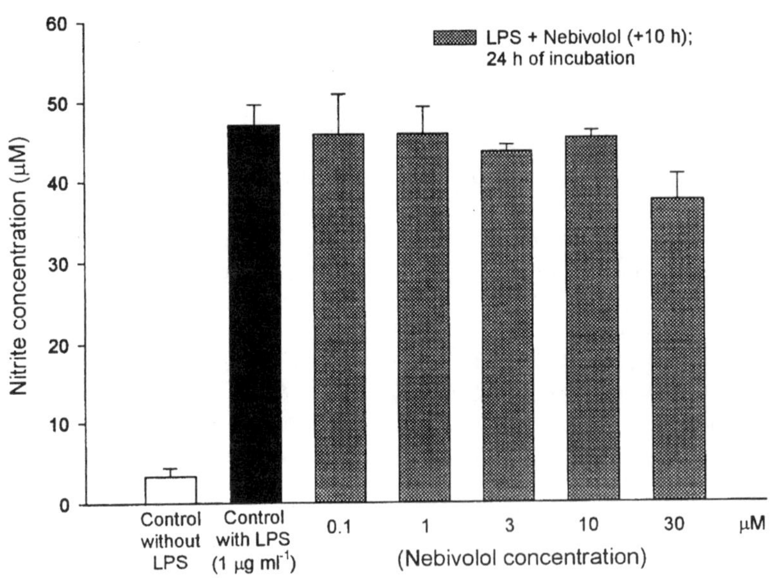

Fig. 6. Influence of nebivolol (0.1 μM – 30 μM) given 15 min prior or 10 hours after LPS (1 μg ml^{-1}) on 24 hours nitrite accumulation in the culture medium of J774.2 macrophages.

the selectivity of nebivolol to activate constitutive NOS-3 was confirmed by a lack of any effect of nebivolol on activation or induction of NOS-2 in cultured mouse macrophage cell line J774.2. Nebivolol - triggered NO production in BAEC was measured on the surface of a single endothelial cell [48]. Nitric oxide production was detected as early as 5 seconds after instillation of nebivolol and was concentration-dependent at a wide range of

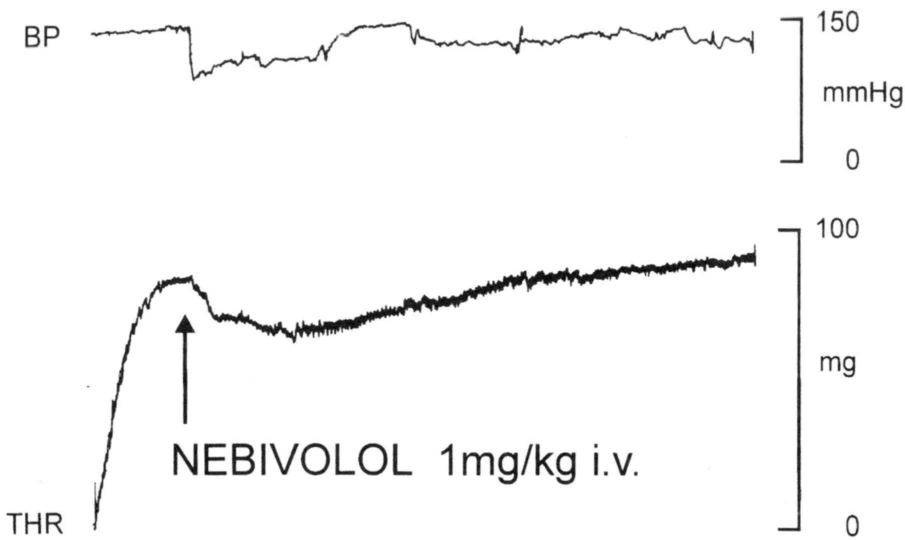

Fig. 7. Original tracings of weight of thrombus (THR) and blood pressure (BP) in a rat with extracorpreal circulation. Effect of nebivolol (1 mg kg^{-1}, i.v) on THR and BP.

concentrations from 0.1 μM to 10 μM. Pre-treatment with either of two NOS inhibitors (L-NAME or L-NNA) prevented the nebivolol-induced NO generation by a single BAEC at the same time confirming the specificity of NO detection by the porphyrinic sensor.

The mechanisms of rising endothelial $[Ca^{2+}]i$ by nebivolol are not clear. It was claimed that vasodilator action of nebivolol engaged involvement of histamine H_2 receptors [49] or serotonin 5-HT_{1A} receptors [50]. Broeders et al. [51] suggest that nebivolol is metabolised to a β_2 adrenoceptor agonist (!), which triggers a rise in $[Ca^{2+}]i$ followed by NO release from endothelial cells. Indeed, in our hands adrenaline (10 μM – 30 μM) produced a significant rise of $[Ca^{2+}]i$ in BAEC [7]. However, an immediate effect of nebivolol on both endothelial rise in $[Ca^{2+}]i$ and release of NO leaves no space for speculation that this drug could have been metabolised in our experiments. Recent discovery of β_3 adrenoceptor localised on rat aortic endothelial cells [52] and in myocardium [53] may offer an explanation of unexpected actions of some adrenergic drugs, like SR 59104A, a selective β_3 adrenoceptor agonist operating in endothelial cells through calcium channels and EDRF(NO) [54]. Still McDonald et al. [55] proposed a differential contribution of classical β and β_3 adrenoceptors to endothelium-dependent vasorelaxant responses. Could it be that one of nebivolol isomers shows a β_3 adrenoceptor stimulating activity on endothelial cells?

Here, we also report that in anaesthetised rats with extracorporeal circulation an intravenous administration of nebivolol produces short-lasting hypotensive and thrombolytic effects, similar to those evoked by exogenous NO-donors like S-nitroso-N-acetylpenicillamine or sodium nitroprusside. Fifteen years ago we reported that another β-adrenoceptor antagonist, sotalol (3 5 mg kg^{-1}), but not the other drugs of this group, would produce thrombolysis in cats similar to nebivolol [56]. It may well be that some other β-adrenolytics share with nebivolol a property of activation of endothelial cells.

Acknowledgements

The State Committee for Scientific Research (grant No 4 PO5A 050 19) and Berlin-Chemie Menarini Group funded this study. Technical help of Barbara Lorkowska and Lucyna Olejniczak are highly appreciated.

References

[1] R.P.Brandes, A.Behra, C.Lebherz, R.H.Boger, S.M.Bode-Boger, A.Mugge. Lovastatin maintains nitric oxide -but not EDHF-mediated endothelium-dependent relaxation in the hypercholesterolemic rabbit carotid artery, *Atherosclerosis*, **142**, (1999) 97–104.

[2] W.H.Kaesemeyer, R.B.Caldwell, J.Huang, R.W.Caldwell. Pravastatin sodium activates endothelial nitric oxide synthase independent of its cholesterol-lowering actions, *J.Am.Coll.Cardiol.*, **33**, (1999) 234–241.

[3] U.Laufs, D.Marra, K.Node, J.K.Liao. 3-Hydroxy-3-methylglutaryl-CoA reductase inhibitors attenuate vascular smooth muscle proliferation by preventing rho GTPase-induced down-regulation of p27(Kip1), *J.Biol.Chem.*, **274**, (1999) 21926–21931.

[4] G.O'Driscoll, D.Green, R.R.Taylor. Simvastatin, an HMG-coenzyme A reductase inhibitor, improves endothelial function within 1 month, *Circulation*, **95**, (1997) 1126–1131.

[5] S.Yusuf, P.Sleight, J.Pogue, J.Bosch, R.Davies, G.Dagenais. Effects of an angiotensin-converting-enzyme inhibitor, ramipril, on cardiovascular events in high-risk patients. The Heart Outcomes Prevention Evaluation Study Investigators [see comments] [published errata appear in *N Engl J Med* 2000;**342**(10):748 and 2000; **342**(18):1376], *New England Journal of Medicine*, **342**, (2000) 145–153.

[6] G.B.Mancini, G.C.Henry, C.Macaya, B.J.O'Neill, A.L.Pucillo, R.G.Carere, T.J.Wargovich, H.Mudra, T.F.Luscher, M.I.Klibaner, H.E.Haber, A.C.Uprichard, C.J.Pepine, B.Pitt. Angiotensin-converting enzyme inhibition with quinapril improves endothelial vasomotor dysfunction in patients with coronary artery disease. The TREND (Trial on Reversing ENdothelial Dysfunction) Study [see comments] [published erratum appears in Circulation 1996 Sep 15;**94**(6):1490], *Circulation*, **94**, (1996) 258–265.

[7] R.J.Gryglewski, S.Chłopicki, W.Uracz, E.Marcinkiewicz. Significance of endothelial prostacyclin and nitric oxide in peripheral and pulmonary circulation, *Med.Sci.Monit.*, **7**(1) (2001) 1–16.

[8] M.Gent. The CAPRIE trial: culmination of the preregistration program for clopidogrel. Clopidogrel versus aspirin in patients at risk of ischaemic events. Clopidogrel versus aspirin in patients at risk of ischaemic events, Seminars in *Thrombosis & Hemostasis*, **25** Suppl 2, (1999) 1–2.

[9] G.J.Hankey, C.L.Sudlow, D.W.Dunbabin. Thienopyridine derivatives (ticlopidine, clopidogrel) versus aspirin for preventing stroke and other serious vascular events in high vascular risk patients, *Cochrane.Database.Syst.Rev.*, (2000) CD001246.

[10] L.A.Harker, J.P.Boissel, A.J.Pilgrim, M.Gent. Comparative safety and tolerability of clopidogrel and aspirin: results from CAPRIE. CAPRIE Steering Committee and Investigators. Clopidogrel versus aspirin in patients at risk of ischaemic events, *Drug Saf*, **21**, (1999) 325–335.

[11] P.Savi, J.M.Pereillo, M.F.Uzabiaga, J.Combalbert, C.Picard, J.P.Maffrand, M.Pascal, J.M.Herbert. Identification and biological activity of the active metabolite of clopidogrel., *Thromb.Haemost.*, **84**, (2000) 891–896.

[12] R.J.Gryglewski, R.Korbut, J.Swies, T.E.Kostka, K.Bieron, J.Robak. Thrombolytic action of ticlopidine: possible mechanisms, *European Journal of Pharmacology*, **308**, (1996) 61–67.

[13] R.J.Gryglewski, R.Korbut, J.Swies, W.Uracz. Effect of ticlopidine on streptokinase-induced thrombolysis in rats, *Wiener Klinische Wochenschrift*, **111**, (1999) 98–102.

[14] R.J.Gryglewski, J.P.Dupin, W.Uracz, J.Swies, J.Madej, G.Hou, D.Gravier, F.Casadebaig. Thrombolysis by thienopyridines and their congeners, *J. Physiol.Pharmacol*, **51**, (2000) 683–693.

[15] L.M.Van Bortel, J.N.de Hoon, M.J.Kool, J.A.Wijnen, C.I.Vertommen, L.G.Van Nueten. Pharmacological properties of nebivolol in man, *Eur.J.Clin.Pharmacol.*, **51**, (1997) 379–384.

[16] Y.S.Gao, T.Nagao, R.A.Bond, W.J.Janssens, P.M.Vanhoutte. Nebivolol induces endothelium-dependent relaxations of canine coronary arteries, *J.Cardiovasc.Pharmacol.*, **17**, (1991) 964–969.

[17] A.J.Bowman, C.P.Chen, G.A.Ford. Nitric oxide mediated venodilator effects of nebivolol, *Br.J.Clin.Pharmacol.*, **38**, (1994) 199–204.

[18] J.R.Cockcroft, P.J.Chowienczyk, S.E.Brett, C.P.Chen, A.G.Dupont, L.Van Nueten, S.J.Wooding,

J.M.Ritter. Nebivolol vasodilates human forearm vasculature: evidence for an L-arginine/NO-dependent mechanism, *J.Pharmacol.Exp.Ther.*, **274**, (1995) 1067–1071.

[19] H.H.Schmidt, S.E.Baeblich, B.C.Zernikow, M.M.Klein, E.Bohme. L-arginine and arginine analogues: effects on isolated blood vessels and cultured endothelial cells, *Br.J.Pharmacol.*, **101**, (1990) 145–151.

[20] B.Ziemianin, R.Olszanecki, W.Uracz, E.Marcinkiewicz, R.J.Gryglewski. Thienopyridines: effects on cultured endothelial cells, *J.Physiol.Pharmacol*, **50**, (1999) 597–604.

[21] G.Grynkiewicz, M.Poenie, R.Y.Tsien. A new generation of Ca2+ indicators with greatly improved fluorescence properties, *J.Biol.Chem.*, **260**, (1985) 3440–3450.

[22] T.Malinski, Z.Taha. Nitric oxide release from a single cell measured in situ by a porphyrinic-based microsensor, *Nature*, **358**, (1992) 676–678.

[23] L.Jia, R.F.Furchgott. Inhibition by sulfhydryl compounds of vascular relaxation induced by nitric oxide and endothelium-derived relaxing factor, *Journal of Pharmacology & Experimental Therapeutics*, **267**, (1993) 371–378.

[24] L.W.Dobrucki, I.Dobrucki, T.Malinski. Kinetics of nitric oxide and superoxide release in the brain during ischaemia/reperfusion. NATO ASI 59th Course: Nitric oxide - basic research and clinical application, Erice, September 7-17, 2000. 2001.

[25] S.Chłopicki, R.J.Gryglewski. Nitric oxide is a major mediator in reactive hyperaemia evoked by a brief coronary occlusion in the guinea pig heart, *Eur.J.Pharmacol.*, **241**, (1993) 117–120.

[26] C.P.Verdon, B.A.Burton, R.L.Prior. Sample pre-treatment with nitrate reductase and glucose-6-phosphate dehydrogenase quantitatively reduces nitrate while avoiding interference by NADP+ when the Griess reaction is used to assay for nitrite, *Anal.Biochem.*, **224**, (1995) 502–508.

[27] C.C.Wu, C.Thiemermann. Biological control and inhibition of induction of nitric oxide synthase, *Methods Enzymol.*, **268**, (1996) 408–420.

[28] R.J.Gryglewski, R.Korbut, A.Ocetkiewicz, J.Stachura. In vivo method for quantitation for anti-platelet potency of drugs, *Naunyn-Schmiedebergs Archives of Pharmacology*, **302**, (1978) 25–30.

[29] R.J.Gryglewski, W.Uracz, J.Swies. Unusual effects of aspirin on ticlopidine-induced thrombolysis, *Thorax*, **55** (2000) Suppl 2:S17-9

[30] Z.G.R.J.Mackiewicz, J.Swies, W.Dabros, W.Uracz. Streptokinase-induced changes in the structure of platelets. Ex vivo study. *Acta Med.Lituanica* **2**, 3–8.

[31] M.Mangrella, F.Rossi, F.Fici, F.Rossi. Pharmacology of nebivolol, *Pharmacol.Res.*, **38**, (1998) 419–431.

[32] H.R.Lu, G.Vandeplassche, L.Wouters, M.Borgers. Beta-blockade in the ischemic reperfused working rabbit heart: dissociation of beta-adrenergic blocking and protective effects, *Arch.Int.Pharmacodyn.Ther.*, **301**, (1989) 165–181.

[33] W.McNeely, K.L.Goa. Nebivolol in the management of essential hypertension: a review [published erratum appears in *Drugs* 1999 Jun;**57**(6):870], *Drugs*, **57**, (1999) 633–651.

[34] L.Van Nueten, A.Schelling, C.Vertommen, A.G.Dupont, J.I.Robertson. Nebivolol vs. enalapril in the treatment of essential hypertension: a double-blind randomised trial, *J.Hum.Hypertens.*, **11**, (1997) 813–819.

[35] B.Waeber. Nebivolol: a beta blocker with vasodilator properties, *Schweiz.Rundsch.Med.Prax.*, **89**, (2000) 631–633.

[36] M.Dawes, S.E.Brett, P.J.Chowienczyk, T.G.Mant, J.M.Ritter. The vasodilator action of nebivolol in forearm vasculature of subjects with essential hypertension, *Br.J.Clin.Pharmacol.*, **48**, (1999) 460–463.

[37] O.Uhlir, I.Dvorak, P.Gregor, I.Malek, L.Spinarova, J.Vojacek, L.Van Nueten. Nebivolol in the treatment of cardiac failure: a double-blind controlled clinical trial, *J.Card Fail.*, **3**, (1997) 271–276.

[38] W.A.Van de, W.Janssens, J.Van Neuten, R.Xhonneux, J.De Cree, H.Verhaegen, R.S.Reneman, P.A.Janssen. Pharmacological and hemodynamic profile of nebivolol, a chemically novel, potent, and selective beta 1-adrenergic antagonist, *J.Cardiovasc.Pharmacol.*, **11**, (1988) 552–563.

[39] R.Xhonneux, L.Wouters, R.S.Reneman, P.A.Janssen. The l-enantiomer of nebivolol potentiates the blood pressure lowering effect of the d-enantiomer, *Eur.J.Pharmacol.*, **181**, (1990) 261–265.

[40] P.J.Pauwels, W.Gommeren, G.Van Lommen, P.A.Janssen, J.E.Leysen. The receptor binding profile of the new antihypertensive agent nebivolol and its stereoisomers compared with various beta-adrenergic blockers, *Mol.Pharmacol.*, **34**, (1988) 843–851.

[41] W.A.Van de, R.Xhonneux, R.S.Reneman, P.A.Janssen. Cardiovascular effects of dl-nebivolol and its enantiomers--a comparison with those of atenolol, *Eur.J.Pharmacol.*, **156**, (1988) 95–103.

[42] P.M.Janssen, O.Zeitz, G.Hasenfuss. Transient and sustained impacts of hydroxyl radicals on sarcoplasmic reticulum function: protective effects of nebivolol, *Eur.J.Pharmacol.*, **366**, (1999) 223–232.

[43] L.Van Nueten, A.G.Dupont, C.Vertommen, H.Goyvaerts, J.I.Robertson. A dose-response trial of nebivolol in essential hypertension, *J.Hum.Hypertens.*, **11**, (1997) 139–144.

[44] S.Taddei, A.Virdis, L.Ghiadoni, I.Sudano, A.Salvetti. Antihypertensive drugs and reversing of endothelial dysfunction in hypertension, *Curr.Hypertens.Rep.*, **2**, (2000) 64–70.

[45] B.R.Brehm, D.Bertsch, J.von Fallois, S.C.Wolf. Beta-blockers of the third generation inhibit endothelin-1 liberation, mRNA production and proliferation of human coronary smooth muscle and endothelial cells, *J.Cardiovasc.Pharmacol.*, **36**, (2000) S401-S403.

[46] R.Troost, E.Schwedhelm, S.Rojczyk, D.Tsikas, J.C.Frolich. Nebivolol decreases systemic oxidative stress in healthy volunteers [In Process Citation], *Br.J.Clin.Pharmacol.*, **50**, (2000) 377–379.

[47] A.Parenti, S.Filippi, S.Amerini, H.J.Granger, A.Fazzini, F.Ledda. Inositol phosphate metabolism and nitric-oxide synthase activity in endothelial cells are involved in the vasorelaxant activity of nebivolol, *J.Pharmacol.Exp.Ther.*, **292**, (2000) 698–703.

[48] P.Wohlfart, T.Malinski, H.Ruetten, U.Schindler, W.Linz, K.Schoenafinger, H.Strobel, G.Wiemer. Release of nitric oxide from endothelial cells stimulated by YC-1, an activator of soluble guanylyl cyclase, *Br.J.Pharmacol.*, **128**, (1999) 1316–1322.

[49] F.Kamali, A.Howes, S.H.Thomas, G.A.Ford, E.Snoeck. A pharmacokinetic and pharmacodynamic interaction study between nebivolol and the H2-receptor antagonists cimetidine and ranitidine, *Br.J.Clin.Pharmacol.*, **43**, (1997) 201–204.

[50] M.Kakoki, Y.Hirata, H.Hayakawa, H.Nishimatsu, Y.Suzuki, D.Nagata, E.Suzuki, K.Kikuchi, T.Nagano, M.Omata. Effects of vasodilatory beta-adrenoceptor antagonists on endothelium-derived nitric oxide release in rat kidney, *Hypertension*, **33**, (1999) 467–471.

[51] M.A.Broeders, P.A.Doevendans, B.C.Bekkers, R.Bronsaer, E.van Gorsel, J.W.Heemskerk, M.G.Egbrink, E.van Breda, R.S.Reneman, Z.R.van Der. Nebivolol: a third-generation beta-blocker that augments vascular nitric oxide release: endothelial beta(2)-adrenergic receptor-mediated nitric oxide production, *Circulation*, **102**, (2000) 677–684.

[52] J.N.Trochu, V.Leblais, Y.Rautureau, F.Beverelli, H.Le Marec, A.Berdeaux, C.Gauthier. Beta 3-adrenoceptor stimulation induces vasorelaxation mediated essentially by endothelium-derived nitric oxide in rat thoracic aorta, *Br.J.Pharmacol.*, **128**, (1999) 69–76.

[53] P.Varghese, R.W.Harrison, R.A.Lofthouse, D.Georgakopoulos, D.E.Berkowitz, J.M.Hare. beta(3)-adrenoceptor deficiency blocks nitric oxide-dependent inhibition of myocardial contractility, *J.Clin.Invest*, **106**, (2000) 697–703.

[54] J.P.Dumas, F.Goirand, M.Bardou, M.Dumas, L.Rochette, C.Advenier, J.F.Giudicelli. Role of potassium channels and nitric oxide in the relaxant effects elicited by beta-adrenoceptor agonists on hypoxic vasoconstriction in the isolated perfused lung of the rat, *Br.J.Pharmacol.*, **127**, (1999) 421–428.

[55] P.H.McDonald, N.L.Cote, F.T.Lin, R.T.Premont, J.A.Pitcher, R.J.Lefkowitz. Identification of NSF as a beta-arrestin1-binding protein. Implications for beta2-adrenergic receptor regulation, *J.Biol.Chem.*, **274**, (1999) 10677–10680.

[56] J.Swies, M.Radomski, A.Dembinska-Kiec, R.Korbut, R.J.Gryglewski, R.Brandt, J.Nowak. Differing adrenolytic effects on cholinergically induced PGI_2 activity in the circulation of cat and man, *Eur.J.Clin.Invest.*, **15**, (1985) 320–326.

Nitric Oxide
R.J. Gryglewski and P. Minuz (Eds.)
IOS Press, 2001

Role of Nitric Oxide in Cardiac Transplant Rejection: Documentation by EPR Spectroscopy and Other Techniques

Galen M. Pieper, Mark B. Adams, Christopher C. Felix and Allan M. Roza
Department of Surgery, Division of Transplant Surgery
and the Biophysics Research Institute, Medical College of Wisconsin,
9200 West Wisconsin Avenue, Milwaukee, WI 53226 USA

Abstract

Northern and Western blot analysis reveal a significant upregulation in inducible nitric oxide synthase (iNOS) gene expression during cardiac transplant rejection. Treatment with various nitric oxide synthase (NOS) inhibitors to decrease nitric oxide (NO) yields conflicting results on prolonging cardiac allograft survival. An alternative approach to intervention with NOS enzyme inhibitors would be pharmacological interventions to limit the action of NO using NO scavengers. Iron-dithiocarbamate-based NO scavengers may have benefits as pharmacological agents to enhance graft survival. In addition, these agents have the potential to be used as diagnostic agents to confirm upregulation of NO using electron paramagnetic resonance (EPR) spectroscopy, EPR imaging or NMR imaging techniques. EPR spectroscopy may also be useful in identifying in allograft tissue biopsies the molecular targeting of excess NO to iron-containing proteins. Binding of NO by these proteins may inhibit their function and may play a role in the development of graft rejection. In untreated cardiac allografts, we observed a progressive increase in iron-nitrosylprotein complex formation prior to rejection. No EPR signals for iron-nitrosylproteins were seen in either isografts or native hearts of allograft recipients. Iron-nitrosylprotein levels were decreased by treatment with the iNOS inhibitor, aminoguanidine, suggesting a role of NO derived from iNOS. A single pulse injection of a lipid-soluble iron-dithiocarbamate-based NO scavenger in untreated allografts on post-operative day #6 revealed an EPR signal of an NO-Fe-dithiocarbamate trapped adduct confirming NO production occurred within the graft tissue. Increases in plasma and urinary levels of nitrate plus nitrite and myocardial iron-nitrosylprotein in untreated cardiac allografts were decreased in recipients treated with the NO scavenger, NOX-100, or by low-dose cyclosporine (CsA). NOX-100 was equieffective to CsA in prolonging graft survival. A short-term combination treatment of NOX-100 plus CsA blocked myocardial iron-nitrosylprotein complex formation and produced a synergistic effect on prolonging graft survival. Long-term combination treatment (days 0-100) yielded permanent graft acceptance with no evidence of iron-nitrosylprotein. These results support a role of NO in cardiac allograft rejection. Furthermore, a therapeutic value of NO scavengers in preventing organ rejection is indicated along with the potential use of the advanced biophysical technique of EPR spectroscopy for diagnosing graft rejection.

1. Biochemical Evidence for a Role of NO in Rejection

Research findings obtained from our laboratory and elsewhere suggest an important role for increased production of nitric oxide (NO) in the etiology of acute cardiac allograft rejection. Indeed, increased concentrations of urinary [1-4] or serum [5] nitrate/nitrite (by-products of NO metabolism) preceded rejection in experimental models of allogeneic cardiac transplantation.

Supplementing these peripheral measurements of NO by-products are studies showing increased mRNA and protein levels for inducible NO synthase (iNOS) in allografts of experimental animals prior to graft rejection [6-8] and by increased iNOS mRNA during cardiac transplant rejection in humans [9]. iNOS protein was present in allografts but absent in isograft controls (Figure 1).

2. Role of NO in Rejection: Pharmacological Intervention

A role of NO in rejection has also been examined using treatment with NOS inhibitors. Indeed, non-selective NOS inhibitors reduced NO production but yielded conflicting results on graft survival [1,10,11]. In contrast, the relatively selective iNOS inhibitor, aminoguanidine, was shown to increase graft survival [5,7]. The role of iNOS in acute rejection was strengthened by studies showing that rejection time was delayed in iNOS knock-out mice [12].

Considering the controversial findings with iNOS inhibitors, we have elected an alternative approach to support a role of NO in cardiac allograft rejection using a NO scavenger [4]. In contrast to NOS inhibitors, NO scavengers would not act to alter NOS enzyme activity but rather to limit the action of excess NO by binding NO. This property may have some advantage. Indeed, iNOS has an important role to play in fighting bacterial [13-15] infections and regulating viral replication [16,17]. In this sense, maintaining the integrity of iNOS enzyme function might be an advantage in immunocompromised patients.

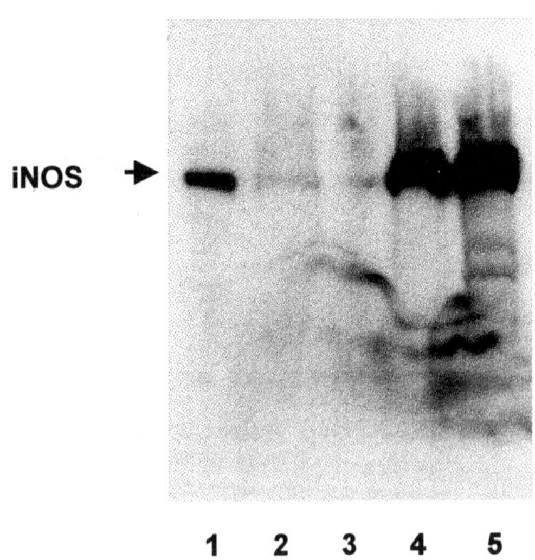

Figure 1. Western blot of iNOS protein at posttransplant day #4 in allograft hearts, Wistar-Furth to Lewis strain (Lanes 4 and 5), but not isograft hearts, Lewis to Lewis strain (Lanes 2 and 3), compared to mouse macrophage iNOS control (Lane 1).

One class of NO scavengers or NO neutralizing agents are the iron-dithiocarbamate derivatives. These agents scavenge NO under *in vitro* and *in vivo* conditions. Theoretically, these agents should potentially limit adverse sequelae arising from excess NO. Indeed, similar NO scavenger analogs of this class have shown benefits in other high-output NO models such as endotoxic shock [18,19] and myocardial ischemia/reperfusion injury [20,21]. Despite these findings, NO scavengers have not been evaluated to date in allograft rejection models. Using the principle of NO scavenging by iron-dithiocarbamate-based derivatives, we evaluated the potential efficacy of a newly-developed NO neutralizer of this class, NOX-100, to improve graft survival and to prevent nitrosylation of myocardial iron-containing proteins [4,22].

3. Principal: NO Scavengers as Neutralizers of NO Action

Dithiocarbamate derivatives bind Fe^{2+} which, in turn, avidly bind NO. The rate constant *in vitro* for the reaction of NO with Fe-proline dithiocarbamate is estimated to be $1.1 \times 10^8 M^{-1}s^{-1}$ [23]. The rate constant of reaction of NO with [Fe-(N-methyl-D-glucamine dithiocarbamate)$_2$] is relatively specific with estimates of approximately 8 orders of magnitude higher than for the reaction with nitrite ion [24].

4. Pharmacological Intervention with NO Scavengers in Acute and Delayed Cardiac Transplant Rejection

On post-operative day #5, plasma nitrate + nitrite concentration was elevated in allograft compared to isograft recipients (Figure 2). This increase was significantly decreased in recipients treated with either NOX-100 or low-dose CsA alone and completely reversed in recipients receiving combination treatment of NOX-100 plus CsA [4].

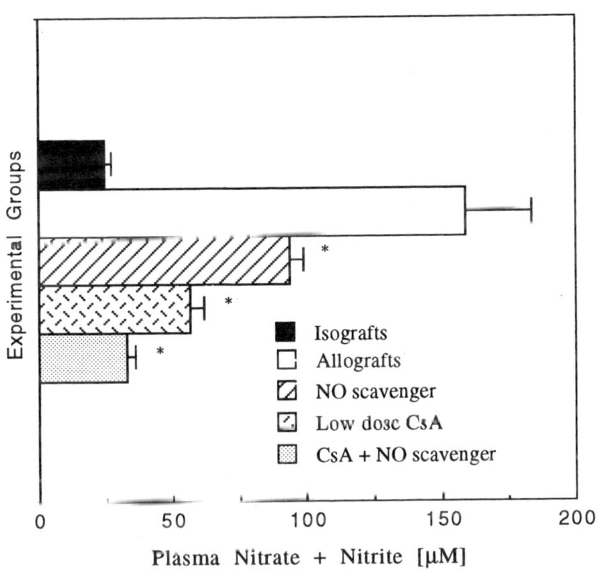

Figure 2. Plasma nitrate + nitrite concentration at post-operative day #5 in isografts, allografts without or with drug treatments. *$P<0.05$ vs. untreated allografts

On post-operative day #6 [25], X-band EPR analysis of an isograft heart revealed background signals for reduced iron-sulfur (Fe-S) complexes at g=2.02 and g=1.94 and a signal characteristic of semiquinone at g=2.004 (Figure 3) . The EPR spectra of isografts were identical to that observed in either normal hearts or native hearts of isograft recipients (not shown). In contrast to isografts, there was a broad EPR signal at g=2.08 and a triplet signal at g =2.014 with hyperfine splitting of 17.5 Gauss. The triplet signal at g=2.014 is characteristic of nitrosylation of heme protein while the upfield component at g=2.08 is designated as arising from nitrosomyoglobin in cardiac grafts [25-27].

The signals for iron-nitrosylprotein complexes on post-operative day #6 were markedly decreased in allograft recipients treated with aminoguanidine suggesting that NO derived from iNOS was responsible for iron-nitrosylprotein complex formation [25]. In addition, treatment with NOX-100 markedly reduced iron-nitrosylprotein formation suggesting that this agent was able to interfere with NO-induced targeting of myocardial protein (Figure 4). In addition to the treatments with NOX-100 alone, treatment with low-dose CsA alone or CsA in combination with NOX-100 completely eliminated EPR signals for nitrosylated heme-protein (Figure 4). In these allograft hearts, there were EPR signals for only the reduced Fe-S complexes and semiquinone which were identical to that seen in isografts or native hearts of allograft recipients.

Treatment with NOX-100 alone increased graft survival equipotent to that achieved by CsA alone (Figure 5). In a model of delayed rejection, short-term combination treatment with NOX-100 (days 1-30) plus CsA (days 1-7) produced a synergistic increase in graft survival (Figure 5).

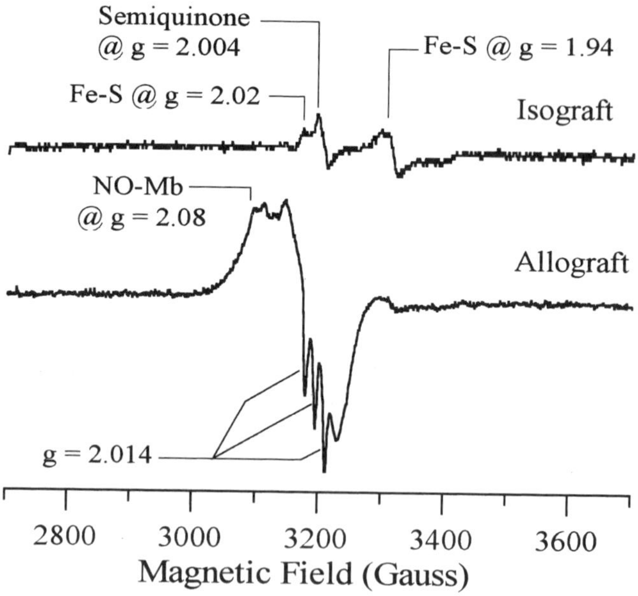

Figure 3. EPR spectra showing background signals in control and isograft hearts.

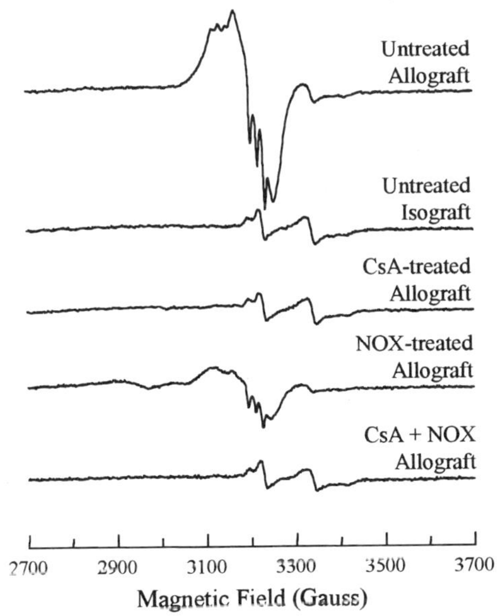

Figure 4. Examples of EPR iron-nitrosylprotein signals in allograft hearts harvested on postoperative day #6 without or with drug treatment.

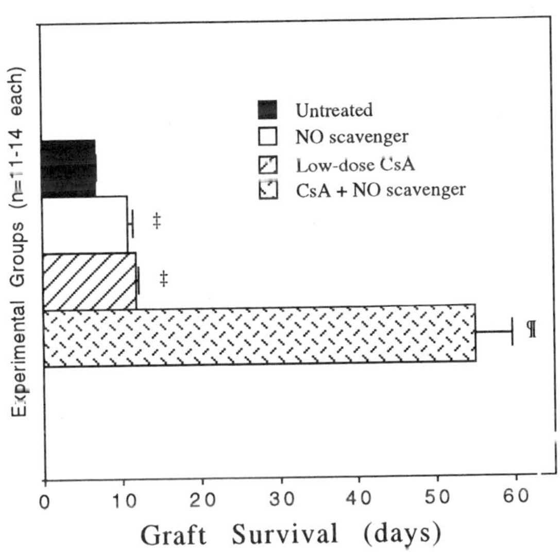

Figure 5. Prolongation of cardiac graft survival in various drug treatment groups. ‡ $P<0.01$ and ¶$P<0.001$ vs untreated allografts.

5. Effects of Long-Term Combination Therapy

In a model of long-term combination treatment, CsA and NOX-100 were given for the first 100 days following transplantation followed by cessation of treatments. At various times during drug treatment, there was no EPR evidence for any iron-nitrosylprotein formation at any time in any sample (not shown). Furthermore, these hearts did not reject even on day #450 (i.e. 350 days after cessation of drug treatment). EPR analysis also revealed no evidence of any iron-nitrosylprotein complex formation.

A subset of animals receiving long-term combination treatment underwent a second cardiac transplant (to the groin) from a Wistar-Furth strain donor rat. These second transplant grafts survived indefinitely without any immunosuppression and exhibited no iron-nitrosylprotein formation. In contrast, a transplant with an ACI strain donor heart promptly rejected within 7 days. Also, a Wistar-Furth skin graft rejected suggesting that this phenomenon was tissue-specific. Thus, by an unknown mechanism, long-term combination treatment was able to produce permanent graft acceptance without the need for immunosuppression.

6. NO Scavengers as Diagnostic Agents

The value of iron-dithiocarbamates is the ability to use them as both therapeutic agents and as pharmacological agents. Addition of NO solution to various buffers containing iron-dithiocarbamates results in a characteristic three-line EPR spectra with g = 2.04 and hyperfine coupling of 12.5 Gauss [18]. These same NO-Fe-dithiocarbamate adducts have been detected by EPR spectroscopy in biological systems. After a pulse injection with Fe-diethyldithiocarbamate (Fe-DETC) in untreated allograft recipients, we have identified a three-line EPR spectra of g=2.04 superimposed on the broader heme-nitrosylprotein EPR signal (Figure 6). These findings document that NO production occurs directly within the graft tissue. Thus, these procedures are feasible to be used for determination of NO in organ rejection by biopsy of tissue. Theoretically, the use of water soluble derivatives such as Fe-N-methyl-D-glucamine dithiocarbamate [$(MGD)_2$-Fe] to trap NO may be useful as noninvasive diagnostic probes of Fe-NO-trapped adducts using EPR spectroscopy of biological fluids such as plasma and urine samples. In fact, this same adduct has been used to diagnose NO formation in septic mice *in vivo.*

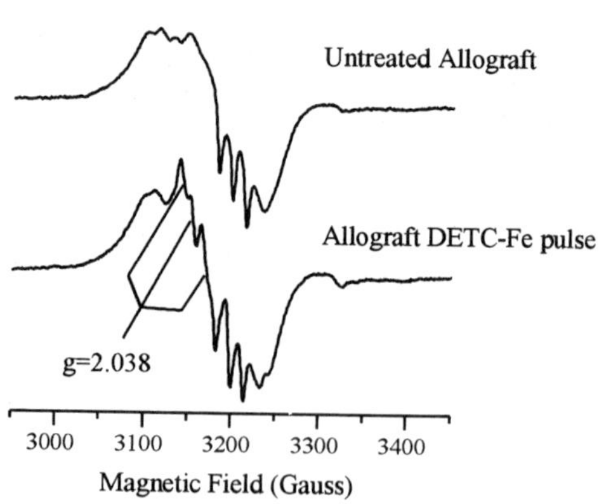

Figure 6. EPR spectra of NO-Fe-DETC adduct indicating *in vivo* trapping of NO within rejecting cardiac graft.

7. NO Scavengers as Potential *In Vivo* Diagnostic Imaging Tools

The feasibility of these agents for imaging were first conducted in studies using an iron complex of *N*-(dithiocarboxy)sarcosine [$(DTCS)_2$-Fe]. Injection of [$(DTCS)_2$-Fe-NO] was used to image the head region of a living rat [28] or the abdominal region of the living mouse [29]. This same derivative has been used in EPR imaging to visualize NO produced by isorbide dinitrate [30] and by induction of iNOS in septic mice [31]. The agent [$(MGD)_2$-Fe] was tested for EPR imaging of NO produced in isolated, ischemic-reperfused rat heart [32]. A variant of this technology was recently devised by Mülsch et al. [33] who used proton-electron double resonance imaging of liver perfused with sodium nitroprusside. Furthermore, [$(MGD)_2$-Fe-NO] is a contrast agent used to improve magnetic resonance imaging [34]. Because NO is believed to be involved in organ rejection based upon the evidence provided above, the development and refinement of these newer advanced biophysical technologies may provide a future additional means of assisting clinicians in the diagnosis of organ transplant rejection.

Acknowledgments

The authors thank Dr. Ching-San Lai, Medinox, Inc. for the generous gift of NOX-100 for these studies. The authors also thank Ms. Gail Moore-Hilton for the surgical assistance and Ms. Karen Hyde for EPR spectra reproductions using SUMSPEC and Grapher (Golden Software, Golden, CO) programs. This work was supported, in part, by the Biomedical Research Center grant #RR01008.

References

[1] N.R. Bastian *et al.*, N^{ω}-Monomethyl-L-Arginine Inhibits Nitric Oxide Production in Murine Cardiac Allografts but does not affect Graft Rejection. *Biochimical et Biophysical Acta* **1226** (1994) 225-231.

[2] D.S. Winlaw *et al.*, Urinary Nitrate Excretion is a Noninvasive Indicator of Acute Cardiac Allograft Rejection and Nitric Oxide Production in the Rat. *Transplantation* **58** (1994) 1031-1036.

[3] S.G. *et al.*, Relative Importance of Cytotoxic T Lymphocytes and Nitric Oxide-Dependent Cytotoxicity in Contractile Dysfunction of Rejecting Murine Cardiac Allografts. *Transplantation* **66** (1998) 413-419.

[4] G.M. Pieper *et al.*, Reduction of Myocardial Nitrosyl Complex Formation by a Nitric Oxide Scavenger Prolongs Cardiac Allograft Survival. *Journal of Cardiovascular Pharmacology* **35** (2000) 114-120.

[5] N.K. Worrall *et al.*, Inhibition of Inducible Nitric Oxide Synthase Attenuates Established Acute Cardiac Allograft Rejection. *Annals of Thoracic Surgery* **62** (1996) 378-385.

[6] X. Yang *et al.*, Induction of Myocardial Nitric Oxide Synthase by Cardiac Allograft Rejection. *Journal of Clinical Investigation* **94** (1994) 714-721.

[7] N.K. Worrall *et al.*, Modulation of In Vivo Alloreactivity by Inhibition of Inducible Nitric Oxide Synthase. *Journal of Experimental Medicine* **181** (1995) 63-70.

[8] M.E. Russell *et al.,* Upregulation and Modulation of Inducible Nitric Oxide Synthase in Rat Cardiac Allografts with Chronic Rejection and Transplant Arteriosclerosis. *Circulation* **92** (1995) 457-464.

[9] N.P. Lewis *et al.,* Induction of Nitric Oxide Synthase in the Human Cardiac Allograft is Associated with Contractile Dysfunction of the Left Ventricle. *Circulation* **93** (1996) 720-729.

[10] D.S. Winlaw *et al.,* Selective Inhibition of Nitric Oxide Production during Cardiac Allograft Rejection causes a Small Increse in Graft Survival. *Transplantation* **60** (1995) 77-82.

[11] L.C. Paul *et al.,* Nitric Oxide Synthase Inhibition is Associated with Decreased Survival of Cardiac Allografts in the Rat. *Transplantation* 62 (1996) 1193-1195.

[12] J . Koglin *et al.,* NOS2 Mediates Opposing Effects in Models of Acute and Chronic Cardiac Rejection. Insights from NOS2-Knockout Mice. *American Journal of Pathology* **153** (1998) 1371-1376.

[13] A.H. Fortier *et al.,* Activation of Macrophages for Destruction of *Francisella tularensis:* Identification of Cytokines, Effector Cells and Effector Molecules. *Infection and Immunity* **60** (1992) 817-825.

[14] M. Denis. Interferon-Gamma-Treated Murine Macrophages Inhibit Growth of Tubercle Bacilli via the Generation of Reactive Nitrogen Intermediates. *Cell Immunity* **132** (1991) 150-157.

[15] J. Mayer *et al.,* Gamma Interferon Induced Nitric Oxide Production Reduces *Chlamydia trachomatis* Infectivity in McCoy Cells. *Infection and Immunity* **61** (1993) 491-497.

[16] F. Chen *et al.,* Nitric Oxide Inhibits HIV Tat-Induced NF-κB Activation. *American Journal of Pathology* **155** (1999) 275-284.

[17] G. Karupiah *et al.,* Inhibition of Viral Replication by Interferon-Gamma-Induced Nitric Oxide Synthase. *Science* **261** (1993) 1445-1448.

[18] C.S. Lai and A.M. Komarov AM. Spin Trapping of Nitric Oxide Produced In Vivo in Septic Shock Mice. *FEBS Letters* **345** (1995) 120-124.

[19] S. Lecour *et al.,* Electron Spin Resonance Detection of Nitric Oxide Generation in Major Organs from LPS-Treated Rats. *Journal of Cardiovascular Pharmacology* **33** (1999) 78-85.

[20] J.L. Zweier *et al.,* Direct Measurement of Nitric Oxide Generation in the Ischemic Heart using Electron Paramagnetic Resonance Spectroscopy. *Journal of Biological Chemistry* **270** (1995) 304-307.

[21] A.M. Komarov *et al.,* EPR Detection of Endogenous Nitric Oxide in Postischemic Heart using Lipid and Aqueous-Soluble Dithiocarbamate-Iron Complexes. *Molecular and Cellular Biochemistry* **175** (1997) 91-97.

[22] A.M. Roza *et al.,* NOX 100, a Nitric Oxide Scavenger, Enhances Cardiac Allograft Survival and Promotes Long-Term Graft Acceptance. *Transplantation* **69** (2000) 227-231.

[23] S.V. Paschenko *et al.,* EPR and Laser Flash Photolysis Studies of the Reaction of Nitric Oxide with Water Soluble NO Trap Fe (II)-Proline-Dithiocarbamate Complex. *Biochemical Biophysical Research Communications* **225** (1996) 577-584.

[24] K. Tsuchiya *et al.,* Nitric Oxide-Forming Reaction Between the Iron-*N*-Methyl-D-Glucamine Dithiocarbamate Complex and Nitrite. *Journal of Biological Chemistry* **275** (2000) 1551-1556.

[25] A.L. Nakanishi *et al.,* Electron Spin Resonance Analysis of Heme-Nitrosyl and Reduced Iron Sulfur-Centered Complexes in Allogeneic, Heterotopic Cardiac Transplants: Effects of Treatment with Pyrrolidine Dithiocarbamate. *Free Radical Biology and Medicine* **25** (1998) 201-207.

[26 M. Cooper *et al.,* Myocardial Nuclear Factor-κB Activity and Nitric Oxide Production in Rejecting Cardiac Allografts. *Transplantation* **66** (1998) 838-844.

[27] E.A. Konorev *et al.,* *S*-Nitrosoglutathione Induces Formation of Nitrosylmyoglobin in Isolated Hearts during Cardioplegic Ischemia- An Electron Spin Resonance Study. *FEBS Letters* **378** (1996) 111-114.

[28] T. Yoshimura *et al.,* In Vivo Electron Paramagnetic Resonance Imaging of NO Bound Iron Complex in a Rat Head. *Chemistry Letters* **8** (1995) 309-310.

[29] T. Ueno and T. Yoshimura. The Physiological Activity and In Vivo Distribution of Dinitrosyl Dithiolato Iron Complex. *Japanese Journal of Pharmacology* **82** (2000) 95-101.

[30] S. Fujii *et al.,* In Vivo Three-Dimensional EPR Imaging of Nitric Oxide Production from Isosorbide Dinitrate in Mice. *American Journal of Physiology* **274** (1998) G857-G862.

[31] T. Yoshimura *et al.,* In Vivo EPR Detection and Imaging of Endogenous Nitric Oxide in Lipopolysaccharide-Treated Mice. *Nature Biotechnology* **14** (1996) 992-994.

[32] J. L. Zweier *et al.,* Electron Paramagnetic Resonance Imaging of the Rat Heart. *Physics Medicine and Biology* **43** (1998) 1823-1835.

[33] A. Mülsch *et al.,* Detection of Nitrosyl-Iron Complexes by Proton-Electron-Double-Resonance Imaging. *Free Radical Biology and Medicine* **27** (1999) 636-646.

[34] H. Fujii *et al.,* In Vivo Imaging of Spin Trapped Nitric Oxide in Rats with Septic Shock: MRI Spin Trapping. *Magnetic Resonance Medicine* **42** (1999) 235-239.

Nitric Oxide
R.J. Gryglewski and P. Minuz (Eds.)
IOS Press, 2001

Role of nitric oxide in the antiarrhythmic effects of ischaemic preconditioning

Ágnes Végh and James Roy Parratt*,

*Department of Pharmacology and Pharmacotherapy, University of Szeged, Albert Szent-Györgyi Medical and Pharmacological Center, Dóm tér 12, H-6721, Szeged, Hungary and *Department of Physiology and Pharmacology, Strathclyde Institute for Biomedical Sciences, University of Strathclyde, 27 Taylor Street, Glasgow G4 ONR, Scotland, UK.*

Abstract

There is a good deal of evidence that nitric oxide, most probably derived from endothelial cells, plays an important role in both the *early* and the *delayed* cardioprotective effects of ischaemic preconditioning, induced either by coronary artery occlusion, rapid cardiac or physical exercise. The evidence for the involvement of nitric oxide comes mainly from pharmacological studies in a canine model of ischaemic preconditioning which demonstrate that inhibition of nitric oxide generation leads to the loss of protection against ischaemia and reperfusion-induced ventricular arrhythmias. These studies suggest that both the *early* and the *delayed* antiarrhythmic effects of preconditioning result from the bradykinin-triggered release of nitric oxide from endothelial cells and that the protective effects of nitric oxide under these conditions are mediated through elevation of myocyte cGMP. This concept suggest that there is a "cross-talk" between coronary vascular cells and cardiac myocytes which leads to cardioprotection. The concept also suggest that under those conditions where endothelial dysfunction is evident, such as hypertension, atherosclerosis, diabetes and ventricular hypertrophy, the susceptibility of the myocardium to ischaemia is increased and that this is due, in part, to the reduced formation of nitric oxide and of other 'endogenous myocardial protective substances' derived from endothelial cells.

Keywords: nitric oxide, ischaemic preconditioning, cardiac pacing, exercise, ventricular arrhythmias, ischaemia-reperfusion.

1. Introduction

In 1980 Furchgott and Zawadzki (1) reported in a seminal paper in Nature that the endothelium releases relaxing substances and later one of these, was identified as nitric oxide (NO; 2). Endothelium-derived nitric oxide has since been recognised as a remarkable modulator of various biological mechanisms, both under physiological and pathophysiological conditions. There is abundant evidence that under physiological conditions the endothelium releases NO and that this is an important regulator of cell to cell interactions. For example, this endothelium-derived NO, by increasing cGMP levels in smooth muscle cells, results in relaxation, inhibits smooth muscle cell proliferation and migration, the aggregation of platelets and the expression of adhesion molecules. More recently, it has also been recognised that endocardial endothelial and coronary vascular endothelial cells communicate with cardiac myocytes (3) and perhaps also with cardiac neurones (4) to modulate function through the release of various endothelium-derived substances which include NO. In this chapter we summarise the evidence that endothelium-derived NO also modifies, in a protective fashion, the life-threatening ventricular arrhythmias that arise soon after the onset of myocardial ischaemia. In particular, we summarise the evidence that nitric oxide plays an important role in the pronounced antiarrhythmic effects of ischaemic preconditioning.

2. The antiarrhythmic effects of ischaemic preconditioning

2.1. The phenomenon of ischaemic preconditioning

A few general words concerning the phenomenon of ischaemic preconditioning might be appropriate at this point. The term was introduced by Murry, Jennings and Reimer in a paper published in 1986 (5). They showed, in anaesthetised dogs, that the extent of myocardial ischaemic damage (necrosis) which resulted from a 40 min occlusion of the left circumflex coronary artery and then by reperfusion, was markedly reduced if that same coronary artery had been occluded briefly (four times for 5 min) just prior to the prolonged ischaemic insult. The phenomenon was termed 'ischaemic preconditioning' and has aroused considerable interest in the last 15 years of cardiovascular research. As the result we now know that the heart, and indeed other organs, are able to withstand severe injury resulting from ischaemia. The stimulus for this protection is a brief period of a similar, that is ischaemic, stress.

There are, at least, four important manifestations of myocardial protection that result from ischaemic preconditioning. These are the reduction in infarct size, the enhanced recovery of contractile function following a period of ischaemia and reperfusion, protection against ischaemia and reperfusion-induced vascular endothelial dysfunction and the suppression of life-threatening ventricular arrhythmias that arise both during ischaemia and reperfusion. The evidence for these various manifestations of myocardial protection resulting from preconditioning has been extensively reviewed (6, 7) and it is not the purpose here to rehearse them again. However, we believe that suppression of ventricular arrhythmias to be a particularly important manifestation of tolerance resulting from brief periods of ischaemia, and the evidence for a possible role of nitric oxide in this will now be reviewed.

2.2. The antiarrhythmic effects of ischaemic preconditioning, induced by short coronary artery occlusions, rapid cardiac pacing and exercise

One of the most important consequences of the abrupt reduction in coronary blood flow that results from coronary artery occlusion, in both humans and experimental animals, is the occurrence of life-threatening ventricular arrhythmias. In the clinical situation, these are responsible for most instances of sudden cardiac death. For example (Figure 1a), in chloralose-urethane anaesthetised dogs, following the abrupt occlusion of the anterior descending branch of the left coronary artery (LAD) for a period of 25 min, there was marked ventricular ectopic activity (VPB) and there was a high incidence, and many separate episodes, of ventricular tachycardia (VT). Ventricular fibrillation (VF) occurred in all these dogs, either during the occlusion period or after reperfusing the ischaemic myocardium. Thus, there were no survivors from such an ischaemia-reperfusion insult. However, if such dogs were subjected to one or two 5 min periods of occlusion 20 min prior to prolonged ischaemia, there was marked arrhythmia suppression (Figure 1b). Compared to the controls, the number of ectopic beats was reduced and only a few dogs exhibited brief periods of VT. Further, in these preconditioned dogs, no VF occurred during the prolonged occlusion, and 50 % of the dogs survived reperfusion. This is really profound protection against these life-threatening ventricular arrhythmias (8, 9).

Suppression of ventricular arrhythmias during myocardial ischaemia can also be achieved by cardiac pacing (10, 11). For example, if dogs were paced at a rate of 300 beats min^{-1} for two times 2 min periods (10) or at a rate of 220 beats min^{-1} for four times 5 min periods (11) through the right ventricle, and then, at various time afterwards, these dogs were subjected to a 25 min coronary artery occlusion, the most marked protection occurred 5 min after the preconditioning stimulus. Increasing the time interval between the pacing stimulus and the occlusion to 15 min, 1h or 6 hours, resulted in attenuation and then complete loss of the protection. However, the antiarrhythmic protection returned several hours later and there was a marked reduction in the occurrence of VF 24 hours after the pacing stimulus (12). This delayed protection against arrhythmias started to fade at 48 hours and was lost at 72 hours (11). Further, the time window of protection against arrhythmias could be prolonged if the preconditioning stimulus had been even as long as 48h after the first pacing stimulus (13). Thus, at these times (48 h and 72 h after the second pacing stimulus) ischaemia-induced ventricular arrhythmias were still markedly reduced (13, 14).

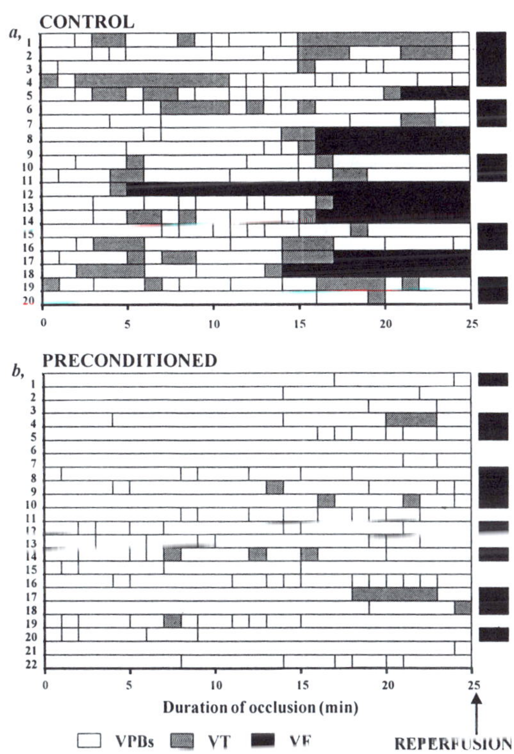

Figure 1. *Panel a.* Ventricular arrhythmias in 20 open-chest anaesthetised dogs subjected to a 25 min occlusion of the LAD, followed by reperfusion. *Panel b.* Ventricular arrhythmias in 22 dogs subjected to ischaemic preconditioning by two 5 min periods of coronary artery occlusion terminating 20 min prior to a 25 min occlusion of the LAD. The severity of ventricular arrhythmias is markedly reduced by preconditioning.

A similar delayed protection against ischaemia and reperfusion-induced ventricular arrhythmias could also be achieved by a single episode of submaximal treadmill exercise (15). For example, if dogs were subjected to a gradually increasing load over a 21 min period of exercise and, then twenty-four hours later, and following anaesthesia, the coronary artery was occluded for 25 min, the number of VBPs, the incidence and the number of episodes of VT and the incidence of occlusion-induced VF, were all markedly reduced. Further, more than two-thirds of the exercised dogs survived the combined ischaemia-reperfusion insult, in contrast to only 10 % in the control non-exercised dogs. Although the time-course of this protection, afforded by exercise, is still not clear the results raise the possibility of being able to keep the heart protected for prolonged periods of time as result of repeated periods of cardiac pacing or of exercise.

3. Mechanisms of antiarrhythmic protection induced by preconditioning, cardiac pacing and exercise: Role of nitric oxide

Although the precise mechanisms of preconditioning have still to be elucidated they seem to involve the generation of endogenous myocardial protective substances, probably derived from the coronary vascular endothelium. This hypothesis has recently been reviewed in detail (16, 17, 18). Firts, and briefly, we started from the assumption that several substances, both deleterious and protective, are released from the myocardium in the very early stages of ischaemia and that these contribute to the generation of those ventricular arrhythmias that arise soon after the onset of the coronary artery occlusion. It is likely that several of these mediators are also released in response to the transient ischaemia which results from the preconditioning stimulus. Further, if preconditioning is really a mechanism through which the heart is able to protect itself against ischaemic stress than it would be reasonable to assume that more than one protective mediator is released. These mediators, acting on different receptors, might induce protection in different ways or there might be a final common pathway. Second, we supposed that these protective substances, were probably derived from vascular endothelial cells or somewhere else within the vascular wall. This supposition came from the fact that it is not only the heart that can be preconditioned but other organs such as the brain, kidney and skeletal muscle as well. The one thing that these organs have in common is, of course, a vasculature. The hypothesis is that the "target" for preconditioning in the heart is the coronary vasculature and, in particular, the endothelium and that mediators, derived from such endothelial cells modify arrhythmogenesis by direct communication, via diffusible mediators, with cardiac cells.

Although there is a good deal of clinical evidence that endothelial dysfunction occurs in patients with hypertension, atherosclerosis, diabetes, ventricular hypertrophy and ischaemic heart disease, the relationship between endothelial dysfunction and arrhythmia severity in such patients is difficult to document. This relationship is also difficult to demonstrate in most experimental models, such as in our own *in vivo* canine model. However, this has been attempted recently by a group in Glasgow (19). They showed that removal of the endothelium in rat isolated perfused hearts resulted in a marked increase in arrhythmia severity (19). This result strongly suggests, that following coronary artery occlusion protective substances are released from the coronary vascular endothelium that modify arrhythmia severity. What are these substances likely to be and what is the evidence that these are involved in the protective effect of ischaemic preconditioning? Endothelial cells generate a variety of potent vasoactive substances (prostanoids, bradykinin, endothelin, nitric oxide) and influence vascular activity by the release of such diffusible vasoactive substances as originally demonstrated in the elegant studies of Furchgott and Zawadzki (1). Further, more recently it has been recognised that coronary vascular (and endocardial) endothelial cells communicate with cardiac myocytes, as well as with vascular smooth muscle cells and that myocardial function is modified by the release of these endothelium-derived mediators (20). We suggest that this vascular endothelial – cardiac myocyte „cross talk" not only modulates cardiac function (contractility) but also modifies, in a protective manner, ischaemia and reperfusion-induced ventricular arrhythmias. We believe that in this cell to cell interaction nitric oxide is particularly involved.

The first piece of evidence for the role for NO (presumably generated as a result of bradykinin release; 21) in the antiarrhythmic effects of preconditioning comes from studies in which dogs were preconditioned in the presence of an inhibitor of the L-arginine-NO pathway, L-NAME (22). Under these conditions, the protective effects of ischaemic preconditioning against arrhythmias were markedly attenuated. Further evidence that NO is involved comes from studies which demonstrated that the non-selective guanylyl cyclase inhibitor, methylene blue, infused locally into the coronary circulation completely prevented the antiarrhythmic effects of preconditioning (23).

The protective role of NO against arrhythmias was also confirmed by studies in which drugs which are able to donate NO, such as nicorandil (24) and isosorbide mononitrate (25), given by intracoronary infusion prior to ischaemia, also markedly reduce the severity of ventricular arrhythmias resulting from coronary artery occlusion and reperfusion.

Although the precise mechanisms of the delayed protection against arrhythmias, induced by cardiac pacing or exercise, remains to be elucidated there is some evidence which suggests that the mechanisms may be similar to those involved in „classical preconditioning". This evidence comes from studies which showed that either dexamethasone (26), which inhibits the induction of cyclooxygenase-2 and iNOS or aminoguanidine (14) and aminoethylisothiourea (13), both more selective inhibitors of iNOS activity, abolished the delayed protective effects of cardiac pacing. Similarly, aminoguanidine markedly attenuated the delayed antiarrhythmic effect resulting from physical exercise (unpublished data).

Another piece of evidence that NO contributes to delayed cardioprotection comes from studies with monophosphoryl lipid A (MLA), a non-toxic derivative of the lipid-A component derived from *Escherichia coli* endotoxin. This also causes delayed protection against arrhythmias 24h later (27). In these dogs MLA given intravenously markedly reduced the severity of arrhythmias that resulted when a coronary artery was occluded 24h later. Interestingly, this protection was only attenuated by the prior administration of aminoguanidine but was completely abolished if aminoguanidine and meclofenamate, an inhibitor of the cyclooxygenase pathway, were given together. These results indicate that, perhaps not just one mediator but several mediators, both of endothelial or myocardial origin, are released and act in concert to overcome ischaemia and reperfusion-induced cardiac injury.

The above evidence supports the hypothesis that coronary vascular, and perhaps also endocardial, endothelial cells play an important role in protecting the heart against the severe, often fatal, consequences of an abrupt reduction in coronary blood flow. Indeed, it is conceivable that ischaemic preconditioning acts by protecting endothelial cells and maintaining the release of endogenous myocardial protective mediators such as bradykinin, prostacyclin and nitric oxide during a subsequent more prolonged ischaemic insult (16, 18). We do not know the precise mechanisms involved in mediator release but our current hypothesis, summarised in Figure 2, suggests that the early release of bradykinin and the subsequent activation of bradykinin B_2 receptors on the surface of endothelial cells may initiate the formation of nitric oxide which then diffuses to the myocyte and sets in progress events which ultimately lead to cellular cardioprotection. One mechanism for this protection might be the elevation of cGMP as a result of the activation of soluble guanylyl cyclase by nitric oxide. This would depress myocardial contractility and reduce energy demand, perhaps by limiting myocardial cAMP levels as a result of the stimulation of the cGMP-dependent phosphodiesterase enzyme. Cyclic GMP could also inhibit L-type calcium channels in

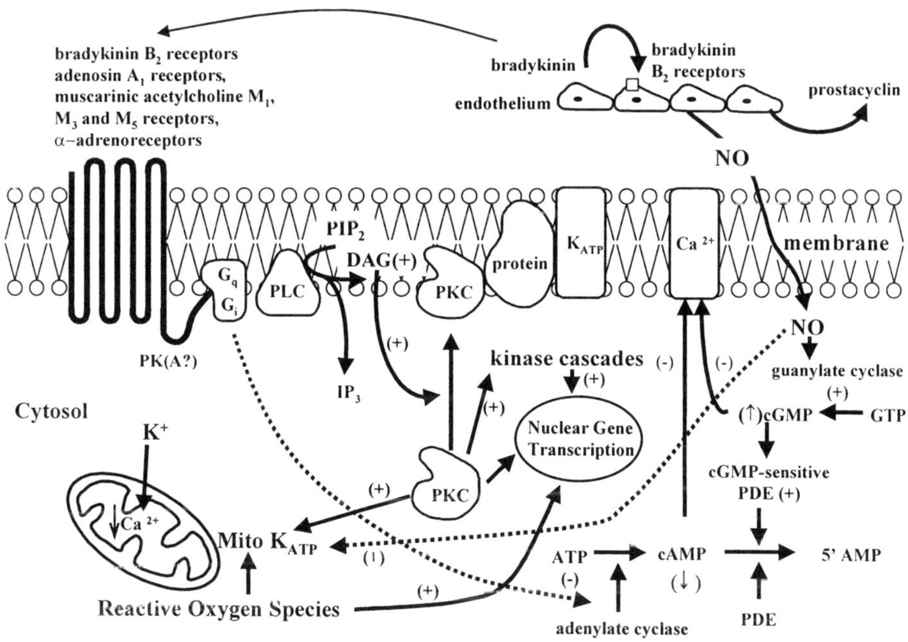

Figure 2. Hypothesis for the role of nitric oxide and of the other endothelium derived endogenous protective mediators in ischaemic preconditioning. Bradykinin released, probably from endothelial cells, acts on B_2 receptors on the endothelial surface to increase the calcium transient within these cells and activate the L-arginine nitric oxide pathway. Nitric oxide then diffuses to the cardiac myocytes, stimulates soluble guanylyl cyclase and raises cGMP. This both reduces cAMP levels and calcium entry through L-type calcium channels as well as directly suppressing myocyte oxygen consumption. Nitric oxide can also modify the opening of mitochondrial K_{ATP} channels which might contribute to cardioprotection by preventing ischaemia/reperfusion-induced mitochondrial damage

myocardial cells through activation of a cGMP-dependent protein kinase.

There is also a strong evidence for the involvement of the adenosine and bradykinin-induced translocation of protein kinase C (PKC) and the subsequent activation of the ATP-dependent potassium channels in the cardioprotective effect of ischaemic preconditioning (6). Moreover, more recent evidence suggests that mitochondrial K_{ATP} channels might be the common end-effector of these various pathways, since both PKC and NO are able to modify the opening of these channels (28). The evidence for this hypothesis comes from studies which demonstrated that the cardioprotective effects of diazoxide, a selective opener of this channel, are similar to those of preconditioning and that the beneficial effects of both preconditioning and diazoxide can be blocked by the selective mitochondrial K_{ATP} channel inhibitor 5-hydroxydecanoate (29, 30). For example, in our own experiments, the local administration of diazoxide resulted in a similar protection against occlusion and reperfusion-induced VF to that obtained by ischaemic preconditioning. Further, 5-hydroxydecanoate (5HD) given either before or after preconditioning abolished the protection resulting from ischaemic preconditioning. Indeed, it was difficult to precondition dogs in the presence of 5HD. Thus, our results support the hypothesis that mitochondrial K_{ATP} channels might also be involved in the antiarrhythmic effects of preconditioning. However, it is still not clear precisely how the opening of these channels contributes to the protection.

As regards the mechanism of the delayed antiarrhythmic effects of pacing (and exercise?), there is some evidence that PKC can also translocate to the nucleus where it then induces the synthesis of cytoprotective stress proteins or protective enzymes such as cyclooxygenase or nitric oxide synthase (31). This may result in the increased formation of NO which perhaps keeps the myocardium in a protected state for a longer period. Whether the increased generation of NO following preconditioning is due to the upregulation of endothelial NOS or to the induction of iNOS is still not clarified but the time course of the protection, with a peak effect one or two days after the preconditioning stimulus, suggest the latter.

Finally, what might the implications of these findings be? Presumably, under those conditions where endothelial dysfunction is evident (hypertension, atherosclerosis, hypercholesterolemia, diabetes, etc.) the ability to generate such endogenous myocardial protective substances, like nitric oxide, may be impaired. This would lead to the attenuation of a major pathway for protection and may explain, in part, the increased susceptibility of such patients to the arrhythmic consequences of acute ischaemia. Thus, one might expect that those drugs which are able to replace the loss of these substaces resulting from such acute and chronic injury (e.g. nitric oxide donors or adenosine agonists), or drugs that enhance the production of these substances (e.g. inhibition of kinin metabolism with ACE and NEP inhibitors with the subsequent increase in NO production) would be cardioprotective. Similarly, regular physical exercise perhaps by increasing endothelial NOS gene expression (32) leads to increased NO production, reduced platelet and leucocyte adherence to the vascular endothelium, inhibition of cardiac sympathetic transmission and modulation of the cAMP/cGMP balance in cardiac myocytes. If NO is indeed an important signalling mechanism in preconditioning, then exercise, even in individuals with elevated cardiovascular risk or established disease, would represent an important means by which benefit is provided in the setting of secondary prevention.

References

1. RF. Furchgott and JN Zawadzki. The obligatory role of endothelial cells in relaxation of arterial smoot muscle by acetylcholine. *Nature* **288** (1980) 373-376.
2. RMJ. Palmer *et al.*, Nitric oxide release account for the biological activity of endothelium-derived relaxing factor. *Nature* **327** (1987) 524-526.
3. JA. Smith *et al.*, Factors released from endothelium of the ferret and pig modulate myocardial contraction. *J Physiol* **439** (1991) 1-14.
4. JL. Ballingand *et al.*, Control of cardiac muscle cell function by an endogenous nitric oxide system. *Proc Natl Acad Sci* **90** (1993) 347-351.
5. CE. Murry *et al.*, Preconditioning with ischemia: a delay of lethal cell injury in ischemic myocardium. *Circulation* **74** (1986) 1124-1136.
6. MS. Marber and DM. Yellon, Ischaemia preconditioning and adaptation. Bios Scientific Publishers, Oxford, 1996
7. CL. Wainwright and JR. Parratt, Myocardial preconditioning. Berlin: Springer, 1996.
8. Á. Végh *et al.*, Protective effects of preconditioning of the ischaemic myocardium involves cyclooxygenase products. *Cardiovasc Res* **12** (1990) 1020-1022.
9. Á. Végh *et al.*, Antiarrhythmic effects of preconditioning in anaesthetised dogs and rats. *Cardiovasc Res* **26** (1992) 486-652.
10. Á. Végh *et al.*, Transient ischaemia induced by rapid cardiac pacing results in myocardial preconditioning. *Cardiovasc Res* **25** (1991) 1051-1053.
11. K. Kaszala *et al.*, Time course of the protection against ischaemia and reperfusion induced ventricular arrhythmias resulting from brief periods of cardiac pacing. *J Mol Cel. Cardiol* **28** (1996) 2085-2095.
12. Á. Végh *et al.*, Cardiac pacing in anaesthetized dogs preconditions the heart against arrhythmias when ischaemia is induced 24h later. *J Physiol Lond* **480** (1994) 89P
13 A.Kis *et al.*, Repeated cardiac pacing extends the time during which canine hearts are protected against ischaemia-induced arrhythmias. Role of nitric oxide. *J Mol Cell Cardiol* **31** (1999) 1229-1241.
14 A. Kis *et al.*, Pacing-induced delayed protection against arrhythmias is attenuated by aminoguanidine, an inhibitor of nitric oxide synthase. *Br J Pharmacol* **127** (1999) 1545-1550.
15. Á. Végh *et al.*, Exercise 24h prior to coronary artery occlusion reduces arrhythmia severity in anaesthetised dogs: role of nitric oxide. *J Physiol* **525** (2000) 14P.
16. JR. Parratt and Á. Végh, Coronary vascular endothelium, preconditioning and

arrhythmogenesis. In: Endothelial modulation of cardiac contraction. Eds: M. Lewis & A. J. Shah, Harwood Academic Publisher, Reading, UK. 1996, pp. 237-254, 1996.

17. JR. Parratt and Á Végh, Preconditioning induces both immediate and delayed protection against arrhythmias resulting from ischemia and reperfusion. In: Advances in Organ Biology. Myocardial Preservation and Cellular adaptation. Eds: E. E. Bittar & D.K. Das,. JAI Press, London, 1998, pp 1-21.

18. J.R. Parratt, Endogenous myocardial protective (antiarrhythmic) substances. *Cardiovasc Res* **27** (1993) 693-702.

19. ZF. Hassanabad *et al.*, Coronary endothelial dysfunction increases the severity of ischaemia-induced ventricular arrhythmias in rat isolated perfused hearts. *Basic Res Cardiol* **93** (1988) 241-249.

20. DL. Brutsaert *et al.*, Effects of damaging the endocardial surface on the mechanical performance of isolated heart muscle. *Circ Res* **62** (1988) 358-366.

21. Á.Végh *et al.*, Prevention by an inhibitor of the L-arginine nitric oxide pathway of the antiarrhythmic effects of bradykinin in anaesthetized dogs. *Br J Pharmacol* **110** (1993) 18-19.

22. Á. Végh *et al.*, Preconditioning of the ischaemic myocardium; involvement of the L-arginine - nitric oxide pathway. *Br J Pharmacol* **107** (1992) 648-652.

23. Á.Végh *et al.*, The local intracoronary administration of methylene blue prevents the pronounced antiarrhythmic effect of ischaemic preconditioning. *Br J Pharmacol* **107** (1992) 910-911.

24. Á. Végh *et al.*, The local intracoronary administration of nicorandil suppresses ventricular arrhythmias in a canine model of myocardial ischaemia. *Eur J Pharmacol* **305** (1996) 163-168.

25. K. György *et al.*, Isosorbide-2-mononitrate reduces the consequences of myocardial ischaemia, including arrhythmia severity: implications for preconditioning. *Cardiovasc Drugs Ther* **14** (2000) 481-488.

26. Á. Végh *et al.*, Prevention by dexamethasone of the marked antiarrhythmic effects of preconditioning induced 20 h after rapid cardiac pacing. *Br J Pharmacol* **113** (1994) 1081-1082.

27. Á. Végh *et al.*, Delayed protection against ventricular arrhythmias by monophosphoryl lipid-A in a canine model of ischaemia and reperfusion. *Eur J Pharmacol* **382** (1999) 81-90.

28. R. Ockaili *et al.*, Opening of mitochondrial K_{ATP} channel induces early and delayed cardioprotective effect: role of nitric oxide. *Am J Physiol* **277** (1999) H2425-H2434.

29. R.M. Fryer *et al.*, Ischaemic preconditioning in rats: role of mitochondrial K_{ATP} channel in preservation of mitochondrial function. *Am J Physiol* **278** (2000) H305-H312

30. C.P. Baines *et al.*, Ischaemic preconditioning depends on interaction between mitochondrial K_{ATP} channels and actin cytoskeleton. *Am J Physiol* **276** (1999) H1361-H1368.

31. G.F. Baxter and D.M. Yellon, Delayed Preconditioning and Adaptive Cardioprotection. Kluwer Academic Publisher, London, 1998.

32. WC. Sessa *et al.*, Chronic exercise in dogs increases coronary vascular nitric oxide production and endothelial cell nitric oxide synthase gene expression. *Circ Res* **74** (1993) 349-353.

Nitric Oxide
R.J. Gryglewski and P. Minuz (Eds.)
IOS Press, 2001

NITRIC OXIDE AND THE DEVELOPING HEART

Bohuslav Ošťádal

Institute of Physiology, Academy of Sciences and Centre for Cardiovascular Research, Vídeňská 1083, 14220 Prague 4, Czech Republic

1. Introduction

Appreciation of the role of NO in the cardiovascular system dates back to the discovery of endothelium–derived relaxing factor (EDRF, 1). In 1987, Moncada (2) and Ignarro (3) independently demonstrated that NO accounted for the biological activity of EDRF. Since then it has been observed that endogenous production of NO is involved in a variety of cardiovascular functions: it is a potent, locally active, vasodilator with a central role in the regulation of vascular smooth muscle tone. In addition, it inhibits leukocyte and platelet adhesion to the endothelium, leukocyte activation and platelet aggregation, and endothelial permeability. It thus optimises blood flow regulation in the microcirculation (4). Under certain pathological conditions, endogenous NO synthesis can considerably increase and cause serious organ damage: excessive NO production may contribute to cardiovascular failure in sepsis. In contrast, other pathological situations have been shown to be associated with a decreased capacity of NO synthesis; these conditions include coronary artery disease, hypertension, diabetes and heart failure (5).

At variance with numerous investigations on the effects of NO on vascular smooth muscle, its actions on the cardiac muscle has only recently attracted scientific interest. Two reviews have comprehensively summarized the role of NO in the regulation of the function of the adult myocardium under physiological and pathological conditions (4, 5). Unfortunately, data dealing with the potential role of NO in the developing heart are almost lacking. This review therefore briefly summarizes the present knowledge on the possible involvement of NO in the regulation of the immature heart.

2. Nitric oxide in the adult myocardium

In the adult heart, NO is generated (i) within cardiac myocytes, (ii) in the endothelial cells of coronary vasculature and (iii) in endocardial endothelial cells. Normal cardiomyocytes selectively express eNOS, while expression of iNOS is detectable only after incubation with various cytokines. These two isoforms of NOS show several major differences. The activity of eNOS is strictly dependent on cytosolic calcium concentration and the enzyme produces NO in nanomolar concentrations. On the other hand, iNOS produces NO in the micromolar concentration range and its activity is not regulated by cytosolic calcium.

The major target of NO in the cardiovascular system is the soluble guanylate cyclase. Activation of this enzyme results in conversion of guanosine trisphosphate (GTP) to the second messenger cyclic guanosine monophosphate (cGMP). This enzyme has been detected in many different cell types including smooth muscle cells and cardiomyocytes. Alternatively, NO can circumvent soluble guanylate cyclase and affect other cellular structures such as transmembrane ion channels or mitochondrial enzymes (5).

It has been repeatedly shown that the activity of myocardial NOSs plays a role in the regulation of myocardial contractile function (6). In summary, physiological concentrations of NO increase contractility and heart rate under basal conditions, while high NO concentrations induce the opposite effects. The results on the inotropic influence of basal endogenous NO production are, unfortunately, not homogenous. Some *in vitro* and *in vivo* studies failed to detect any influence on myocardial contraction or obtained even opposite results. The reason for this apparent discrepancy is unknown: it might be related to species differences or experimental conditions (5). Exogenous and endogenous NO has been shown to accelerate myocardial relaxation. The lusitropic effect of NO is small, as already noted for the direct inotropic effect. Moreover, NO has vascular and myocardial antiproliferative potential and can act as a bifunctional regulator of cell apoptosis (6).

Myocardial expression of iNOS can be stimulated by treatment with cytokines or E.coli lipopolysacharide (7). Subsequent to iNOS expression, a decrease in basal myocardial contractility, a diminished response to beta-adrenergic stimulation and higher rate of cell deaths have been observed. Expression and activity of iNOS may be involved in the development of contractile dysfunction of the ventricular myocardium in different diseases, such as endotoxic shock (together with hypotension and vascular hyporeactivity), heart failure, rejection after transplantation, dilated cardiomyopathy and myocarditis (4).

Recent research has implicated NO as both a trigger and mediator of the cardioprotective effect of ischaemic preconditioning (8). Free NO rapidly reacts with other radicals to form toxic side products. However, free radicals can also trigger ischaemic preconditioning since the administration of antioxidants has been found to prevent both early and late protection (9). Thus a detrimental component of NO can paradoxically also trigger subsequent protection.

3. Why developmental approach?

The importance of the developmental approach for experimental and clinical cardiology is indisputable. Clinical-epidemiological data have namely clearly shown that the risk factors of serious cardiovascular diseases, such as atherosclerosis and ischaemic heart disease, are present already during the early phases of ontogenetic development (10). Some of them, including excessive food intake and an increased level of cholesterol, operate after birth, whereas genetic factors are present even before birth. Thus atherosclerosis and ischaemic heart disease can no longer be considered the diseases of the fifth and higher decades of life, but their origin and consequences may be essentially influenced by risk factors acting already during early development. Moreover, a developmental approach offers new possibilities in the experimental studies of pathogeny, prevention and therapy of serious cardiovascular diseases. Accordingly, it is not surprising that the interest of theoretical and clinical cardiologists in the developmental approach keeps increasing.

There is, however, much controversy concerning the mechanisms of developmental variations in cardiovascular responsiveness; the obvious discrepancies may be partly due to different experimental procedures used for studying different animal species with a different time course of maturation. In this connection it is worth noting that ontogenetic conclusions are often based on observations from 2-3 developmental stages, such as "fetal", "newborn", and "adult"; the possible developmental changes between selected points thus remain undetermined. Precise knowledge of individual developmental periods that are critical for cardiac ontogeny is thus crucial for the validity and reproducibility of each developmental study (11). Developmental changes in the possible role of NO may serve as a typical example.

4. Development of the sources of NO in the cardiac muscle

As has been mentioned above, there are three sources of nitric oxide in the adult myocardium: endothelium of coronary vasculature, endocardial endothelium, and cardiomyocytes. Although the mechanisms of cardiac endothelial modulation of myocardial performance have not yet been fully elucidated, several laboratories have provided evidence that NO, resulting from the activity of cNOS and iNOS in endocardial and coronary endothelial cells or within cardiomyocytes themselves directly modulates myocardial contractility by activation of the guanylate cyclase–cGMP pathway (8). All cardiac sources of NO undergo significant, closely related changes during ontogenetic development.

Developmental pattern of the myocardium

To summarize the structural, metabolic and functional development of the myocardium would be a topic of a separate review. We therefore focus only on the few decisive events in cardiac ontogeny. The key stages in the development of the vertebrate myocardium are (Fig. 1): (i) formation of the primitive tubular heart, similar to those of the invertebrates; (ii) trabecular formation, first in the ventricle, later in the atrium; this is the final stage for some fish and amphibians; (iii) development of the compact layer, which coincides with coronary vascularization from the epicardium and (iv) development of three-layered architecture of the compact myocardium, best present in the left ventricle of homeotherms and enabling efficient pumping at high systemic pressure (11, 12).

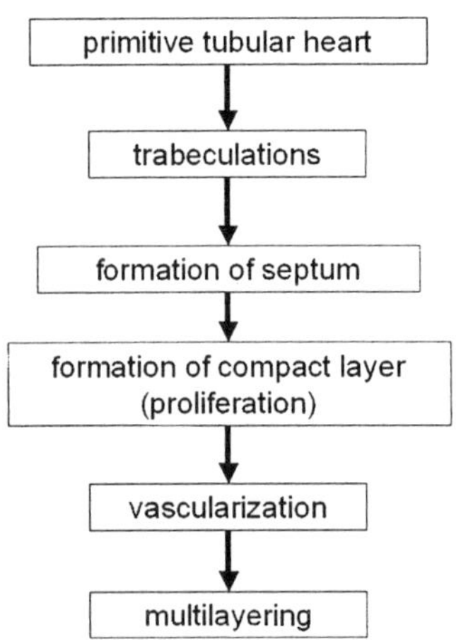

Figure 1
Developmental pattern of the myocardium (according to ref. 12).

Structural development of the myocardium is accompanied by metabolic and functional maturation of cardiac cells. Fetal metabolism is primarily anaerobic, and this adaptive property is retained during the neonatal period. The immature heart mainly depends on glycolysis because the capacity to use fatty acids is impaired due to delayed maturation of enzymes associated with mitochondrial fatty acid transport and metabolism. Postnatal development of enzyme activities depends on the degree of maturation of individual species at birth. Significant ontogenetic changes affect the systems responsible for calcium handling. The contraction of neonatal myocardium where the sarcoplasmic reticulum is not fully developed depends to a large extent on the influx of calcium across the sarcolemma via the calcium channels. During further development, the ability of the sarcoplasmic reticulum to accumulate calcium increases, and there is a progressive calcium release from the sarcoplasmic reticulum (13). In this connection it is necessary to mention that the immature heart is significantly more tolerant to oxygen deprivation than is the adult myocardium. Unfortunately, the mechanisms of the higher resistance have not yet been satisfactorily clarified (14), the possible role of NO cannot be excluded (15).

Development of endocardial endothelium

The development of endocardial endothelium has been recently summarised by Brutsaert and coworkers (16). In vertebrates, the endocardial endothelium has a different ontogenetic origin than the endothelium of coronary vessels. During early development, the proendocardial cells originate from the splanchnic mesoderm. After a process of directional migration, these cells form small cell clusters that merge into a plexus, which subsequently forms a single endocardial tube. In the tubular heart, the endocardial and myocardial layer are separated by a thick layer of extracellular material, termed the cardiac jelly. During further development, the tubular heart elongates and the myocardium develops into an outer compact and inner spongy zone. Subsequently, endocardial endothelial cells invade the inner spongy myocardium, thereby disturbing the smooth endocardial surface (17).

During the stage of trabeculation, endocardial endothelial evaginations reach deep into the myocardium and form a plexus of sinusoids, which communicate with the ventricular lumen and thus establish primitive nutrient vessels prior to the development of a true coronary circulation. These primitive channels are believed to be precursors of Thebesian vessels. The importance of the endocardial endothelium at this stage for normal development of ventricular muscle has recently been shown by Fishman et al. (18). In human hearts, the final structural organisation of the endocardium is attained by the eighth month of fetal life. Remodelling of the endocardium continues during the first years after birth; in atria, the endocardium is much thicker than in ventricles (50 –300 vs. 600-900 μm) and reaches maximum in the first decade of human life.

Development of coronary vessels

The development of the vascular bed in the heart of homeotherms has been summarized by Rakusan (19), Tomanek (20) and Ostadal et al. (11) (Fig.2). Recent studies have shown that capillaries form via vasculogenesis, i.e. from precursor cells, which migrate to the epicardium from the liver region and then form tubes. These structures first appear 13 days after conception in rats and during the fourth week of pregnancy in humans. The vascular tubes mature and become associated with pericytes; these vessels subsequently grow by angiogenesis, a process which consists of disruption of the basement membrane and migration and proliferation of endothelial cells to form new channels. The capillaries are first connected to the venous system, the formation of the arterial system is a later event. The fate of primitive sinusoidal vessels lined by endocardial endothelium is not yet clarified; they probably regress and do not participate in the development of the coronary system. The major coronary arteries form by coalescence of microvessels that grow toward and penetrate the aorta, in humans as early as after 44 days of pregnancy. The stimulus for the angioblast invasion is probably hypoxia, most pronounced in the outermost subepicardial layers. The mediators are fibroblasts and vascular endothelial growth factors.

The development of arterial vasculature continues after birth. The early postnatal period is characterised by a rapid rate of coronary capillary formation. The ultrastructural differentiation of capillary endothelium in rats is completed by the end of the second postnatal week but the development of various enzymes characteristic of capillary endothelium (ATPase, alkaline phosphatase, and dipeptidyl peptidase IV) is only completed after weaning. The main quantitative changes of the coronary capillary bed are a decrease in the muscle fiber-to-capillary ratio and a decrease in capillary density. It follows that the capillary domain, i.e. the area supplied by a single capillary increases with increasing cardiac mass and is similar in both rats and humans (21).

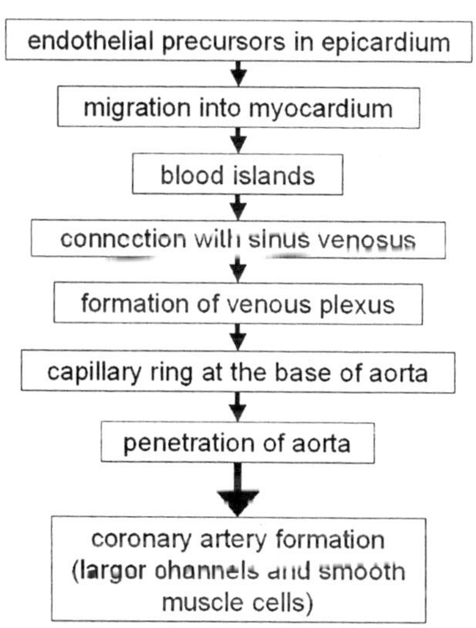

Figure 2
Mechanism of the development of coronary arteries (acording to ref. 20)

It may be concluded that changes in the heart size during ontogenetic development of homeotherms are accompanied by the gradual transformation of the avascular spongious musculature into a compact myocardium supplied through coronary vessels (Fig. 3). Ontogenetic development of the myocardial blood supply can be thus divided into three periods: (i) lacunar-up to the development of coronary arteries the myocardium is entirely spongious and supplied from the ventricular cavity lined by endocardial endothelium, (ii) transient – from the beginning of the arterial stem development to the time when the definite coronary vasculature is formed, the myocardium may be supplied both from the lumen and from the developing coronary bed;

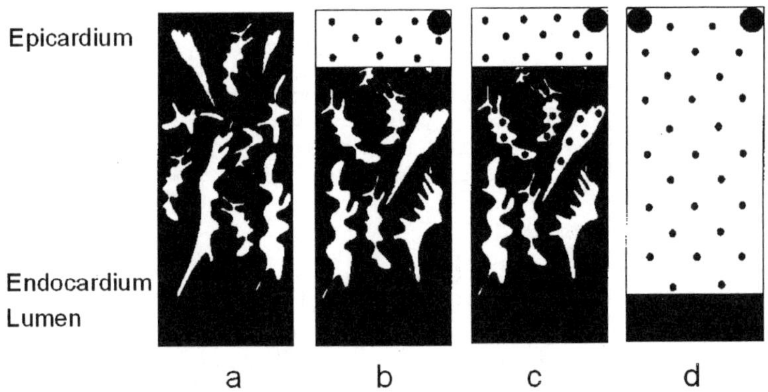

Figure 3
Different types of myocardial blood supply. a: Spongious musculature supplied from ventricular lumen. b: Inner spongious layer is covered by an outer compact musculature with vascular supply. c: As in b, but capillaries are present also in some trabecels of spongylike musculature. d: Compact musculature supplied from coronary vessels (according to ref. 11).

the amount of endothelial cells of vascular origin increases; these two stages are typical also for poikilotherm animals; (iii) coronary – when the development of coronary vessels is completed the blood supply is effected mostly from coronary arteries (11).

Unfortunately, there are no data available on the cardiac endothelium–cardiomyocyte interactions in the developing heart. The ontogenetic approach offers stimulating possibilities for this analysis: for example, there are precisely defined phylogenetic and ontogenetic developmental periods where only the endocardial endothelium is present, periods with different sensitivity of myofilaments to intracellular calcium etc.

5. Role of NO during cardiomyogenesis

As it has been mentioned above, data on the role of NO in the developing heart are very scarce. Bloch et al. (22) have demonstrated that iNOS and eNOS isoforms are expressed early during mouse and rat embryonic heart development whereas they are down-regulated in the last quarter of the fetal life. Indeed, prior to birth only a weak expression of eNOS can be detected and iNOS expression was hardly detectable. These changes were found to be associated with a switch in the signaling cascades involved in the regulation of intracellular calcium. In addition to the classical way of action of NO through intracellular signaling cascades via activation of soluble guanylyl cyclase and cGMP, NO-mediated gene regulation by direct interaction with DNA binding domains has been proposed (23).

Bloch et al. (22) further investigated whether NO plays a role in cardiomyogenesis, using *in vitro* embryonic stem cell-derived cardiomyocytes. The NOS expression pattern in these cells paralleled the one detected *in vivo*. They found that continuous incubation of cardiomyocytes with NOS inhibitors resulted in a pronounced differentiation arrest; this effect could be reversed by the NO donor. It seems, therefore, that NO generation is required for cardiomyogenesis since NOS inhibitors prevent the maturation of terminally differentiated cardiomyocytes.

6. NO and protection of the developing heart

Degree of cardiac hypoxic/ischaemic injury depends not only on the intensity and duration of the hypoxic stimulus but also on the level of cardiac tolerance to oxygen deprivation. It is, therefore, not surprising that the interests of many experimental and clinical cardiologists during the past 40 years has been focused on the question of how cardiac tolerance to oxygen deprivation might be increased. Whereas a substantial amount of data is available concerning protection of the adult myocardium, much less is known about this phenomenon in the developing heart. In this regard there are two main possibilities: long lasting adaptation to chronic hypoxia or short lasting adaptation, i.e. preconditioning.

Adaptation to chronic hypoxia

Adaptation to chronic hypoxia is characterized by a variety of functional changes that help to maintain homeostasis with minimum energy expenditure. Such adjustment may protect the heart under conditions that require enhanced work and consequently increased metabolism. In addition to such protective effects, adaptation to chronic hypoxia may also exert adverse influences on the cardiopulmonary system, including pulmonary hypertension and right ventricular hypertrophy, which may result in congestive heart failure. It was reported already in late 1950s that the incidence of myocardial infarction is lower in people who live at high altitude. Epidemiological observations are consistent with the experimental results, using acute ischaemia or hypoxia for testing the myocardial resistance in animals adapted to simulated high altitude in a barochamber.

It has been demonstrated that adaptation to chronic hypoxia increases cardiac resistance to acute hypoxic injury in both adult and immature hearts (24,25). However, the exact cardioprotective mechanisms of adaptation to chronic hypoxia are still unknown. Recently, Asemu et al (26) suggested, that activation of mitochondrial ATP-dependent potassium channels might be involved in the cardioprotection of adaptation to chronic hypoxia. Mechanisms that may lead to activation of mitochondrial K_{ATP} channels in chronic hypoxia-induced cardioprotection are still unknown. Although several possibilities can be referred in the literature, direct evidence is missing.

Baker et al. (27) have found that exposure of immature rabbit hearts to chronic hypoxia increased mRNA levels for eNOS, as well as release of nitrite, nitrate, and tissue cGMP content. Moreover, NO donor GSNO increased recovery of post-ischaemic function in normoxic hearts but not in hearts chronically hypoxic from birth. These results suggest that NO production is maximal in hypoxic hearts, and that no further protection would be expected in the presence of NO donor. Conversely, NOS inhibitors abolished the cardioprotective effect of hypoxia. Similar effect of NOS inhibitor was observed in neonatal rat hearts, adapted to high altitude hypoxia (28). Baker et al. (27) proposed that increased production of NO in chronically hypoxic animals leads to activation of sarcolemmal and mitochondrial K_{ATP} channels via soluble guanylyl cyclase, causing accumulation of cGMP and activation of cGMP-dependent protein kinase. Thus, increased NO production in the hypoxic hearts may be an adaptive response as other proposed mechanisms, like polycythaemia and increased hemoglobin level (29).

Ischaemic preconditioning

It is now 14 years since the cardioprotective phenomenon termed "ischaemic preconditioning" has been formally recognized. In their study, Murry et al (30) reported that infarct size produced by sustained coronary occlusion could be markedly reduced if they first "preconditioned" the heart with the several episodes of brief ischaemia. Whereas extensive data are available on ischaemic preconditioning in the adult myocardium, information on whether this protective phenomenon also occurs in immature heart is inadequate. It has been shown (14), that classical ischaemic preconditioning, at least in rats is not present at birth and that the enhanced postischaemic recovery of contractile function only develops at the end of the first postnatal week.

The precise mechanisms of ischaemic preconditioning in the adult myocardium are still unclear and the same is valid for the immature heart. Among the long list of possible mediators, NO has been suggested to play an important role in the cardioprotective effect of ischaemic preconditioning. It has been namely shown, that the early and late phases of this phenomenon can be blocked by the NO pathway inhibitors (31, 32, 8). The precise mechanism is not known, but the finding that NO can potentiate the ATP sensitive K channel (33) raises the distinct possibility that NO could modulate the possible end effector of ischaemic preconditioning – mitochondrial K_{ATP} channels (8). Recently, Baker et al (34) have shown that also preconditioning in immature rabbit hearts is associated with activation of K_{ATP} channels, possibly those localized in mitochondria, similarly as in the cardioprotective effect of adaptation to chronic hypoxia. Nevertheless, further studies on the postnatal development of this protective phenomenon would be desirable.

7. NO and the fetal and neonatal pulmonary circulation

Finally, we would like to mention ontogenetic changes in the role of NO in the pulmonary circulation. Transition of a fetus to postnatal life is characterized by a rapid decrease in pulmonary vascular resistance and a 10-fold increase in pulmonary blood flow at birth (35). Endothelial cells modulate this pulmonary vasodilation by release of NO, a potent pulmonary vasodilator. A significant increase in expression of eNOS mRNA and NO production have been demonstrated at birth and during neonatal period. Moreover, the postnatal decline of pulmonary vascular resistance can be attenuated by the treatment with NOS inhibitors. Basal NO synthesis in the pulmonary circulation has thus a characteristic ontogenetic development: it is high in the fetus, culminates at the time of birth, declines postnatally, and is relatively small during adulthood (36).

8. Conclusions

Whereas a lot of data are available concerning the role of NO in the adult myocardium, much less is known on the physiology and pathophysiology of NO in the developing heart. Three sources of NO in the cardiac muscle (endocardial endothelium, vascular endothelium and cardiomyocytes) change substantially during ontogenetic development. Both iNOS and eNOS are expressed already during early stages of cardiomyogenesis; their role is, however, not yet fully understood but it may be suggested that NO is required for maturation of cardiomyocytes. Another very stimulating field is the involvement of NO in the mechanisms of the two most potent cardioprotective phenomena: adaptation to chronic hypoxia and ischaemic preconditioning. And last but not least, NO modulates postnatal development of the right heart by rapid decrease in pulmonary vascular resistance at birth. Developmental approach offers new stimulating possibilities for the analysis of the role of NO in the regulation of cardiac function as well as for the analysis of the very complex interactions of endothelial and myocardial cells. We do hope that developmental biologists might come to a new appreciation of the amazing experimental potential of the developing cardiovascular system.

References

[1] R.F. Furchgott and J.V . Zawadski, The obligatory role of endothelial cells in the relaxation of arterial smooth muscle by acetylcholine, *Nature* 288 (1980) 373-376.

[2] R.M.J. Palmer *et al.*, Nitric oxide release accounts for the biological activity of endothelium-derived relaxing factor, *Nature* 327 (1987) 524-526.

[3] L.J. Ignarro *et al.*, Endothelium-derived relaxing factor produced and released from artery and vein is nitric oxide, *Proc Natl Acad Sci USA* 84 (1987) 9265-9269.

[4] A.M. Shah, Inducible nitric oxide synthase and cardiovascular disease, *Cardiovasc Res* 45 (2000) 148-155.

[5] G. Kojda and K. Kottenberg, Regulation of basal myocardial function by NO, *Cardiovasc Res* 41 (1999) 514-523.

[6] R.A. Kelly *et al.*, Nitric oxide and cardiac function, *Circ Res* 79 (1996) 363-380.

[7] R. Schulz *et al.*, Induction and potential biological relevance of a calcium-independent nitric oxide synthase in the myocardium, *Br J Pharmacol* 105 (1992) 575-580.

[8] R.D. Rakhit *et al.*, Nitric oxide, nitrates and ischeamic preconditioning, *Cardiovasc Res* 43 (1999) 621-627.

[9] J.-Z. Sun *et al.*, Evidence for an essential role of reactive oxygen species in the genesis of late preconditioning against myocardial stunning in conscious pigs, *J Clin Invest* 97 (1996) 562-776.

[10] Z. Fejfar, Prevention against ischemic heart disease" a critical review. In:M.F. Oliver (ed.), Modern Trends in Cardiology, Butterworth-Heinemann, Boston, 1975, pp.465-499.

[11] B. Ošťádal *et al.*, Development of cardiac sensitivity to oxygen deficiency: comparative and ontogenetic aspects, *Phys Rev* 79 (1999) 635-659.

[12] D. Sedmera *et al.*, Developmental patterning of the myocardium, *Anat Record* 258 (2000) 319-337.

[13] H. Tanaka and K. Shigenobu, Effect of ryanodine on neonatal and adult rat heart: developmental increase in sarcoplasmic reticulum function, *J Mol Cell Cardiol* 21 (1989) 1305-1313.

[14] I. Ošťádalová *et al.*, Tolerance to ischaemia and ischaemic preconditioning in neonatal rat heart. *J Mol Cell Cardiol* 30 (1998) 857-865.

[15] S. Brovkovych and T. Malinski, The role of nitric oxide in ischemia of the developing heart, Exp Clin Cardiol 5 (2000) 35.

[16] D.L. Brutsaert *at al.*, The cardiac endothelium: functional morphology, development, and physiology, *Progress Cardiovasc Dis* 39 (1996) 239-262.

[17] J.M. Icardo and F.J. Manasek, Fibronectin distribution during early chick embryo heart development, *Dev Biol* 95 (1983) 19-30.

[18] M.C. Fishman and D.Y.R. Stainier, Cardiovascular development. Prospects for genetic approach, *Circ Res* 74 (1994) 757-763.

[19] K. Rakusan, Coronary angiogenesis. Mechanisms, methods, and consequences. *Adv Organ Biol* 7 (1999) 1-23.

[20] R.J. Tomanek, Vascularization of the heart during prenatal and perinatal growth, *Adv Organ Biol* 7 (1999) 111-127.

[21] K. Rakusan *et al.*, Morphometry of human coronary capillaries during normal growth and the effect of age in left ventricular pressure-overload hypertrophy, *Circulation* 86 (1992) 38-46.

[22] W. Bloch *et al.*, Nitric oxide synthase expression and role during cardiomyogenesis, *Cardiovasc Res* 43 (1999) 675-684.

[23] S.A. Henderson *et al.*, Nitric oxide reduces early growth response-1 gene expression in rat lung macrophages treated with interferon-gamma and lipopolysaccharide, *J Biol Chem* 269 (1994) 25239-25242.

[24] B. Ošťádal *et al.*, Ontogenetic differences in cardiopulmonary adaptation to chronic hypoxia, *Phys Res* 44 (1995) 41-45.

[25] J.E. Baker *et al.*, Tolerance of the developing heart to ischemia: impact of hypoxemia from birth. *Am J Physiol* 268 (1995) H1165-H1173.

[26] G. Asemu *et al.*, Adaptation to high altitude hypoxia protects the rat heart against ischemia-induced arrhythmias. Involvement of mitochondrial K_{ATP} channel, *J Mol Cell Cardiol* 41 (1999) 1821-1831.

[27] J.E. Baker *et al.*, Adaptation of hearts to chronic hypoxia increases tolerance to subsequet ischemia by increased nitric oxide production. In: Hudetz and Bruley (eds.), Oxygen transport tissue to XX. Plenum Press, New York, 1998.

[28] I. Ošťádalová *et al.*, Adaptation to chronic hypoxia and ischaemic preconditioning in neonatal rat heart, *Exp Clin Cardiol* 5 (2000) 43.

[29] B. Ošťádal and F. Kolář, Cardiac ischemia: From Injury to Protection. ISBN 0-7923-8642-6. Kluwer Academic Publishers, Boston, Dordrecht, London, 1999.

[30] C. E. Murry *et al.*, Preconditioning with ischemia: a delay of lethal cell injury in ischemic myocardium, *Circulation* 74 (1986) 1124-1136.

[31] A. Vegh *et al.*, Preconditioning of the ischaemic myocardium; involvement of the L arginine nitric oxide pathway, *Br J Pharmacol* 107 (1992) 648-652.

[32] R. Bolli *et al.*, Evidence that late preconditioning against stunning in conscious rabbits is triggered by the generation of nitric oxide, *Circ Res* 81 (1997) 42-52.

[33] A. Shinbo and T. Ijima, Potentiation by nitric oxide of the ATP-sensitive K current induced by K channels openers in guinea-pig ventricular cells, *Br J Pharmacol* 120 (1997) 1568-1574.

[34] J.E. Baker *et al.*, Adaptation to chronic hypoxia confers tolerance to subsequent myocardial ischemia by increased nitric oxide production, *Ann N Y Acad Sci* 30 (1999) 236-253

[35] M.E. Villaneuva *et al.*, Decreased gene expression of endothelial nitric oxide synthase in newborns with persistent pulmonary hypertension, *Pediatric Res* 44 (1998) 338-343.

[36] V. Hampl and J. Herget, Role of nitric oxide in the pathogenesis of chronic pulmonary hypertension, *Phys Rev* 88 (2000) 1337-1372.

Nitric Oxide
R.J. Gryglewski and P. Minuz (Eds.)
IOS Press, 2001

Oxidized Lipids and Platelet Activation: Interaction with Nitric Oxide

Pietro MINUZ, Stefania GAINO, Valeria ZULIANI[1], Rosamaria TOMMASOLI[1], Alessandro LECHI, Clara LECHI SANTONASTASO[1].

Departement of Biomedical and Surgical Sciences and Department of Morphological and Biomedical Sciences[1], University of Verona, Policlinico GB Rossi, Piazzale LA Scuro, 37134 Verona, Italy

Abstract: Oxidized low density lipoprotein (LDL) and products of lipid peroxidation such as the F_2-isoprostane 8-iso-prostaglandin (PG) $F_{2\alpha}$ are weak platelet agonists, able to induce shape change and platelet adhesion, but not platelet aggregation. However both oxidized LDL and 8-iso-PG$F_{2\alpha}$ are able to prevent in vitro the antiaggregatory effects of endothelial cells releasing nitric oxide (NO) or NO-donors. These effects appear at least partly dependent on antagonism towards NO-induced reduction in platelet intracellular calcium. Activation of kinases and cytoskeleton-associated proteins may also explain the activatory effects of oxidized lipid and LDL.

A number of clinical conditions related to athero-thrombosis are characterized by an intermediate phenotype consisting of altered endothelial function, increased oxidative stress and platelet activation. Hypercholesterolemia and diabetes mellitus are typical conditions in which this syndrome is observed. Reduction in nitric oxide (NO) bio-activity has been extensively demonstrated in these diseases and this thought to be the patho-physiological determinant of endothelial dysfunction[1,2]. It has been hypothesized that increase in oxidative stress might be one of the mechanisms responsible for alteration in nitric oxide release and activity. In fact it has been observed that alteration in endothelium-dependent vasodilatory activity, detected in humans by means of the intra-arterial infusion of acethylcholine, can be reversed by the co-administration of vitamin C, thus suggesting that oxidation underlies the reduction in NO bio-activity[3,4].

Platelet activation may also derive from increased oxidative stress[5]. This has been demonstrated *in vivo* by showing increased urinary excretion of thromboxane A_2 metabolites in patients with diabetes mellitus and hypercholesterolemia, accompained by the parallel increase in urinary excretion of 8-isoprostaglandin $F_{2\alpha}$, index of lipid peroxidation[6,7]. There is evidence from functional and mechanistic studies that oxidized lipids antagonize the antiplatelet activity of NO.

1. F_2 -isoprostanes, biolgically active lipid peroxides

Products of lipid oxidation are generated *in vivo* in clinical conditions characterized by increased oxidative stress that may account for the alteration in the equilibrium between pro and antiaggregatory factors.

We focused our attention on the possible role of F_2-isoprostanes as platelet activators. These compounds are prostaglandin isomers synthesized *in vivo* mostly independently of the activity of cyclo-oxygenase, through a free radical-catalyzed peroxidation of arachidonic acid in the biological membranes[8]. F_2-isoprostanes generation occurs also during copper and cell-induced oxidation of LDL[9,10] and we observed that also mildly oxidized LDL, obtained by leaving native LDL in the light at room temperature contain appreciable amounts of 8-epi-$PGF_{2\alpha}$ (1 ng/mg protein).
F2-isoprostanes are currently considered reliable markers of lipid peroxidation and used to estimate oxidative stress *in vivo*[11]. However, at least some of these compound have been shown to have biological activities. In fact, 8-iso-$PGF_{2\alpha}$ is able to induce platelet shape-change and induce platelet aggregation in the presence of minimal doses of strong platelet agonists[12]. It is also a vasoconstricor and possesses a variety of additional biological activities, such as broncocostriction and neutrophil adhesion[13,14]. Confirmation of the potential role of 8-iso-$PGF_{2\alpha}$, and possibly of other related compounds, as platelet activators *in vivo* comes from the observation that 8-iso-$PGF_{2\alpha}$ modulates platelet function when infused in experimental animals[15].

2. 8-iso-$PGF_{2\alpha}$ and the antiplatelet activity of NO

We investigated some aspects of platelet function focusing our attention on early and critical events of platelet activation such as platelet adhesion, the expression of adhesion molecules and platelet secretion. We also studied the effects of 8-iso-$PGF_{2\alpha}$ in the presence of endothelial cells releasing NO and in the presence of NO-donors to mimic the antiaggregatory environment in which circulating platelets are located. We hypothesized that platelets could be primed and partially activated by F_2-isoprostanes and that this activatory signal could counteract the effects of NO.
To investigate the effects of 8-iso-$PGF_{2\alpha}$ on platelet adhesion we used fibrinogen coated microwells. In this experimental model platelet adhesion is entirely dependent on fibrinogen binding to its platelet receptor GP-IIb/IIIa, since completely prevented by anti-GP-IIb/IIIa monoclonal antibodies, is accompanied by cell spreading and requires several minutes to be completed. This model may not be representative of the events that have been identified occur in the bloodstream, where engagement of the vonWillebrand receptor appears to initiate platelet activation[16]. However the use of immobilized ligands allows to explore the interaction between soluble platelet agonists and antagonists. Using immobilized fibrinogen it has been shown that platelet adhesion is increased by platelet agonists and reduced by NO[17].
NO is an inhibitor of platelet adhesion. This effect has been observed *in vitro* under different experimental conditions[18]. It also inhibits the expression of adhesion molecules on platelet surface, as shown *in vivo* in the human coronary circulation where the expression of P-selectin (also an index of platelet secretion, since stored in the α–granules in resting platelets) and GP-IIb/IIIa was found reduced in the presence of the NO-donor S-nitrosoglutathione[19]. We observed that *in vitro* NCX-4016, an acetylsalicylic acid analogue able to gradually release NO, dose-dependently reduced platelet P-selectin and GP-IIb/IIIa expression in thrombin-stimulated platelets with a parallel reduction in platelet adhesion to immobilized fibrinogen[20].
Using these methods to investigate platelet adhesion we found that a dose-dependent increase in the number of cells adhering to immobilized fibrinogen and an increase in the

expression of GP-IIb/IIIa was induced by 8-epi-$PGF_{2\alpha}$ in the presence the NO-donor NOR-3.
An interesting aspect of the profile of activity of 8-iso-$PGF_{2\alpha}$ on platelets is that it is unable to induce platelet secretion (assessed by measuring the expression of P-selectin) and platelet aggregation at any concentration[21]. However when 8-iso-$PGF_{2\alpha}$ is added to threshold doses of a strong platelet agonists such as thrombin, irreversible platelet aggregation is induced[12].
To further investigate the functional interaction between 8-iso-$PGF_{2\alpha}$ and NO, we tested the effects of 8-epi-$PGF_{2\alpha}$ on platelet aggregation using a previously described experimental model of platelet-endothelial cell interaction[22]. At least *in vitro,* the antiaggregatory effects of endothelial cells is almost completely mediated by the release of NO and prostacyclin. The effects NO can be investigated by using endothelial cells treated with cyclo-oxygenase inhibitors[23]. If endothelial cells and platelets are stimulated by a strong agonist, such as thrombin, platelet activation is prevented by the release of antiaggregatory factors from endothelial cells. When cyclo-oxygenase is inhibited, endothelial cells maintain a residual antiaggregatory activity that is dependent entirely on NO release, since can be completely reversed by monomethyl-L-arginine, inhibitor of NO synthase[23]. Using thrombin as a stimulus, NO released from endothelial cells inhibited platelet aggregation and this inhibitory effect was reduced in the presence of 8-epi-$PGF_{2\alpha}$ (Fig. 1). Similarly the nitric-oxide donor NOR-3 reduced the aggregatory effects of thrombin in aspirin-treated platelets and this activity was counteracted by the addition of 8-epi-$PGF_{2\alpha}$[21].

3. Intraplatelet signaling of 8-iso-$PGF_{2\alpha}$

These functional experiments indicate that 8-epi-PGF_2 is a weak platelet agonist. Nevertheless it is able to antagonize the antiaggregatory effects of endothelial-derived NO and that of NO-donors. It appears plausible that intracellular interaction with NO-dependent signaling systems may occur.
We tested this hypothesis by measuring free intracellular calcium with the fluorescent indicator FURA-2. In fact, a number of experimental data indicate that NO reduces free intracellular calcium in platelets[24]. When washed platelets were stimulated with 8-epi-$PGF_{2\alpha}$ 1 μmol/L a very limited and transient increase in intracellular calcium was observed[21]. When 8-iso-$PGF_{2\alpha}$ was added to thrombin, intracellular calcium increased in a dose-dependent manner from 1 nmol/L to 1 μmol/L, mainly through the release of calcium from intracellular stores. Incubation with a NO-donor reduced platelet intracellular calcium and this effect was reversed by low nanomolar concentrations of 8-epi-$PGF_{2\alpha}$[21].
Antagonistic regulation of intracellular calcium appears therefore to be a mechanism by which 8-epi-$PGF_{2\alpha}$ functionally interacts with NO. However, additional mechanisms may be involved.
There are evidences that most of the effects of 8-iso-$PGF_{2\alpha}$ is mediated through the activation of thromboxane/endoperoxide (TP) receptors. However the signaling system engaged by 8-iso-$PGF_{2\alpha}$ to induce platelet activation is somewhat different from that engaged by thromboxane A_2. It appears that 8-iso-$PGF_{2\alpha}$ does not operate through the system that mediates platelet secretion and aggregation leading to a limited increase in intracellular calcium[21]. Part of the activatory effects of 8-epi-$PGF_{2\alpha}$ in the platelets depend on the engagement of non-receptor tyrosine-kinase pathways. Shape-change is associated

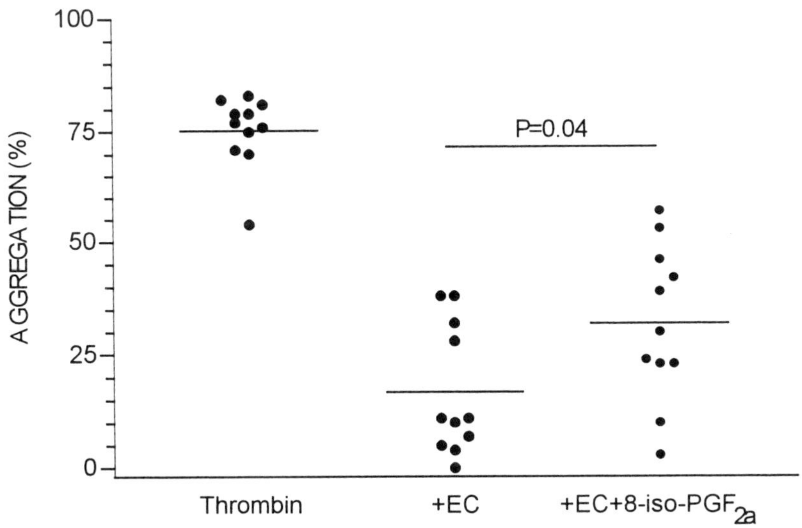

Figure 1: Endothelial cells releasing nitric oxide (2-7.5 x 10^5 cells/ml) prevent platelet aggregation induced by 0.3 U/ml thrombin. The F2 isoprostane, 8-epi-$PGF_{2\alpha}$ 1μmol/L, partially reversed the antiaggregatory effect of nitric oxide. The experiments were performed using washed platelets (1 x 10^8 cells/ml). Platelet aggregation was monitored for 3 minutes using a turbidimetric aggregometer.

with the phosphorylation of p72 Syk and the cytoskeleton-associated protein cortactin. Both tyrosine phosphorylation of cortactin and p72 Syk occur very early during platelet activation and are kinetically correlated with the occurrence of shape-change[25].
We observed that biphosphorylation of the stress and mitogen activated (MAP) kinase p38 is also induced by 8-epi-$PGF_{2\alpha}$ (personal observations). This kinase phosphorylation occurs upstream cytoskeleton reorganization, since not altered by cytochalasin D, and this is functionally associated to activation events, such as increased platelet adhesion to immobilized fibrinogen and shape-change, as demonstrated by the inhibitory effects of the specific p38 MAP-kinase inhibitor SB203580 (personal observations).

4. Oxidized low density lipoprotein (LDL) and platelet activation

Experimental data accumulated in last ten years indicate that oxidized low density lipoprotein (LDL) induces platelet activation. Platelet secretion, sensitization to agonists and increased platelet aggregability have been observed to be induced in a dose dependent manner by oxidized LDL and also by native LDL[26,27].
Early observations indicated protein kinase C as the intracellular target for the activatory effects of oxidized LDL. More recently, additional kinases been observed activated in platelet exposed to LDL. It has been shown that the p38 MAP kinase and 125 Syk become phosphorylated in the presence of high concentrations of native LDL[28,29]. These event are associated with a transient increase in intracellular calcium. Activation of p38 MAP kinase and increased intracellular calcium appear to be responsible for phospholipase A_2 phosphorylation and cyctoskeleton reorganization[29]. Moreover, mildly oxidized LDL has been shown to induce platelet shape change via myosin light-chain and moesin phosphorylation. The activation of these cytoskeleton-associated enzymes occurs

downstream Rho-kinase phosphorylation and is not accompanied with any significant increase in intracellular calcium[30]. All these data strongly suggest that LDL-induced platelet activation is linked to cytoskeleton reorganization.

5. Reduction in the antiplatelet activity of NO by oxidized LDL

We performed some experiments to investigate the role of oxidation in LDL-induced platelet activation. We also evaluated any potential antagonistic role of oxidized LDL towards NO. Our intention was to simulate *in vitro* the vascular microenvironment in which circulating platelets, endothelial cells and oxidized LDL interact.
Using the experimental model of platelet-endothelial cells interaction, we studied the effects of mildly oxidized LDL, obtained by leaving native LDL in the light at room temperature for at least seven days[23]. Under these condition the lipid component of LDL becomes oxidized and peroxides such as linoleate hydroperoxides and isoprostanes are generated. These LDL preparations markedly reduced the anti-aggregatory activity of NO released from endothelial cells. This was not observed using native LDL indicating that oxidation was responsible [23].

6. Mechanisms of LDL-NO interaction

There are many possible mechanisms by which oxidized lipids may interfere with the anti-aggregatory activity of NO. If oxidized LDL are placed in contact with endothelial cells, NO activity may be affected through the reduction in NO-synthase. However, there is evidence that in our experimental conditions LDL activity was exerted on platelets rather than endothelial cells: the effect of oxidized LDL was almost immediate, was not reproduced by the preincubation of endothelial cells with oxidized LDL and, finally, was obtained also by using NO-donors instead of endothelial cells[23]. In fact we observed that a number of compounds, able to gradually release nitric oxide, can mimic the activity of endothelial derived nitric oxide on platelet function[31]. The antiaggregatory effects of NO-donors was evident when platelet were stimulated with a strong platelet agonist at a concentration able to induce a sub-maximal aggregation[32]. The addition of mildly oxidized LDL in this platelet preparation reduced the effects of NO (Fig.1).
To further investigate the role of oxidation in the biochemical mechanism of action of mildly oxidized LDL we used antioxidants. The addition of ascorbic acid to a platelet preparation did not modify the effects of thrombin. However, in the presence of a NO-donor the proaggregatory effects of ox-LDL were counteracted by ascorbic acid (Fig.1).
These results concerning the interaction of mildly oxidized LDL with NO are consistent with the hypothesis that reactive lipid peroxides generated within LDL may lead to the formation of nitroperoxides[33]. An alternative hypothesis that can be formulated is that NO-dependent signaling system in the platelets is altered by oxidized LDL. As observed in different experimental models, lipid components of oxidized LDL may be responsible of cell activation and the reduction in NO bio-activity[34,35]. However, this has not yet been demonstrated in platelets.
To investigate the mechanisms by which ox-LDL affect the bio-activity of nitric oxide in the platelets we measured cGMP. We observed that platelet cGMP increase induced by a NO-donor (NOR-3, 1-5 μmol/L) was completely prevented by mildly oxidized LDL but not

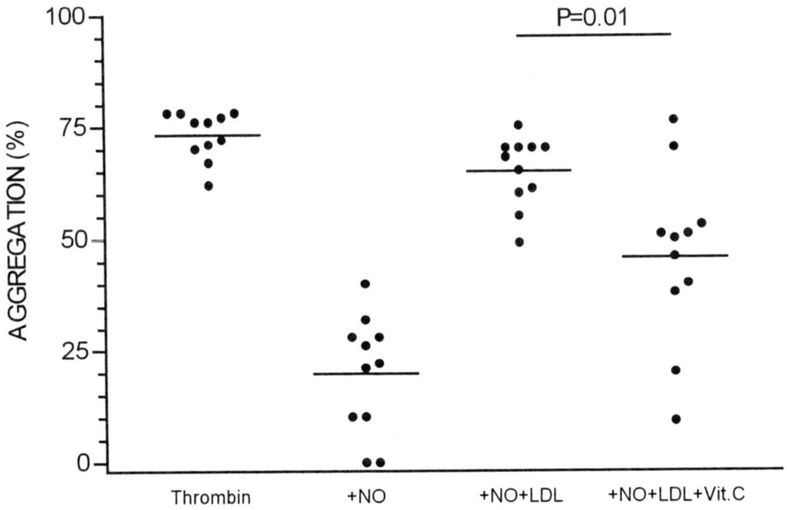

Figure 2: The antiaggregatory activity of NO-donors (NOR3, 1-10 μmol/L), is antagonized by mildly oxidized LDL (200 μg protein/ml). The addition of vitamin C (60 μg/mL) partially prevents the effects of LDL. Washed platelet ($1x10^8$ cells/mL) were stimulated with thrombin (0.03 U/mL). Each set of experiments was performed in parallel using the same cell and LDL preparations. Platelet aggregation was monitored using a 4 channel turbidimetric aggregometer.

native LDL (personal observation). We also evaluated free intracellular calcium. Using the fluorescent indicator FURA-2 to measure changes in intracellular calcium, we observed that neither native or mildly oxidized-LDL affected calcium concentrations in resting and thrombin-stimulated platelets. Nevertheless, mildly oxidized LDL was able to counteract the reduction in intracellular calcium caused by NO[32].

There is also evidence that oxidized LDL are active also on NO generated within the platelets[36]. A concentration-dependent increase in thrombin-induced platelet aggregation and serotonin release was observed to be induced by oxidized LDL. This was associated with reduction in cGMP in washed platelets, indicating a direct effect on the intra-platelet generation of NO. This was demonstrated to be related to a diminishing NO synthase expression as well as a decreased uptake of L-arginine[37].

In conclusion the activity of products of lipid oxidation may explain the increase in platelet activation observed in clinical conditions characterized by increased oxidative stress. This appears to be related partly to direct platelet stimulation and partly to inhibition of intra-platelet bio-activity of NO.

References

1. Arcaro G, Zenere BM, Travia D, Zenti MG, Covi G, Lechi A, Muggeo M: Non-invasive detection of early endothelial dysfunction in hypercholesterolaemic subjects. Atherosclerosis 114:247, 1995

2. De Vriese AS, Verbeuren TJ, Van de Voorde J, Lameire NH, Vanhoutte PM: Endothelial dysfunction in diabetes. Br J Pharmacol 130:963, 2000

3. Levine GN, Frei B, Koulouris SN, Gerhard MD, Keaney JF, Vita JA: Ascorbic acid reverses endothelial vasomotor dysfunction in patients with coronary artery disease. Circulation 93:1107, 1996

4. Ting HH, Timimi FK, Haley EA, Roddy MA, Ganz P, Creager MA: Vitamin C improves endothelium-dependent vasodilation in forearm resistance vessels of humans with hypercholesterolemia. Circulation 95:2617, 1997

5. Keaney JF, Loscalzo J: Diabetes, oxidative stress, and platelet activation. Circulation 99:189, 1999

6. Davi G, Alessandrini P, Mezzetti A, Minotti G, Bucciarelli T, Costantini F, Cipollone F, Bon GB, Ciabattoni G, Patrono C: In vivo formation of 8-Epi-prostaglandin $F_{2\alpha}$ is increased in hypercholesterolemia. Arterioscler Thromb Vasc Biol 17:3230, 1997

7. Davi G, Ciabattoni G, Consoli A, Mezzetti A, Falco A, Santarone S, Pennese E, Vitacolonna E, Bucciarelli T, Costantini F, Capani F, Patrono C: In vivo formation of 8-iso-prostaglandin $F_{2\alpha}$ and platelet activation in diabetes mellitus: effects of improved metabolic control and vitamin E supplementation [see comments]. Circulation 99:224, 1999

8. Roberts LJ, Morrow JD: Measurement of F(2)-isoprostanes as an index of oxidative stress in vivo. Free Radic Biol Med 28:505, 2000

9. Gopaul NK, Nourooz-Zadeh J, Mallet AI, Anggard EE: Formation of F_2-isoprostanes during aortic endothelial cell-mediated oxidation of low density lipoprotein. FEBS Lett 348:297, 1994

10. Lynch SM, Morrow JD, Roberts LJ, Frei B: Formation of non-cyclooxygenase-derived prostanoids (F_2-isoprostanes) in plasma and low density lipoprotein exposed to oxidative stress in vitro. J Clin Invest 93:998, 1994

11. Morrow JD: The isoprostanes: their quantification as an index of oxidant stress status in vivo. Drug Metab Rev 32:377, 2000

12. Pratico D, Smyth EM, Violi F, FitzGerald GA: Local amplification of platelet function by 8-Epi prostaglandin $F_{2\alpha}$ is not mediated by thromboxane receptor isoforms. J Biol Chem 271:14916, 1996

13. Wilson SH, Best PJ, Lerman LO, Holmes DR, Richardson DM, Lerman A: Enhanced coronary vasoconstriction to oxidative stress product, 8-epi-prostaglandin$F_{2\alpha}$, in experimental hypercholesterolemia. Cardiovasc Res 44:601, 1999

14. Fontana L, Giagulli C, Minuz P, Lechi A, Laudanna C: 8-Iso-PG$F_{2\alpha}$ induces ss(2)-integrin-mediated rapid adhesion of human polymorphonuclear neutrophils : A link between oxidative stress and Ischemia/Reperfusion injury. Arterioscler Thromb Vasc Biol 21:55, 2001

15. Audoly LP, Rocca B, Fabre JE, Koller BH, Thomas D, Loeb AL, Coffman TM, FitzGerald GA: Cardiovascular responses to the isoprostanes iPF(2α)-III and iPE(2)-III are mediated via the thromboxane A(2) receptor in vivo. Circulation 101:2833, 2000

16. Kulkarni S, Dopheide SM, Yap CL, Ravanat C, Freund M, Mangin P, Heel KA, Street A, Harper IS, Lanza F, Jackson SP: A revised model of platelet aggregation. J Clin Invest 105:783, 2000

17. Shahbazi T, Jones N, Radomski MW, Moro MA, Gingell D: Nitric oxide donors inhibit platelet spreading on surfaces coated with fibrinogen but not with fibronectin. Thromb Res 75:631, 1994

18. Sneddon JM, Vane JR: Endothelium-derived relaxing factor reduces platelet adhesion to bovine endothelial cells. Proc Natl Acad Sci U S A 85:2800, 1988

19. Langford EJ, Brown AS, Wainwright RJ, de Belder AJ, Thomas MR, Smith RE, Radomski MW, Martin JF, Moncada S: Inhibition of platelet activity by S-nitrosoglutathione during coronary angioplasty. Lancet 344:1458, 1994

20. Lechi C, Andrioli G, Gaino S, Tommasoli R, Zuliani V, Ortolani R, Degan M, Benoni G, Bellavite P, Lechi A, Minuz P: The antiplatelet effects of a new nitroderivative of acetylsalicylic acid--an in vitro study of inhibition on the early phase of platelet activation and on TXA_2 production. Thromb Haemost 76:791, 1996

21. Minuz P, Andrioli G, Degan M, Gaino S, Ortolani R, Tommasoli R, Zuliani V, Lechi A, Lechi C: The F_2-isoprostane 8-epiprostaglandin $F_{2\alpha}$ increases platelet adhesion and reduces the antiadhesive and antiaggregatory effects of NO. Arterioscler Thromb Vasc Biol 18:1248, 1998

22. Mollace V, Salvemini D, Sessa WC, Vane JR: Inhibition of human platelet aggregation by endothelium-derived relaxing factor, sodium nitroprusside or iloprost is potentiated by captopril and reduced thiols. J Pharmacol Exp Ther 258:820, 1991

23. Minuz P, Lechi C, Gaino S, Bonapace S, Fontana L, Garbin U, Paluani F, Cominacini L, Zatti M, Lechi A: Oxidized LDL and reduction of the antiaggregating activity of nitric oxide derived from endothelial cells. Thromb Haemost 74:1175, 1995

24. Geiger J, Nolte C, Walter U: Regulation of calcium mobilization and entry in human platelets by endothelium-derived factors. Am J Physiol 267:C236, 1994

25. Gallet C, Rosa JP, Habib A, Lebret M, Levy-Toledano S, Maclouf J: Tyrosine phosphorylation of cortactin associated with Syk accompanies thromboxane analogue-induced platelet shape change. J Biol Chem 274:23610, 1999

26. Naseem KM, Goodall AH, Bruckdorfer KR: The effects of native and oxidised low density lipoproteins on platelet activation. Biochem Soc Trans 21:140S., 1993

27. Hackeng CM, Huigsloot M, Pladet MW, Nieuwenhuis HK, van Rijn HJ, Akkerman JW: Low-density lipoprotein enhances platelet secretion via integrin-αIIβ3-mediated signaling. Arterioscler Thromb Vasc Biol 19:239, 1999

28. Hackeng CM, Pladet MW, Akkerman JW, van Rijn HJ: Low density lipoprotein phosphorylates the focal adhesion-associated kinase p125(FAK) in human platelets independent of integrin αIIβ3. J Biol Chem 274:384, 1999

29. Hackeng CM, Relou IA, Pladet MW, Gorter G, van Rijn HJ, Akkerman JW: Early platelet activation by low density lipoprotein via p38MAP kinase. Thromb Haemost 82:1749, 1999

30. Retzer M, Siess W, Essler M: Mildly oxidised low density lipoprotein induces platelet shape change via Rho-kinase-dependent phosphorylation of myosin light chain and moesin. FEBS Lett 466:70, 2000

31. Lechi C, Gaino S, Tommasoli R, Zuliani V, Bonapace S, Fontana L, Degan M, Lechi A, Minuz P: In vitro study of the anti-aggregating activity of two nitroderivatives of acetylsalicylic acid. Blood Coagul Fibrinolysis 7:206, 1996

32. Zuliani V, Tommasoli R, Gaino S, Degan M, Cominacini L, Davoli A, Lechi C, Lechi A, Minuz P. Oxidized low density lipoprotein (LDL) and platelet intracellular calcium: interaction with nitric oxide. Platelets 1998;9:269-72.

33. Carr AC, Zhu BZ, Frei B: Potential antiatherogenic mechanisms of ascorbate (vitamin C) and alpha-tocopherol (vitamin E). Circ Res 87:349, 2000

34. Chin JH, Azhar S, Hoffman BB: Inactivation of endothelial derived relaxing factor by oxidized lipoproteins. J Clin Invest 89:10, 1992

35. Witztum JL, Berliner JA: Oxidized phospholipids and isoprostanes in atherosclerosis. Curr Opin Lipidol 9:441, 1998

36. Freedman JE, Loscalzo J, Barnard MR, Alpert C, Keaney JF, Michelson AD: Nitric oxide released from activated platelets inhibits platelet recruitment. J Clin Invest 100:350, 1997

37. Chen LY, Mehta P, Mehta JL: Oxidized LDL decreases L-arginine uptake and nitric oxide synthase protein expression in human platelets: relevance of the effect of oxidized LDL on platelet function. Circulation 93:1740, 1996

Nitric Oxide
R.J. Gryglewski and P. Minuz (Eds.)
IOS Press, 2001

EFFECT OF OXIDIZED LDL ON NITROTYROSINES PRODUCTION IN ENDOTHELIAL CELLS

L. Cominacini, A. Fratta Pasini, U. Garbin, A. Rigoni, A. Davoli, V. Lo Cascio , A. Pastorino.
Clinica di Medicina Interna D - University of Verona - Verona (Italy)

ABSTRACT

The reaction product of NO with superoxide (O_2^-), is the peroxide of nitrous acid (peroxynitrite), a non radical product with of a strong oxidative power. Peorxynitrite oxidizes tyrosines to nitrotyrosines which can be detected immunogenically. Using a novel Elisa method for direct detection of nitrotyrosines, in this study we examined the effect of oxidized LDL on nitrotyrosines production in endothelial cells. In our experimental conditions the incubations of bovine aortic endothelial cells (BAECs) with different concentrations of oxidized LDL generated a dose- and time-dependent increase in nitrotyrosines. This increase is likely to be related to a receptor-dependent, abnormally increased intracellular production of reactive oxygen species (ROS) and in particular of O_2^- , which is known to inactivate NO in a chemical reaction during which the cytotoxic radical peroxynitrite is formed. In conclusion the results of this study show that one of the pathophysiological consequences of ox-LDL incubation with endothelial cells may be the production of nitrityrosines derived from the formation of peroxynitrite as a result of the inactivation of NO through an increased cellular production of O_2^-.

INTRODUCTION

Nitric oxide (NO) is an endogenous vasodilator secreted mainly by endothelial cells and active to relax smooth muscle cells by activating cGMP synthesis. Having odd number of electrons it is a free radical and therefore reactive with other compound with odd number of electrons (1). In particular NO can react in a very fast, "diffusion limited" event with superoxide (O_2^-) a second long living radical secreted by many cells as a metabolite or as lateral product of oxygen metabolism (2). The reaction product of NO with of O_2^-. is the peroxide of nitrous acid (peroxynitrite), a non radical product with of a strong oxidative power (3). In biological milieu peroxynitrite decomposes: homolitically to hydroxy radicals and nitrogen dioxide, or heterolitically to a range of reactive molecules possibly including also the nitronium cation (4). Alternatively recent opinions consider peroxynitrite in equilibrium between a cis configuration, fairly stable, and a trans configuration less energetically favoured but provided of the reactivity proper hydroxy-radical and nitronium cation. (3).Nitronium cation or trans-peroxynitrite are very powerful nitrating agents of aromatic rings and can nitrate tyrosine, giving rise to 3-nitrotyrosine. This reaction is catalysed by a glutamic acid vicinal to tyrosine resulting in a fast event lasting not more than few minutes (5). This reaction can modify consistently the properties of many proteins also in vivo and a number of diseases has been shown to involve the presence of nitrated proteins, including atherosclerosis (6), Altzeimer's diseases (7), celiac diseases (8), asthma (9), renal failure and septic shock (10), carbon monoxide poisoning (11) and also cigarette smocking (12).

Oxidized (ox)- LDL has been observed to induce abnormalities in endothelial function which may be relevant for the progression of atherosclerotic lesions (13). In particular functional alterations of the endothelial cells may be involved in the reduction of vasodilation, in response to stimuli which induced NO release, in isolated arteries exposed to ox-LDL (14). Recently, an endothelial receptor for ox-LDL, called lectin-like ox-LDL receptor-1 (LOX-1) was cloned from cultured bovine aortic endothelial cells (BAECs) (14). It has been suggested that ox-LDL uptake through this receptor may be involved in endothelial activation or dysfunction in atherogenesis (15). In this context we recently reported that ox-LDL binding to LOX-1 determined a significant increase in the generation of reactive oxygen species (ROS) in endothelial cells (16) and that this increase is responsible of the reduced NO availability in endothelial cells (17).

In this study we investigated the effect of incubation of oxidized LDL wih BAECs on the formation of nitrotyrosines,

MATERIALS AND METHODS

LDL isolation

Whole blood, obtained by venipuncture from healthy volunteers after 12 h of fasting, was collected into Vacutainer tubes (Becton Dickinson, Meylan, France) containing ethylenodiamine tetraacetic acid (EDTA) (1 mg/ml), and processed for LDL separation within 1 day by sequential flotation in NaBr solution (18) containing 1 mg/ml EDTA.

LDL oxidation and modification

Cu^{2+} modified LDL (1.7 mg protein/ml) was prepared by exposure of LDL to 5 μM $CuS0_4$ for 18 h at 37°C, as previously described (19, 20). The extent of LDL oxidation was determined by thiobarbituric acid-reactive substances (TBARS), as reported (20). Protein was measured by the Pierce BCA protein assay reagent (21).

Cell cultures

BAECs were isolated and cultured as previously described (22). Cells used for experiments were at passage levels between 2 and 4. Cell survival was monitored according to the method of Landegren (23). Confluent BAECs in 24-well plates were incubated in DMEM (Sigma, St. Louis, Mo, USA) for different times (18-48 h) with increasing concentrations (100-250 μg protein/ml) of native and ox-LDL and the nitrated proteins were measured in cell extracts.

Nitration of proteins

Bovine serum albumine (BSA) (Sigma) was nitrated by several consecutive additions of peroxinitrite (Sigma) and purified by gel-filtration as described by Beckman (19)
The extent of nitration was in the range of 2-4 mol/mol protein as was determined by absorbance at 438 nm pH9 using a molar coefficient of 4300 M^{-1} cm^{-1}. In some cases it was confirmed by mass spectrometry (18).

ELISA assay of nitroproteins

The assay was a modification of the blotting assay used by Crown and Ischiropoulos (15) adapted to work with plate-bound nitroproteins.
0.1 ml protein from cell extracts in Tris/saline pH 10 and NO_2BSA at different dilutions (from 5,92 to 1.82 nM) were added to plate wells and allowed to bind to the plate for the night at +4°C. Then the remaining active sites of the wells were blocked with 0.3 ml of Tris/Milk pH 7.4 (Sigma) for 2 hr at room temperature. After washing with Tris/0.05% tween20, 0.1 ml of immunoaffinity purified policlonal anti-nitrosine rabbit IgG 1/1000 (Sigma) in Tris pH 7.4 was added for 1 hr at room temperature. After washing, anti-rabbitIgG coupled with alkaline phosphatase was added for 1 hr at room temperature. After washing with DEA buffer pH 10, p-nitrophenylphosphate (1 mg/ml in DEA pH 10) was added. The colour was allowed to develop at 37°C for 40 min and then the absorbance read at 405 nm with UV plate reader (BioRad). The standard curve was fitted with a nonlinear fitting program using a saturation mathematical model (20) and the concentration of the unknown determined with the same program and espressed as molar nitrotyrosine equivalent to the nitro-BSA.

Statistical analysis

Statistical analysis was performed by analysis of variance and subsequent by post hoc analysis, using the "SYSTAT" program and statistical software manual (SYSTAT inc., Evanston, IL) for Macintosh.

RESULTS

The calibration curve by using nitrated BSA as a standard is represented in figure 1. The calibration curve is of the saturation kind and can be easily fitted by the conventional non-linear fitting programs using a saturation kinetics equation (27).

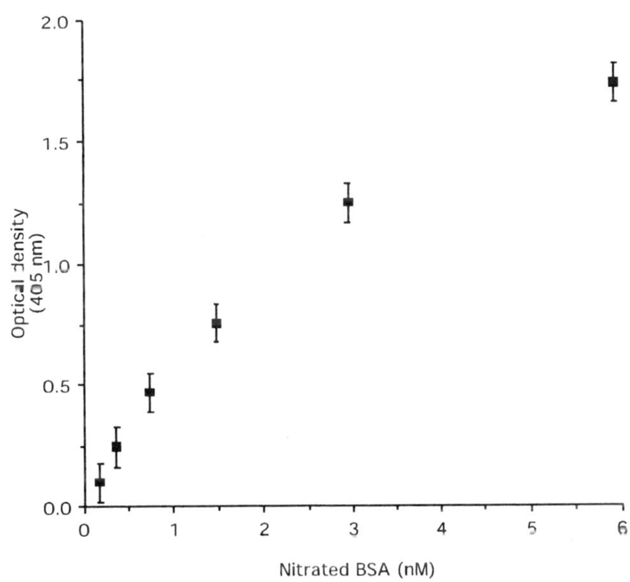

Figure 1. Calibration curve for the measuremey of nitrityrosines by using nitrated bovine serum albumine (BSA). . The calibration curve was fitted by the conventional non-linear fitting programs using a saturation kinetics equation (27).

Accurate measurement of plate-bound protein showed that the protein of tissue extracts and BSA bind with the same yield to the plate so that the standard analytical signal was directly correlated to that of tissue extracts and the correction for different binding yields was not necessary.
In our experimental conditions the incubations of BAECs with 200 μg oxidized LDL determined a tme dependent increase in nitrated proteins evaluated in cell extracts. In particular, when expressed as % increase of the control (no addition of oxidized LDL), the increase in nitrotyrosines was about 100% after 19 hr, 150% after 25 hr and 180% after 45 hr ($p<0.01$). The time-dependent increase of nitrotyrosines determined after incubation of oxidized LDL with BAECs is represented in figure 2. From the figure it is evident that native LDL produced no effect on the concentration of nitrated proteins.

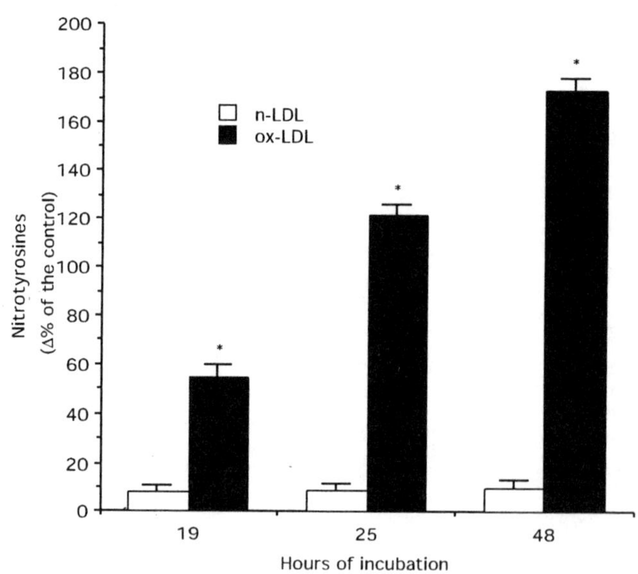

Figure 2. Time-dependent increase of nitrotyrosines determined after incubation of oxidized LDL with BAECs. The values of nitrotyrosines are expressed as % variation (Δ) of the control. *$p<0.01$ versus control (no addition of native (n) or oxidized (ox) LDL.

The incubation of BAECs with increasing amounts of ox-LDL for 24 hr, dose-dependently, increased nitrated proteins evaluated in cell extracts ($p<0.001$) while n-LDL did. The dose-dependent increase of nitrotyrosines obtained after incubation of oxidized LDL with BAECs is indicated in figure 3.

DISCUSSION

Using a novel Elisa method for direct detection of nitrotyrosines, in this study we examined the effect of oxidized LDL on nitrotyrosines production in endothelial cells.

In our experimental conditions the incubations of BAECs with different concentrations of oxidized LDL generated a dose- and time-dependent increase in nitrotyrosines.

Dual increases in superoxide and NO are the prerequisite conditions for increasing peroxynitrite formation (2). Peorxynitrite oxidizes tyrosine to nitrotyrosine which can be detected immunigenically. As for NO generation, several lines of evidences suggest that NO is produced under basal conditions by endothelial cells (28). Oxidized LDL has been recently demomstrated to increase reactive oxygen species (ROS) by binding to oxidized LDL receptor-1 (LOX.) in BAECs

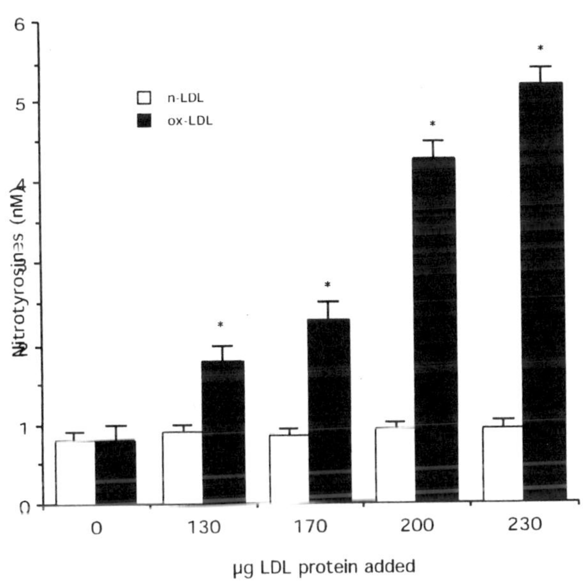

Figure 3. Dose-dependent increase of nitrotyrosines obtained after incubation of oxidized LDL with BAECs. *p<0.01 versus concentration 0 of LDL

(16) and it has been suggested that the generation of O_2^- after ox-LDL exposure may be related t the increased activity of NADPH oxidase and/or mitochondrial complex 1 (17).
Oxidative inactivation of NO is regarded as an important cause of its decreased biological activit (29) The vascular release of superoxide is sharply increased in atherosclerotic arteries (30, 31) an superoxide is known to inactivate NO in a chemical reaction during which the cytotoxic radica peroxynitrite is formed (32, 33). The presence of peroxynitrite-derived nitrotyrosines has recentl been demonstrated in human atherosclerotic lesions (34).
In this context we recently demonstrated that ox-LDL, by binding to LOX-1, induced a significant dose-dependent decrease in intracellular NO concentration both in basal and stimulated condition through an increased production of superoxide and therefore through an augmented formation o peroxynitrite (17).The decrease of intracellular NO concentration was prevented by preincubatin BAECs with different antioxidants known to work as radical scavengers and with anti LOX-monoclonal antibody (17). These data support the conclusion that the incubation of ox-LDL wit BAECs is associated with a receptor-dependent, abnormally increased intracellular production o ROS and in particular of O_2^-. NO decrease therefore may be secondary to O_2^- formation whicl is known to inactivate NO in a chemical reaction during which the cytotoxic radical peroxynitrite i formed (32, 33). The fact that in this study the incubation of oxidized LDL with BAECs i associated with a time- and dose-dependent increase in cell nitrotyrosines which can derive onl from the activity of peroxynitrite, agrees with this conclusion.
In conclusion the results of this study show that one of the pathophysiological consequences of ox LDL binding to LOX-1 may be the production of nitrityrosines derived from the formation o peroxynitrite as a result of the inactivation of NO through an increased cellular production of O_2^-.

REFERENCES

1. M. Feelisch, The biochemical patways of nitric oxide formation from nitrovasodilators appropriate choiche of exogenous NO donor and aspects of preparation and handling of acueou NO solutions, *J Cardiovacs Pharmacol* 3 (1991):25-33.

2. J.S. Beckman and W.H. Koppenol,.Nitric oxide, superoxide, and peroxynitrite: the good, th bad, and the ugly, *Am J Physiol* 271 (1996)1424-1437.

3. W. H.Koppenol, J.J. Moreno, W.A. Pryor, H. Ischiropoulos, J.S.Beckman, Peroxinitrite, cloaked oxidant formed by nitric oxide and superoxide *Chem Res Toxicol* 5 (1992) 834-842.

4. A. van der Vliet, D. Smith, C.A. O'Neill, H. Kaur, V. Darley-Usmar, C.E. Cross and B Halliwell,.Interaction of peroxynitrite with human plasma and its constituents: oxidative damage an antioxidant depletion. *Biochem J* 303 (1994) 295-301.

5. O. Curcuruto, L. Rovatti, A.M. Pastorino, M. Hamdan,Peptide nitration by peroxynitrite Characterization by liquid chromatography/tandem mass spectrometry. *Rapid Commun Mas. Spectrom* 13 (1999) 56-163.

6. J,S, Beckman, Y.Z. Ye, P.G. Anderson, J. Chen, M.A. Accavitti, M.M. Tarpey, C.R. White..Extensive nitration of protein tyrosines in human atherosclerosis detected by immunohistochemistry. *Bio. Chem Hoppe-Seyler* 375 (1994):81-88.

7. M.A. Smith, P.L. Richey Harris, L.M. Sayre, J.S. Beckman, G. Perry, Widespread peroxinitrite-mediated damage in Alzheimer's disease. *J Neurosci* 17 (1997):2653-2657.

8. J.C.A. ter Steege, L. Koster-Kamphuis, E.L.van Straaten, P.Forget, W. Buurman, Nitrotyrosine in plasma of celiac disease patients as detected by a new sandwich ELISA. *Free Radical Biol. Med* 25 (1998):953-963.

9. D. Saleh, P. Ernst, S.Lim, P.J.Barnes, A. Giaid, .Increased formation of potent oxidation peroxynitrite in airways of asthmatic patientsis associated with induction of nitric oxide synthase: effect of inhaled glucocorticoids. *FASEB j* 12 (1998):929-937.

10. N. Fukuyama, Y. Takebayashi, M. Hida, H. Ishida, K. Ichimori, H. Nakazawa, Clinical evidence of peroxynitrite formation in chronic renal failure. *Free Radical Biol Med* 22 (1997) 771-774.

11. H. Ischiropoulos, M. Beers, S.T. Ohnishi, D.Fisher, S.E. Garner, S.R. Thom, Nitric oxide production and perivascular tyrosine nitration in brain after carbon monoxide poisoning in the rat. *J Clin Invest* 97 (1996) 2260-2267.

12. J. Khan, D.M. Brennand, K.R. Bruckdorfer,M. Jacobs, Nitro-tyrosine-like immunoreactivity in human plasma of cigarette smockers and non-smockers *Br J Pharmacol* 116: 81 (1995) 328-335.

13. J.L. Witztum, The oxidation hypothesis of atherosclerosis. Lancet. 344 (1994) :793-795.

14. G. Noll, T.F. Luscher, Influence of lipoproteins on endothelial function. Thromb Res 74 (Suppl. 1) 81994):S45-S54.

15. T. Sawamura, N. Kume, T. Aoyama. H. Moriwaki, H. Hoshikawa Y. Alba, T. Tnaka, S. Miwa Y. Katsura, T. Kita, T. Masaki, An endothelial receptor for oxidized low-density lipoprotein. Nature. 386 (1997):73-77.

16. L. Cominacini, A. Fratta Pasini, U. Garbin, A. Davoli, L. Tosetti, M. Campagnola, A. Rigoni A.M. Pastorino, V. Lo Cascio, T. Sawamura, Oxidized low-density lipoprotein binding to LOX-1 in endothelial cells induces the activation of NF-kB through an increased production of intracellular reactive oxygen species. J Biol Chem 275 (2000)12633-12638.

17. L. Cominacini, A. Rigoni, A. Fratta Pasini, U. Garbin, A. Davoli, M. Campagnola, A Pastorino, V. Lo Cascio, T. Sawamura, The binding of oxidized low-density lipoprotein (ox-LDL) to ox-LDL receptor-1 in endothelial cells reduces the intracellular concentration of nitric oxide through an increased production of superoxide. J Biol Chem 2001 (in press).

18. R.J. Havel, M.A.Eder, J.M.Bragdon, The distribution and chemical composition of ultracentrifugally separated lipoprotein in human serum. J Clin Invest. 134 (1955)1345-1353.

19. L. Cominacini, U. Garbin, A. Davoli, R. Micciolo, O. Bosello, G. Gaviraghi, A simple test for predisposition to LDL oxidation based on fluorescence development during copper-catalyzed oxidative modification. J Lipid Res. 32 (1991):349-358.

20. L. Cominacini U. Garbin. A. De Santis, M. Campagnola, A. Davoli, A, Fratta Pasini, G Faccini, E. Pasqualini, L. Bertozzo, R. Micciolo, A.M. Pastorino, V. Lo Cascio, Mechanisms involved in in the in vitro modification of low density lipoprotein by human umbilical vein endothelial cells and copper ions. J Lipid Med. 13 (1996)19-33.

21. P.K. Smith, R.I. Krohn, G.T. Hermanson, A. Mallia, F.H. Gartner, A.C. Provenzano Measurement of protein using bicinchoninic acid. Anal Biochem. 105 (1985) 293-299.

22. T. Murase, N. Kume, R. Korenaga , J. Ando, T. Sawamura, T. Masaki, T, Kita, Fluid shear stress transcriptionally induces lectin-like oxidized LDL receptor-1 in vascular endothelial cells Circ Res 83 (1998) 328-333.

23. U. Landegren U, Measurement of cell number by means of the endogenous enzyme hexosaminidase: application to detection of lymphokines and cell surface antigens. J Immuno Methods. 67 (1984) 379-385.

24. J.S. Beckman, J. Chen, H. Ischiropoulos, J. Crow, Oxidative chemistry of peroxynitrite *Method Enzymol.* 233 (1994) 229-240.

25. J. Khan, D.M. Brennand, N. Bradley, B. Gao, R. Bruckdorfer, M. Jacobs, .3-nitrotyrosine in the proteins of human plasma determined by ELISA method. *Biochem J* 330 (1998) 795-801.

26. J. Crow, J. and H. Ischiropoulos, Detection and quantitation of nitrotyrosine residues in proteins: in vivo marker of peroxinitrite. *Method Enzymol* 269 (1996)185-194.

27. R.J. Leatherbarrow, *GraFit Erithacus Software Ltd. , Staines, U. K.* Version 3.0, 1992

28. R.M.J. Palmer, D.S. Ashton, S. Moncada S. Vascular endothelial cells synthesize nitric oxide from L-arginine. Nature 336 (1988) 664-666.

29. R.H. Boger, S.M.Bode-Boger, J.C. Frolich, The L-arginine-nitric oxide pathway: role in atherosclerosis and therapeutic implications. Atherosclerosis 127 (1996) 1-11.

30. A. Mugge, R. Brandes, R.H. Boger, A. Dwenger, S. Bade-Boger, S. Kienke, J.C. Frolic, P.R Lichtlen,. Vascular release of superoxide radicals is enhanced in hypercholesterolemic rabbits. J Cardiovasc Pharmacol 24 81994) 994-998.

31. Y. Ohara, T.E. Peterson, D.G. Harrison, Hypercholesterolemia increases endothelial superoxide anion production. J Clin Invest 91 (1993) 2546-2551.

32. R.J. Ignarro, Biosynthesis and metabolism of endothelium-derived nitric oxide. Annu Rev Pharmacol Toxicol 30 (1990) 535-560.

33. R. E. Huie S Padmaja, The reaction of NO with superoxide. Free Radical Res Comms 18 (1993) 195-199.

34. J.S. Beckman, Y.Z. Ye, P.G.Anderson. J. Chen, M.A. Accavitti, M.M. Tarpey, C.R. White Extensive nitration of protein tyrosines in human atherosclerosis detected by immunoistochemistry Biol Chem Hoppe-Seyler 375 (1994) 81-88.

Nitric Oxide and Circulatory Shock: An Update

Christoph Thiemermann
Professor of Pharmacology and Chairman, Department of Experimental Medicine & Nephrology, The William Harvey Research Institute, St. Bartholomew's Hospital Medical College, Charterhouse Square, London EC1M 6BQ, United Kingdom. Fax: 0044 171 251 1685; E-Mail: c.thiemermann@mds.qmw.ac.uk

1. Introduction

The medical syndrome of shock can be defined as a *"progressive failure of the circulation to provide blood and oxygen to vital organs of our body"*. In clinical practice, the key symptom of shock is a severe fall in blood pressure, which is often associated with the dysfunction or failure of several important organs including lung, kidney, liver and brain. The most common cause of shock is bacteraemia resulting in systemic infection and ultimately shock (septic shock). Other causes of shock include severe haemorrhage (haemorrhagic shock), trauma (traumatic shock), failure of the heart to maintain a sufficient cardiac output (cardiogenic shock), interruption of the inervation of blood vessels (neurogenic shock) and severe allergic reactions (anaphylactic shock). Despite improvements in intensive care medicine, the mortality of septic shock remains very high. Thus, there is still a great need for new approaches to improve therapy and outcome of patients with shock. This brief review provides an update of the role of NO in the pathophysiology of shock. The results of recent, phase III clinical trial aimed at elucidating the effects of an inhibitor of NO formation in patients with septic shock are discussed.

Septic shock, regardless of its aetiology, is defined as sepsis (systemic response to infection) with hypotension despite adequate fluid replacement resulting in impaired tissue perfusion and oxygen extraction [1]. The definition of septic shock is independent of the presence or absence of a multiple organ dysfunction syndrome (MODS), which is defined as impaired organ function such that homeostasis cannot be maintained without intervention [2]. Primary MODS is a direct result of a well-defined insult to a specific organ. Secondary MODS occurs as a consequence of an exaggerated host response, termed systemic inflammatory response syndrome (SIRS). SIRS may occur in response to infection, multiple trauma, haemorrhage, ischaemia and immune-mediated organ injury [2]. Current therapeutic approaches for septic shock include antimicrobial chemotherapy, volume replacement, inotropic and vasopressor support, oxygen therapy and mechanical ventilation as well as haemodialysis and haemofiltration. These have, however, failed to make a substantial impact on the high mortality associated with septic shock [3] and, hence, septic shock remains the major cause of death in non-coronary intensive care units with an estimated mortality ranging between 50 and 80%. As shock is also by far the most common cause of prolonged admission to an intensive care unit, the clinical and socio-economical importance of this illness is substantial. Thus, there is still a great need to explore the pathophysiological events leading to circulatory failure, tissue ischaemia and MODS in septic shock.

2. Current knowledge: Role of NO in the pathophysiology of shock: experimental studies

Nitric oxide (NO) is generated from L-arginine by a family of enzymes collectively called NO synthases (NOS). The synthesis of NO from L-arginine and molecular oxygen involves the following the generation of N^G-hydroxy-L-arginine and water and subsequently the oxidation of N^G-hydroxy-L-arginine in the presence of molecular oxygen to form NO, L-citrulline and water (Figure 1). NO is generated from L-arginine in many mammalian cells by at least three different isoforms of NOS. The NOS in endothelial cells (eNOS or NOS III) and neuronal cells (nNOS or NOS I) are expressed constitutively, and both enzymes require an increase in intracellular calcium for its activation. Activation of macrophages and many other cells with pro-inflammatory cytokines or endotoxin results in the expression of a distinct isoform of NOS (inducible NOS; iNOS or NOS II), the activity of which is functionally independent of changes in intracellular calcium. When generated, NO diffuses to adjacent cells where it activates soluble guanylate cyclase, resulting in the formation of cGMP, which in turn mediates many of the effects of NO. Thus, NO has many biological functions in the cardiovascular, nervous and immune systems [4]. For instance, activation of eNOS by shear stress results in a continuous release of NO (active vasodilatation) which regulates blood pressure and organ blood flow, and inhibits the adhesion of platelets and polymorphonuclear cells to the endothelium. The generation of large quantities of NO by activated macrophages (iNOS) has bactericidal and tumouricidal effects and, hence, an important role in the host defence system [4–9].

Hypotension: Endotoxaemia is associated with an enhanced formation of NO due to the early activation of eNOS [10] and the delayed induction of iNOS activity in e.g. macrophages, vascular smooth muscle, hepatocytes, cardiac myocytes etc. [11,12]. This overproduction of NO may contribute to hypotension, vascular hyporeactivity to vasopressor agents [13], myocardial dysfunction, organ injury and possibly MODS. The circulatory failure associated with shock of various aetiologies is characterised by severe hypotension (peripheral vasodilatation), hyporeactivity of the vasculature to vasoconstrictor agents, myocardial dysfunction, maldistribution of organ blood flow and reduced tissue oxygen extraction. There is now good evidence that an enhanced formation of NO contributes to several of these pathophysiological features of septic shock. For instance, an enhanced formation of NO due to activation of eNOS (acute phase of shock) and particularly following the induction of iNOS in the vascular wall (late phase of shock)

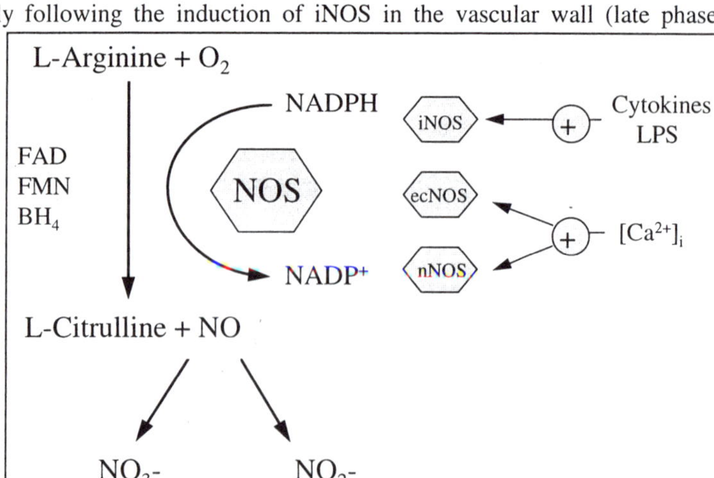

Figure 1. Synthesis and metabolism of NO

importantly contributes to the *hypotension* in animals and man with septic shock. Interestingly, endotoxin does not cause hypotension in mice in which the gene for iNOS has been inactivated ("iNOS knock-out" mice) [14]. In addition, selective inhibition of iNOS activity (with 1400W or L-NIL) abolishes the delayed hypotension caused by endotoxin in rodents [15]. Our hypothesis [10] that an enhanced formation of endogenous NO importantly contributes to the hypotension associated with endotoxic shock, is now supported by numerous studies (in various different species from rodents to man) using different pharmacological (e.g. prevention of iNOS expression, inhibition of iNOS activity with non-selective or iNOS-selective inhibitors, use of agents which scavenge NO etc.) or molecular biological approaches (e.g. gene-targeting of the iNOS gene). Similarly, an enhanced formation of NO by iNOS also contributes to the circulatory failure associated with (1) Gram-positive shock [16,17]; (2) the delayed hypotension in animals and patients exposed to immunotherapy with interleukin-2, (3) haemorrhagic shock [18] and (4) possibly to the hyperdynamic circulatory failure associated with liver cirrhosis. Although the circulatory failure associated with anaphylactic shock, traumatic shock or burns also results in an increase in the plasma levels of nitrite, it is unclear whether this is merely a surrogate markers of these disorders or cause of the associated haemodynamic alterations [19].

Vascular Hyporeactivity: The peripheral vascular failure in animals and man with septic shock also results in a progressive attenuation of the pressor effects afforded by norepinephrine and other vasoconstrictor agents (epinephrine, vasopressin, angiotensin II, serotonin, histamine, calcium, potassium). This phenomenon, which has also been termed *"vasoplegia"* also contributes to the therapy-refractory hypotension in septic shock. Clearly, the hyporeactivity of blood vessels obtained from animals exposed to endotoxic or hemorrhagic shock (for several hours) to catecholamines is largely — but not exclusively — due to an enhanced formation of NO secondary to the induction of iNOS. In endotoxemia, a NO-mediated *vascular hyporeactivity* occurs in conductance, resistance as well as venous vessels [20]. An enhanced formation of NO by eNOS (early phase) and iNOS (late phase) also contributes to the hyporeactivity to catecholamines in rats with haemorrhagic shock [18].

Myocardial Dysfunction: The question as to whether an enhanced formation of NO contributes to the *myocardial dysfunction* associated with shock is still controversial [21,22]. Large amounts of exogenous NO (authentic NO or NO-donors) reduces cardiac contractility by a cGMP-dependent mechanism it is unclear whether the amounts of NO generated under physiological or even pathophysiological conditions are sufficient to exert this effect. Clearly, isolated (rat) cardiac myocytes or papillary muscle cells exposed to TNFα or IL-1 express iNOS; and NOS inhibitors attenuate the impairment of contractility in isolated myocytes obtained from animals with endotoxic shock. Although the iNOS activity in the heart of animals exposed to endotoxaemia (for up to 6 h) is relatively small when compared with the activity found in other organs, longer periods of septic shock may well result in a more pronounced expression of iNOS protein and activity. It has been difficult to demonstrate that NOS inhibition in animals with septic shock results in an improvement in cardiac contractility or cardiac output, as most studies have employed non-selective NOS inhibitors of even eNOS-selective inhibitors (L-NAME), which cause a reduction in cardiac output by reducing myocardial blood flow (eNOS inhibition). Thus, further studies are warranted aimed at evaluating the effects of iNOS-selective NOS inhibitors on cardiac function in septic shock.

Defect in tissue oxygen extraction: Circulatory shock often results in a marked *defect in tissue oxygen extraction* resulting in tissue hypoxia and an increased venous oxygen concentration. As the local generation of large amounts of NO e.g. by activated macrophages serves to kill bacteria or tumour cells as part of the host defence, it is not surprising that the generation of NO by iNOS in other cells is cytotoxic (suicide mechanism). Indeed, large amounts of NO cause an autoinhibition of mitochondrial respiration by inhibiting several key enzymes in the mitochondrial respiratory chain (NADH-ubiquinone reductase, succinate-ubiquinone oxidoreductase) or in the Krebs' cycle (e.g. cis-acconitase) resulting in a shift in glucose metabolism from aerobic to anaerobic pathways [8,23]. NO (like other radicals and oxidants) also causes DNA strand breakage which triggers a futile, energy-consuming repair cycle by activating the nuclear enzyme poly (ADP-ribose) polymerase (PARP). Activation of PARP results in the rapid depletion of the intracellular concentration of NAD^+ (its substrate) slowing the rate of glycolysis, electron transfer and ATP formation which ultimately results in cell death ("PARP suicide hypothesis") [24]. NO and superoxide anion generate peroxynitrite anions [25], which also cause DNA strand breaks and activate PARP. There is some evidence that inhibitors of PARP activity reduce the organ injury associated with endotoxin [26,27] and haemorrhagic shock [28,29]. It still unclear, whether peroxynitrite rather then oxygen-derived radical contributes to the activation of PARP in conditions associated with ischaemia-reperfusion and shock. Nevertheless, the generation of large amounts of NO by iNOS may contribute to the defect in oxygen extraction and ultimately cell hypoxia and death by causing (i) maldistribution of regional blood flow (reduced oxygen supply), (ii) formation of a diffusion barrier for oxygen within the vascular wall (reduced oxygen transport), (iii) inhibition of the generation of ATP (reduced oxygen utilisation), and (iii) excessive and futile consumption of ATP. In concert with the severe hypotension (reduced perfusion pressure), these effects of the local overproduction of NO may importantly contribute to the organ injury and dysfunction associated with septic shock.

Host Defence: The recent generation of mice deficient in iNOS (iNOS mutant or "iNOS knock-out" mice) has helped to shed a further light into the physiological and/or pathophysiological importance of the generation of NO by iNOS [14,30]. Most importantly, these studies provide strong evidence that the enhanced formation of NO by iNOS (e.g. in macrophages) importantly contributes to *host defence* against infection. For instance, MacMicking and colleagues demonstrate that iNOS-deficient mice failed to restrain the replication of Listeria monocytogenes *in vivo* or lymphoma cells *in vitro*. Moreover, iNOS-deficient mice, but not wild-type or heterozygous mice, were highly susceptible to infection with the protozoa parasite *Leishmania major* and exhibited reduced non-specific inflammatory responses to carageenin [30]. The importance of NO as a mediator of host defence is also highlighted by studies demonstrating that non-selective inhibitors of NOS activity increase mortality in mice challenged with live bacteria (infection) rather than endotoxin (endotoxaemia). In contrast, the hypotension (see above) and early mortality caused by endotoxin is reduced in iNOS deficient mice, while the

Table 1
Possible effects of administration of NOS inhibitors in septic shock

Beneficial	Adverse
• Increased blood pressure	• Excessive vasoconstriction
• Restores responsiveness to pressor agents	• Pulmonary hypertension
• Cardiac output return to baseline values	• Fall in cardiac output
• Decreased production of peroxynitrite	• Increased platelet adhesiveness
• Attenuation of inhibition of mitochondrial respiration	• Increased neutrophil adhesion
• Improved organ function	• Worsened organ function
• Improved survival	• Reduction in survival

degree of liver injury is unaltered [14]. Taken together, these studies support the view that an enhanced formation of NO by iNOS defends the host against infectious agents and tumour cells, while an excessive induction of iNOS in other tissues (e.g. vasculature) may cause shock and possibly tissue injury.

The formation of NO also exerts beneficial effects in endotoxic shock including vasodilatation, prevention of platelet and leukocyte adhesion, improvement of microcirculatory blood flow and augmentation of host defence. Thus, it is not surprising that inhibition of NOS activity exerts both beneficial as well as detrimental effects (Table 1). The following paragraphs provide a brief review of effects and side effects of NOS inhibitors in animals and man with septic shock.

3. Inhibition of NOS activity: Effects and Side Effects in animals with septic shock

L-NMMA (and other analogues of L-arginine): Inhibition of iNOS activity can be achieved with competitive inhibitors of NOS enzyme activity. The discovery of *L-arginine-analogues,* which inhibit NOS activity including N^G-methyl-L-arginine (L-NMMA) provided the first tool to explore beneficial or side effects of NOS inhibitors in animals and men with septic shock. The subsequent discovery of the NOS inhibitors N^G-nitro-L-arginine (L-NA) and its methyl ester (L-NAME) [31], which in contrast to L-NMMA were cheap and readily available, stimulated numerous studies aimed at evaluating the role of NO in septic shock by using (high doses of) L-NAME. This was somewhat unfortunate, as L-NAME is a more potent inhibitor of eNOS than iNOS activity. Thus, L-NAME caused many adverse effects resulting from the inhibition of eNOS activity including excessive vasoconstriction including a fall in cardiac output, pulmonary hypertension and enhanced adhesion of platelets and neutrophils to the endothelium. Thus, it was not surprising that high doses of L-NAME or L-NMMA increased mortality in mice and rabbits with endotoxic shock. The hypothesis that the basal release of NO by eNOS has an important role in the regulation of regional blood flow (beneficial effects of NO), while the excessive generation of NO by iNOS "in the wrong place at the wrong time" contributes to some aspects of the pathophysiology of shock (harmful effects of NO), has stimulated the search for selective inhibitors of iNOS activity

Aminoguanidine was the first relatively selective inhibitor of iNOS activity discovered [32]. Although aminoguanidine is a more potent inhibitor of iNOS than eNOS activity *in vitro* and *in vivo* [33,34], aminoguanidine is not a very potent inhibitor of iNOS activity (IC_{50}: approximately 100–150 μM). In addition, aminoguanidine is not a very specific inhibitor of iNOS activity, as this guanidine has many other pharmacological properties including inhibition of (1) histamine decarboxylase, (2) polyamine catabolism, (3) the formation of advanced glycosylation end products, and of (4) catalase activity [34]. Interestingly, aminoguanidine also prevents the expression of iNOS protein by a hitherto unknown mechanism [35]. Thus, aminoguanidine (like aminoethyl-isothiourea, see below) has to be regarded as an agent which has affects iNOS activity by two distinct mechanisms, namely prevention of the expression of iNOS protein and activity and inhibition of iNOS activity.

S-substituted isothioureas (ITUs) are non-amino acid analogues of L-arginine and also potent inhibitors of iNOS activity with variable isoform selectivity [36–38]. The most potent isothioureas were those with only short alkyl chains on the sulphur atom and no substituents on the nitrogen atoms. For instance, S-ethyl-ITU is a potent competitive inhibitor of human iNOS, eNOS and bNOS activity. In contrast to S-ethyl-ITU, aminoethyl-ITU and S-methyl-ITU are more selective inhibitors of iNOS than of eNOS activity *in vitro* and *in vivo* [38]. Interestingly, aminoethyl-ITU is metabolised to mercaptoethyl-guanidine (and possibly the isothiourea, 2-aminothiazoline), which may represent the active principle of aminoethyl-ITU. Indeed, mercaptoalkyl-guanidines, in particular mercaptoethyl- and mercaptopropyl-guanidines, are the most potent (guanidino) inhibitors of iNOS activity [38].

Amidines: S-substituted ITUs and guanidines contain the amidine function (-CH(=NH)NH_2), a feature which they have in common with O-substituted isoureas and amidines themselves. Indeed, amidines including 2-iminopiperidine, butyramine, 2-aminopyridine, propioamidine and acetamidine inhibit NOS activity. Interestingly, both 2-iminopiperidine and butyramidine were more potent inhibitors of iNOS activity than L-NMMA in murine macrophages [39]. Recently, an analogue of acetamidine termed 1400W [N-(3-(aminomethyl)benzyl)acetamidine] has been reported to be a slow, tight binding inhibitor of human iNOS. The inhibition by 1400W of the activity of human iNOS was (1) extremely potent, (2) dependent on the co-factor NADPH and (3) either irreversible or extremely slowly reversible. Most notably, 1400W was approximately 5000-fold more potent as an inhibitor of iNOS activity than of eNOS activity. Moreover, the inhibition by this agent of the activity of eNOS or nNOS activity was reversible by L-arginine, while iNOS inhibition was not. In a rat model of vascular injury caused by endotoxin, 1400W was 50-fold more potent as an inhibitor of iNOS than eNOS activity [40]. Thus, 1400W appears to be the most potent and selective inhibitor of iNOS activity known to date and, hence, will be an ideal tool to elucidate the role of NO from iNOS in shock and other diseases associated with the induction of iNOS. We have recently discovered that 1400W (3-10 mg/kg/h) attenuates both iNOS activity as well as the delayed hypotension caused by endotoxin in the rat in a dose-related fashion. Most notably, infusion of 10 mg/kg/h of 1400W (starting 2 h after endotoxin) abolished both the increase in nitrite+nitrate as well as the delayed hypotension caused by LPS. These beneficial haemodynamic effects of 1400W did, however, not translate into a protection of organs such as the kidney, liver and pancreas against the multiple organ injury caused by endotoxin in the rat. These results suggest that an enhanced formation of NO from iNOS contributes to the circulatory failure, but not the organ dysfunction associated with endotoxic shock [15].

4. Inhibition of NOS activity: Effects and Side Effects in patients with septic shock

Although our understanding of the role of NO in animal models of circulatory shock has improved substantially over the past years, our knowledge regarding the biosynthesis and importance of NO in the pathophysiology of patients with shock (of various aetiologies) is still very limited. Indeed, a Medline search covering the time period from 1987 to November 1995 revealed that only 8 to 14% of all of the publications which included the key word 'nitric oxide' also included the key word 'human' [41]. What, then, is the evidence that septic shock in man is associated with an enhanced formation of NO? Elevated plasma and urine levels of nitrite/nitrate have been reported in adults and children with severe septic shock as well as in patients with burns injuries who subsequently developed sepsis. Moreover, elevated plasma levels of nitrite/nitrate occur in patients receiving IL-2 chemotherapy. In contrast, there is also evidence that the plasma levels of nitrite/nitrate are lower in patients after trauma, surgery and in patients with HIV infections [41]. Interestingly, the increase in iNOS activity in leukocytes obtained from patients with sepsis appears to correlate with the number of failing organs, but not with blood pressure. Taken together, these studies support the conclusion that septic shock in man is associated with an enhanced formation of NO. It should, however, be stressed that the increase in the plasma levels of nitrite/nitrate elicited by endotoxin, cytokines or bacteria in rodents (10-

fold) is substantially higher than the observed increases in the plasma levels of these metabolites of NO in other animal species (pig, sheep etc) or humans. Moreover, our understanding of (i) the biosynthesis of NO, (ii) the regulation of and the mechanism involved in the expression of iNOS, and (iii) the role of NO in MODS in shock are largely based on animal experiments of endotoxic shock in rodents. In contrast, we know relatively little about the role of NO in patients with septic and other forms of circulatory shock.

There is evidence that endotoxin and cytokines (when given in combination) causes the expression of iNOS as well as the formation of NO in various human cells (primary or cell lines) including hepatocytes, mesangial cells, retinal pigmented epithelial cells and lung epithelial cells [8,41]. Interestingly, IL-1 causes the hyporeactivity of human hand veins *(in situ)* to the constrictor effects elicited by exogenous or endogenous norepinephrine, and this vascular hyporeactivity is largely attenuated by L-NMMA suggesting that it is mediated by NO (P.Vallance, personal communication). Early reports of beneficial haemodynamic effects of L-NMMA in humans with septic shock [42] stimulated a phase I, multi-center, open-label, dose-escalation (1, 2.5, 5, 10 or 20 mg/kg/h for upto 8 hours) study using L-NMMA (546C88) in 32 patients with septic shock. In this study, L-NMMA sustained blood pressure and enabled a reduction in vasopressor (norepinephrine) support. The cardiac index fell (possibly due to an increase in peripheral vascular resistance) and left ventricular function was well maintained. Moreover, L-NMMA increased oxygen extraction, while pulmonary shunt was not worsened [44]. This first study was followed by a placebo-controlled multi-center study involving 312 patients with septic shock, which evaluated the effects of L-NMMA on the resolution of shock at 72 h (primary endpoint). The severity of illness according to the SAPS II score was similar between placebo and the L-NMMA group. Infusion of L-NMMA enhanced mean arterial blood pressure and systemic vascular resistance index and decreased cardiac output (from elevated towards normal levels). L-NMMA had no effect on left ventricular systolic work index indicating the fall in cardiac output was not due to an impairment in cardiac contractility. In patients treated with L-NMMA, there was a transient increase in mean pulmonary artery pressure. Interestingly, L-NMMA did not affects the thrombocytopenia or renal dysfunction caused by sepsis. Most notably, 41% of patients treated with L-NMMA, but only 21% of patients treated with placebo, recovered from shock within 72 h. There was a strong trend for a reduction in mortality (at day 14) in patients treated with L-NMMA.

In June 1997, Glaxo Wellcome started a Phase III clinical trial evaluating the effect of 546C88 (targinine, L-NMMA) in patients with septic shock. The company has stopped the trial in spring 1998, because of concerns about a higher mortality in the treated group than in the placebo group at an interim analysis. The trial, which involved 177 centers from 26 countries, started in June 1997. It had enrolled 797 patients at the time of suspension. The interim analysis included data from 522 patients, 309 of whom had received 546C88. The data and safety monitoring committee reported a trend towards increased mortality in the active treatment group and thus recommended stopping the trial because of patients safety concerns. The trial will not be resumed, and it is unlikely that development of the drug will continue *(SCRIP, 1998; No 2330, p21).*

5. Acknowledgements

The author is a Senior Fellow of the British Heart Foundation (FS/96018).

References

[1] Parillo JE: Septic shock in humans. Advances in the understanding of pathogenesis, cardiovascular dysfunction and therapy. Ann Intern Med 1990;113:227–242.
[2] Baue AE: The multiple organ or systems failure syndrome. In: Pathophysiology of shock, sepsis and organ failure. Schlag G. & Redl H (eds), Springer:Berlin, 1993:1004–1108.
[3] Nathanson C, Hoffmann WD, Suffredini EF, Eichacker PQ, Danner RL: Selected treatment strategies for septic shock based on proposed mechanism of pathogenesis. Ann Intern Med 1994;120:771–783.
[4] Moncada S, Higgs A: The L-arginine-nitric oxide pathway. N Eng J Med 1993;329:2202–2212.
[5] Moncada S, Palmer RMJ, Higgs EA: Nitric oxide: physiology, pathophysiology and pharmacology. Pharmacol Rev 1991;43:109–142.
[6] Nathan C: Nitric oxide as a secretory product of mammalian cells. FASEB J 1992;6:3051–3064.
[7] Dinerman JL, Lowenstein CJ, Snyder SH: Molecular mechanism of nitric oxide regulation: potential relevance to cardiovascular disease. Circ Res 1993;73:217–222.
[8] Morris SM, Billiar TR: New insights into the regulation of inducible nitric oxide synthase. Am J Physiol 1994;266:E829–839.
[9] Szabo C, Thiemermann C: Regulation of the expression of the inducible isoform of nitric oxide synthase. Adv Pharmacol 1995;34:113–154.
[10] Thiemermann C, Vane JR: Inhibition of nitric oxide synthesis reduces the hypotension induced by bacterial lipopolysaccharide in the rat. Eur J Pharmacol 1990;82:591–595.
[11] Knowles RG, Salter M, Brooks SL, Moncada S: Anti-inflammatory glucocorticosteroids inhibit the induction by endotoxin of nitric oxide synthase in the lung, liver and aorta of the rat. Biochem Biophys Res Commun 1990; 72:1042–1048.
[12] Szabo C, Mitchell JA, Thiemermann C, Vane JR: Nitric oxide-mediated hyporeactivity to noradrenaline precedes the induction of nitric oxide synthase in endotoxin shock. Br J Pharmacol 1993;108:786–792.
[13] Julou-Schaeffer G, Gray GA, Fleming I, Schott C, Parratt JR, Stoclet JC: Loss of vascular responsiveness induced by endotoxin involves the L-arginine pathway. Am J Physiol 1990;259:H1038–1043.
[14] MacMicking JD, Nathan C, Hom G, Chartrain N, Fletcher DS, Trumbauer M, Stevens K, Xie QW, Sokol K, Hutchinson N: Altered responses to bacterial infection and endotoxic shock in mice lacking inducible nitric oxide synthase. Cell 1995; 82:641–650.
[15] Wray GM, Miller CG, Hinds CJ, Thiemermann C: Selective inhibition of the activity of inducible nitric oxide synthase prevents the circulatory failure, but not the organ injury/dysfunction, caused by endotoxin. Shock 1998;9:329–335.
[16] DeKimpe SJ, Kengatharan M, Thiemermann C, Vane JR: The cell wall components peptidoglycan and lipoteichoic acid from Staphylococcus aureus act in synergy to cause shock and multiple organ failure. Proc Natl Acad Sci USA 1995;92:10359–10363.
[17] Kengatharan KM, DeKimpe S, Robson C, Foster SJ, Thiemermann C: Mechanism of Gram-positive shock: Identification of peptidoglycan and lipoteichoic acid moeities essential in the induction of nitric oxide synthase, shock, and multiple organ failure. J Ex Med 1998;188:305–315.
[18] Thiemermann C, Szabo C, Mitchell JA, Vane JR: Vascular hyporeactivity to vasoconstrictor agents and haemodynamic decompensation in haemorrhagic shock is mediated by nitric oxide. Proc Natl Acad Sci USA 1993;90:267–271.
[19] Szabo C, Thiemermann C: Role of nitric oxide in haemorrhage, traumatic and anaphylactic shock, and thermal injury. Shock 1994;2:145–155.
[20] Fleming I, Gray GA, Julou-Schaeffer G, Parratt, JR, Stoclet JC: Incubation with endotoxin activates the L-arginine pathway in vascular tissue. Biochem Biophys Res Commun 1990;171:562–568.
[21] Kumar A, Parillo JE: Nitric oxide and the heart in sepsis. In: Role of nitric oxide in sepsis and ARDS. Fink MP & Payen D H (eds), Springer:Berlin, 1995:73–99.
[22] Lefer AM: Cellular actions of nitric oxide in the cardiovascular system. In: Role of nitric oxide in sepsis and ARDS. Fink MP & Payen D H (eds), Springer:Berlin, 1995:114–124.
[23] Thiemermann C: The role of L-arginine:nitric oxide pathway in circulatory shock. Adv Pharmacol 1994;28:45–79.
[24] Schraufstatter I, Hinshaw D, Hyslop P, Spragg R, Cochrane C: Oxidant injury of cells. DNA strand breaks activate polyadenosine diphosphate ribose polymerase and lend to depletion of nicotinamide adenine dinucleotide. J Clin Invest 1986;77:1312–1319.
[25] Beckman JS, Beckman TW, Chen J, Marshall PA, Freeman B: Apparent hydroxyl radical production by peroxynitrite: Implications for endothelial injury from nitric oxide and superoxide. Proc Natl Acad Sci USA 1990;87:1620–1624.

[26] Zingarelli B, O'Conner M, Wong H, Salzman AL, Szabo C: Peroxynitrite-mediated DNA breakage activates polyadenosine diphosphate ribosyl sythetase and causes cellular energy depletion in macrophages stimulated with bacterial lipopolysaccharide. J Immunol 1996;156:350–358.
[27] Wray GM, Hinds CJ, Thiemermann C: Effects of inhibitors of poly(ADP-ribose) synthetase on hypotension and multiple organ dysfunction caused by endotoxin. Shock 1998;10:13–19.
[28] Szabo A, Hake P, Salzman AL, Szabo C: 3-Aminobenzamide, an inhibitor of poly (ADP-ribose) synthetase, improves haemodynamics and prolongs survival in porcine model of hemorrhagic shock. Shock 1998;10:347–353.
[29] McDonald M, Mota-Filipe H, Thiemermann C: Effects of inhibitors of poly (ADP-ribose) synthetase on the organ injury and dysfunction caused by haemorrhagic shock. Br J Pharmacol 1999; *in press*.
[30] Wei X, Charles IG, Smith A, Ure J, Feng GJ, Huang FP, Xu D, Muller W, Moncada S, Liew FY: Altered immune responses in mice lacking inducible nitric oxide synthase. Nature 1995:375:408–411.
[31] Moore PK, Oluyoumi AO, Babbedge RC, Wallance P, Hart SL: L-NG-Nitro-L-arginine methyl ester exhibits antinociceptive activity in the mouse. Br J Pharmacol 1991;102:198–202.
[32] Corbett JA, Tilton RG, Chang, K, Hasan KS, Ido Y, Wang JL, Sweetland MA, Lancaster JR, Williamson JR, McDaniel ML: Aminoguanidine, a novel inhibitor of nitric oxide formation, prevents diabetic vascular dysfunction. Diabetes 1992;41:552–558.
[33] Misko TP, Moore WM, Kasten TP, Nickols DA, Corbett JA, Tilton RG, McDaniel ML, Williamson, JR, Currie MG: Selective inhibition of the inducible nitric oxide synthase by aminoguanidine. Eur J Pharmacol 1993;233:119–125.
[34] Wu CC, Chen SJ, Szabo C, Thiemermann C, Vane JR: Aminoguanidine attenuates the delayed circulatory failure and improves survival in rodent models of endotoxic shock. Br J Pharmacol 1995;114:1666–1672.
[35] Thiemermann C, Ruetten H, Wu CC, Vane JR: The multiple organ dysfunction syndrome caused by endotoxin in the rat: Attenuation of liver dysfunction by inhibitors of nitric oxide synthase. Br J Pharmacol 1995;116:2845–2851.
[36] Garvey PE, Oplinger JA, Tanoury GJ, Sherman PA, Fowler M, Marshall S, Marmon MF, Paith JE, Furfine ES: Potent and selective inhibition of human nitric oxide synthases. Inhibition by non-amino acid isothioureas. J Biol Chem 1994; 269:26669–26676.
[37] Szabo C, Southan G and Thiemermann C: Beneficial effects and improved survival in rodent models of septic shock with S-methyl-isothiourea sulfate, a novel, potent and selective inhibitor of inducible nitric oxide synthase. Proc Natl Acad Sci USA 1994;91:12472–12476.
[38] Southan GJ, Szabo C Selective pharmacological inhibition of distinct nitric oxide synthase isoforms. Biochem Pharmacol 1996;51: 383–394.
[39] Southan GJ, Szabo C, O'Conner MP, Salzman AC, Thiemermann C: Amidines are potent inhibitors of constitutive and inducible nitric oxide synthases: Preferential inhibition of the inducible isoform. Eur J Pharmacol 1996;291:311–318.
[40] Garvey EP, Oplinger JA, Furfine ES, Kiff RJ, Laszlo F, Whittle BJ, Knowles RG: 1400W is a slow, tight binding, and highly selective inhibitor of inducible nitric oxide synthase *in vitro* and *in vivo*. J Biol Chem 1997;272:4959–4963.
[41] Preiser JC, Vincent JL: Nitric oxide involvement in septic shock: Do human beings behave like rodents? In: Vincent JL (ed), 1996 Yearbook of Intensive Care and Emergency Medicine. Springer, Berlin 1996:358–365.
[42] Petros A, Lamb G, Leone A, Moncada S, Bennett D, Vallance P: Effects of a nitric oxide synthase inhibitor in humans with septic shock. Cardiovasc Res 1994;28:34–39.

Nitric Oxide
R.J. Gryglewski and P. Minuz (Eds.)
IOS Press, 2001

Role of Inducible Nitric Oxide Syntethase and Reactive Oxygen Species in Zymosan-Induced Shock

Salvatore Cuzzocrea*
Institute of Pharmacology, University of Messina, Messina, Italy

Abstract: The aim of the current article is to overview the recent developments in the field of zymosan-induced shock research, as it relates to the roles of nitric oxide (NO) in the pathogenesis of this condition. The first part of the review focuses on the roles of peroxynitrite, a reactive oxidant produced from the reaction of NO and superoxide. The second part of the review deals with the novel findings related to the recently identified regulatory roles of the inducible isoform of nitric oxide synthase (iNOS) in the expression of pro-inflammatory mediators in zymosan-induced shock. [1] The role of peroxynitrite: Immunohistochemical and biochemical evidence demonstrate the production of peroxynitrite in zymosan-induced shock. Peroxynitrite can initiate a wide range of toxic oxidative reactions. These include initiation of tyrosine nitration, lipid peroxidation, direct inhibition of mitochondrial respiratory chain enzymes, inactivation of glyceraldehyde-3phosphate dehydrogenase, inhibition of membrane sodium/potassium ATP-ase activity, inactivation of membrane sodium channels, and other oxidative modifications of proteins. All these toxicities are likely to play a role in the pathophysiology of zymosan-induced shock. A combined anti-inflammatory agent, aminoguanidine, which selectively inhibits iNOS and scavenges peroxynitrite, prevents the delayed the organ failure associated with zymosan-induced shock. Peroxynitrite is a potent trigger of DNA single strand breakage, with subsequent activation of the nuclear enzyme poly (ADP ribose) synthetase (PARS), leading to eventual severe energy depletion of the cells, and necrotic-type cell death. Pharmacological inhibition of PARS, with 3-aminobenzamide or nicotinamide, prevents organ failure and prolongs survival time in rodent models of zymosan-induced shock. [2] Novel signalling roles of induced NO in zymosan-induced shock. Although the severity and duration of shock may dictate the timing and extent of NOS expression, it is now evident that the up-regulation of NOS can take place during sustained shock. Accumulated data indicate that NOS expressed during shock contributes to organ failure. Pharmacological inhibition of iNOS with Aminiguanidine or genetic inactivation of NOS (iNOS knockout mice) results in a marked reduction of lung, intestine and liver injury produced by zymosan-induced shock. Thus, induced nitric oxide, in addition to being a "final common mediator" of zymosan-induced shock, is essential for the up-regulation of the inflammatory response in zymosan-induced shock. Furthermore, a picture of a pathway is evolving that contributes to tissue damage both directly via the formation of peroxynitrite, with its associated toxicities, and indirectly through the amplification of the inflammatory response.

* Corresponding author: Salvatore Cuzzocrea, Ph. D. Institute of Pharmacology, School of Medicine, University of Messina, Torre Biologica – Policlinico Universitario Via C. Valeria – Gazzi – 98100 Messina Italy. Tel.: + 39 090 2213644 Fax: +39 090 694951; E-mail: salvator@www.unime.it

1. Introduction. Overview of nitric oxide synthases and their roles in Zymosan-induced Multiple Organ Failure

The free radical nitric oxide (NO) is synthesized from the guanidino group of L-arginine by a family of enzymes termed NO synthases (NOS). Three isoforms have been described and cloned: endothelial cell NOS (ecNOS, or type 3), brain NOS (bNOS, nNOS, or type 1), and inducible macrophage type NOS (iNOS, or type 2). All of the NOS isoforms can be inhibited-to a variable degree-with Nr-substituted L-arginine analogs, e.g., N G -methyl-L-arginine (L-NMA). The formation of NO is linked to the incorporation of molecular oxygen into the molecule. All forms of NOS contain four prosthetic groups: flavin-adenine dinucleotide; flavin mononucleotide; tetrahydrobiopterin; and a heme complex, iron protoporphyrin IX. Recently, zinc was identified as an additional prosthetic group of NOS. NOS forms NO and L-citrulline in two steps, the first being the formation of N G -hydroxy-L-arginine, and the second being a three-electron oxidation of this intermediate species.

All NOS isoforms are dependent on NADPH and calmodulin. In iNOS, calmodulin is present in a tightly bound form; thus WOS produces NO in a sustained manner in the presence of adequate substrate. For the biochemistry and regulation of NO synthases, the reader is referred to specialized reviews [1–3]. Many of the biological actions of NO are mediated through the guanylyl cyclase/cGMP system. NO, a lipophilic small molecule, diffuses to adjacent cells and readily enters the cytosol where it activates soluble guanylyl cyclase by binding to the iron on its heme component, thereby moving the iron out of the plane of the porphyrin ring. Increased levels of cyclic GMP (cG I MP) trigger a reduction of calcium concentration by enhancing extrusion of calcium and its sequestration into intracellular stores. The decrease in intracellular calcium concentration is responsible for the NO-mediated relaxation of vascular and non-vascular smooth muscle, inhibition of platelet adherence and aggregation, inhibition of neutrophil chemotaxis, and signal transduction in the central and peripheral nervous systems [4–6]. It is now well established that NO also has a number of cGMP-independent actions. The cytotoxic effects of NO (in high local concentrations) involve the inhibition of key mitochondrial Fe-S enzymes, including NADH:ubiquinone oxidoreductase, NADH:succinate oxidoreductase, and aconitase [7]. cGMP-independent activation by NO of other enzymes, such as cyclooxygenase, has also been described This action may be related to the reaction of NO with the iron-heme center at the active site of the enzyme [8]. NO inhibits the activity of cytochrome P-450 enzymes [9]. NO may modulate gene transcription and translation: in endothelial cells, it activates c-fos [10] whereas in neurons it potentiates the effect of calcium on gene expression linked to the c-fos promoter [11].

EcNOS is a calcium-calmodulin dependent, constitutively expressed enzyme [12]. The regulation of ecNOS in endothelial cells is complex, involving post-translational modifications (such as palmitoylation) and protein-protein interactions, for example, with calmodulin, caveolin-1, and heat shock protein 90. At this time, the roles of calcium and calmodulin are the best understood. Agents that increase intracellular calcium, by activating the influx of extracellular calcium or releasing calcium from intracellular stores, elicit endothelium- and NO dependent relaxations *in vivo* and *in vitro*. Such endothelium dependent vasodilators include acetylcholine, ATP, bradykinin, serotonin, norepinephrine, and platelet activating factor [4–6]. Important from a physiological point of view is the finding that pulsatile flow is a potent stimulant of endothelium-derived NO production. Administration of NOS inhibitors reduces blood flow to most organs. Other important functions of endothelium-derived NO include the regulation of adhesion and activation of circulating blood cells. For example, NO inhibits the adhesion and activation of platelets and the adhesion of neutrophil granulocytes. Inhibition of endothelial NO production may induce microvascular disruption, implicating a role for NO in the control of vascular permeability. This latter phenomenon has important implications for the pathophysiology of multiple organ failure, where interference with ecNOS activity has deleterious effects. For example, a series of studies has demonstrated that NG-nitro-L-arginine or its methyl ester L-NAME, as well as high doses of the isoform nonselective L-NMA, exacerbate organ injury associated with shock [13–16].

Shock is very sensitive to inhibition of the constitutive NO synthesis, perhaps because endotoxin shock is associated with a "baseline" impairment of the endothelial NO generation and reductions in endothelium-dependent relaxations [17–19]. Thus, any additional inhibition of the endothelial NO synthesis results in severe detrimental consequences.

Many inflammatory conditions are associated with production of comparatively large amounts of NO, produced by iNOS, with consequent cytotoxic effects. iNOS, first identified in macrophages, can be expressed in essentially any cell type. Although constitutive expression of iNOS has been localized to the kidney, the intestine, and the bronchial epithelia, iNOS is expressed typically in response to immunological stimuli and produces nanomoles, rather than picomoles, of NO. Once produced in high local concentrations, NO may act as a cytostatic and cytotoxic molecule for fungal, bacterial, helminthic, and protozoal organisms as well as tumor cells. Zymosan and a variety of pro-inflammatory cytokines also induce the expression of iNOS in a number of non-hematopoietic cells, including fibroblasts, glial cells, and cardiac myocytes, as well as vascular and non-vascular smooth muscle cells [7]. iNOS is active as a dimer consisting of two 131 kDa NOS units. iNOS produces large amounts of NO for prolonged periods. The expression of iNOS is regulated both at the level of transcription and at the level of iNOS mRNA stability. The mechanism of iNOS induction involves de novo transcription and the biosynthesis of new protein. Cis-regulatory elements in the murine iNOS promoter have been identified that bind transactivating elements induced by lipopolysaccharide and zymosan [20–22]. Induction of iNOS can be inhibited by numerous agents, including glucocorticoids, thrombin, macrophage deactivation factor, tumor growth factor beta, platelet-derived growth factor, IL-4, IL-8 and IL-10. The same immunological stimuli that induce iNOS also induce GTP cyclohydrolase, an enzyme that produces BH_4 and thus supplies iNOS with its co-factor [23]. Induction of iNOS is associated with the induction of arginosuccinate synthetase, which may supply iNOS with its substrate from intracellular sources by turning on the "re-cycling" of L-arginine from L-citrulline [24]. In some cell type, the same stimuli that induce iNOS also up-regulate the membrane transport system for L-arginine, thereby supplementing intracellular L-arginine from extracellular sources.

Induction of iNOS may have either toxic or protective effects. Factors that appear to dictate the consequences of iNOS expression include the type of insult, the tissue type, the level and duration of NOS expression, and probably the redox status of the tissue. Much attention has focused on the toxicity of iNOS. For example, induction of iNOS in endothelial cells produces endothelial injury [25].

Induction of iNOS has been shown to inhibit cellular respiration in macrophages and vascular smooth muscle cells; these processes can lead to cell dysfunction and cell death. Such processes, when occuring within vascular smooth muscle cells, play a key role in the and other species [26] which, in turn, can act as NO donors, activating guanylyl cyclase [27,28]. Such a mechanism has been proposed in relation to bradykinin-induced cell signaling in pulmonary blood vessels [29]. Currently, little information is available regarding these "physiological" roles of peroxynitrite, while the evidence for the roles of peroxynitrite in pathophysiological conditions is expanding.

Although there are a number of experimental difficulties related to delineation of the actual role of peroxynitrite in shock and other pathophysiological conditions, theoretical considerations strongly favor the production of peroxynitrite when NO and superoxide are produced simultaneously, because the reaction of these two species is nearly diffusion-controlled. In fact, the reaction of superoxide with NO is the only reaction that outcompetes the reaction of superoxide with superoxide dismutase [30,31]. Although chemical considerations favor the production of peroxynitrite, the actual demonstration of the presence or production of peroxynitrite in pathophysiological conditions is far from straightforward. Peroxynitrite rapidly oxidizes the fluorescent probe dihydrorhodamine 123 to rhodamine 123 *in vitro* [32]. The production of peroxynitrite can be evidenced as increased oxidation of dihydrorhodamine 123 to rhodamine 123 in the plasma [33]. Caution should be exercised with this method: oxidation of dihydrorhodamine can be triggered by oxidants other than peroxynitrite (hydroxyl radical, for example). However, a NOS inhibitor inhibitable component of an increased oxidation of dihydrorhodamine can be taken as a relatively specific evidence of an effect of peroxynitrite [33,34]. Using this method of detection, a marked, and NOS inhibitor inhibitable increase in the peroxynitrite-dependent conversion of dihydrorhodamine 123 to rhodamine has been demonstrated in zymosan-induced multiple organ failure [Cuzzocrea…35]. The finding that peroxynitrite is produced during zymosan or LPS shock is not surprising, in light of the previous evidence for the overproduction of oxygen derived free radicals.

Nitrotyrosine formation, and its detection by immunostaining, was initially proposed as a relatively specific means for detecting the "footprint" of peroxynitrite [36]. Recent evidence, however, indicates that certain other reactions can also induce tyrosine nitration; for example, the reaction of nitrite with hypochlorous acid, and the reaction of myeloperoxidase (and certain other peroxidases) with hydrogen peroxide can lead to the formation of nitrotyrosine [37,38]. The physiological or pathophysiological. relevance of this reaction remains to be further clarified. More recent reviews take an increased nitrotyrosine staining as an indication of "increased nitrosative stress", rather than a specific marker of peroxynitrite [39]. The formation of nitrotyrosine has recently been demonstrated in the lung, intestine, liver and aortas of rats subjected to zymosan-induced shock; the staining was abolished by treatment of the animals with the iNOS inhibitor and peroxynitrite scavenger [40].

Thus, multiple lines of evidence strongly suggest that peroxynitrite is produced in zymosan-induced shock. Specific peroxynitrite scavengers that could help further delineating the role of peroxynitrite in circulatory shock are not available. Agents that (e.g., glutathione, melatonin; see above). Uric acid, a putative scavenger of peroxynitrite, is sometimes used as a probe for peroxynitrite [32,41–43]. However, uric acid can interfere with a number of other oxidants [44,45], and caution should therefore be applied when interpreting the results obtained with uric acid. Therefore, the evidence implicating the role of peroxynitrite in a given pathophysiological condition can only be indirect. A simultaneous protective effect of superoxide neutralizing strategies and NO synthesis inhibition, coupled with the demonstration of peroxynitrite in the particular pathophysiological condition, can be taken as a strong indication for the role of peroxynitrite. However, it is likely that additional interactions of oxygen- and nitrogen-derived free radicals also contribute to the inflammatory cell injury.

2. Mechanisms of action of peroxynitrite in zymosan-induced shock

According to *in vitro* experiments, the ratio of superoxide and NO determines the apparent reactivity of peroxynitrite: excess NO reduces the oxidation elicited by peroxynitrite [33,46,47]. The oxidant reactivity of peroxynitrite is mediated by an intermediate with the biological activity of hydroxyl radical, which is not hydroxyl radical per se, but rather peroxynitrous acid or its activated isomer [30]. *In vitro* experiments have demonstrated that peroxynitrite is highly reactive. Peroxynitrite induces the oxidation of sulfhydryl groups and thioethers and the nitration and hydroxylation of aromatic compounds, such as tyrosine, tryptophan, and guanine. These reactions, when occurring during the reaction of peroxynitrite with various enzymes of the cell, can markedly suppress the catalytic activity of these enzymes. For instance, peroxynitrite has been shown to inhibit manganese superoxide dismutase, tyrosine hydroxylase, membrane sodium/potassium ATP ase, membrane sodium channels glyceraldehyde-3-phosphate dehydrogenase, mitochondrial and cytosolic aconitase, and a number of critical enzymes in the mitochondrial respiratory chain, as well as NOS [as overviewed in 30,31]. While peroxynitrite inactivates many enzymes, the catalytic activity of some enzymes is actually enhanced by peroxynitrite-a primary example being cyclooxygenase [48]. In addition to the interactions of peroxynitrite with proteins, an important interaction of peroxynitrite occurs with nucleic acids. Two main types of reactions have been described: DNA base modifications and DNA single strand breakage [see for review 49]. It is likely that the above-mentioned protective effects of oxyradical geavengets, NOS inhibitors, and peroxynitrite scavengers in zymosan-induced shock are, at least in part, related to prevention of the cytotoxic effects triggered by peroxynitrite. However, the relative contribution of the various above-mentioned cytotoxic mechanisms (lipid peroxidation, protein modification, mitochondrial inhibition) remain to be dissected by further studies.

In addition to direct cytotoxic effects, an indirect pathway of peroxynitrite-induced cellular injury has also been proposed. The generation of peroxynitrite, either intracellularly or extracellularly, has been shown to trigger DNA single strand breakage and activation of the nuclear enzyme poly (ADP-ribose) synthetase (PARS). The importance of this. pathway is reviewed in detail elsewhere [50,51]. Briefly, DNA single strand breakage is the obligatory trigger of activation of PARS. When activated, PARS catalyzes the cleavage of NAD into ADP-ribose and nicotinamide. PARS covalently attaches ADP-ribose to various nuclear proteins, such as histones and PARS itself. Activation of PARS can rapidly deplete NAD, slowing the rate of glycolysis, electron transport, and ATP formation, resulting in cell dysfunction and cell death. Pharmacological inhibition of PARS activity has been shown to protect against cell damage in response to exogenously or endogenously produced peroxynitrite. The above-described metabolic derangements generally lead to cell death via the necrotic pathway. This is the pathway affected by pharmacological inhibitors of PARS [51–53]. On the other hand, peroxynitrite (lower levels, longer time of exposure) can also lead to cell death via the apoptotic pathway. PARS, however, does not play a role in this latter process, since inhibition of PARS does not appear to prevent peroxynitrite-induced apoptosis [51-56].

Recent investigations tested the effects of pharmacological inhibitors of PARS in rodent model of zymosan-induced shock shock. In a rat model of zymosan shock, the rats treated with the PARS inhibitor 3-aminobenzamide (3-AB) or nicotinamide (NIC) showed a significant improvement in survival [57]. Similar to the protective effect of 3-AB, the NO synthase inhibitors, L-NMA and L-NAME, and a cell permeable superoxide dismutase mimetic also improved survival times and improved blood pressure during zymosan-induced shock [58]. The theoretical advantage of PARS inhibitors as agents for the zymosan-induced MOF include: (1) no obvious hemodynamic effects in control animals, (2) no side effects in control or shocked animals, and (3) the fact that the PARS inhibitor targets a delayed process of cell death, all permit the treatment to be effective even with a delayed start of administration.

Another distal mechanism of peroxynitrite-mediated cytotoxicity is the activation of the pro-apoptotic caspase cascade. Two independent groups have reported that caspase-3 becomes activated in cells exposed to peroxynitrite: the activation occurs at 4–6 h after exposure to bolus dose of peroxynitrite associated with the induction of apoptosis [51,52,59]. The activation of caspases may be related to the release of cytochrome c from the mitochondria induced by peroxynitrite. Caspases, in turn lead to apoptotic DNA fragmentation and inibition of caspase-3 suppresses peroxynitrite-induced apoptosis. Recent studies have utilized specific caspase-inhibitor tetrapeptides in *in vivo* studies. In such studies, protection against myocardial and neuronal reperfusion injury has been observed [60,61]. Future studies are required to establish whether inhibition of caspases exerts protective effects in zymosan-induced shock. However, the finding that caspase activity can be efficiently inhibited by S-nitrosylation [62], a reaction requiring conversion of NO to a product with NO' reactivity, indicates that the chemical fate of NO (ONOO- vs. NO+) is one factor that dictates toxicity vs. protection. In zymosan-induced shock, peroxynitrite formation is favored.

3. NOS as an amplifier of the inflammatory cascade in zymosan-induced shock

NO and peroxynitrite can have signalling functions that might regulate cellular responses following shock. One such response is inflammation. It is well known that zymosan initiates inflammatory cascades characterized by increased cytokine production, leukocyte adhesion molecule expression, and neutrophil infiltration into tissues [63]. Neutrophils are considered an important effector of tissue damage in zymosan-induced shock. To determine if induced NO participated in pro-inflammatory signaling following resuscitation, two approaches were undertaken [64]. First, mice subjected to zymosan-induced shock were treated with the iNOS inhibitor aminoguanidine. Second, the responses of iNOS knockout mice to zymosan-induced shock were compared to responses of wild-type mice with the same genetic background [64]. In both instances, the lack of iNOS was associated with a marked reduction in the neutrophil infiltatration in the liver, intestine and lung, as well as in the peritoneal cavity [64]. Therefore genetic or pharmacological inhibition of iNOS was associated with a dramatic reduction of liver, intestine and lung injury measured by plasma transaminase levels, wet-to-dry ratios, respectively [65]. Subsequent studies have revealed a similar protective effect of iNOS inhibition on bowel damage and maintenance of small bowel motility as well as preventing the gut permeability following zymosan shock [66]. Similar protective effects were also observed using antioxidants [67].

These studies indicated that NO may be a key mediator in zymosan shock. With the many known targets for NO and peroxynitrite, this finding is perhaps not surprising. However, a potentially unique aspect of NO-mediated signaling in zymosan shock is the associated redox stress. It is possible under conditions of reduced antioxidant capacity that NO or a reaction product, such as peroxynitrite, activates intracellular redox-sensitive signaling pathways. Further, nitrogen-based radicals most likely represent only one of the oxidant species that participate in oxidant signaling in shock.

The role of iNOS-derived NO (or peroxynitrite) as an amplifier of the inflammatory response is now also supported by the following observations: inhibition of iNOS suppresses TNF production in the delayed phase of allergic encephalomyelitis [68]; inhibition of iNOS suppresses interleukin-1, collagen and stromelysin production in arthritis [69,70]; inhibition of iNOS suppresses interferon-gamma production in a murine model of leishmaniasis [71]; and the expression of certain chemokines is suppressed in the absence of iNOS in zymosan-induced peritonitis [72]. Thus, a picture of a pathway is evolving that contributes to tissue damage both directly via the formation of peroxynitrite, with its associated toxicities, and indirectly through the amplification of the inflammatory response.

Although the current review is not aimed at a comprehensive overview of the role of iNOS in various pathophysiological conditions, it is important to note that the role of iNOS is widely dependent on the inflammatory condition. The participation of iNOS ranges from an amplifier and terminal mediator of tissue injury (e.g., in various forms of shock, see above), to cytoprotective or anti-inflammatory roles. In certain diseases, iNOS can play multiple, sometimes opposing, roles. A prime example for this is allergic encephalomyelitis, where studies with pharmacological inhibitors and knockout approaches concluded that iNOS is a pathogenetic factor in the induction phase of EAE, but has an inhibitory role in the progression phase of EAE [73–76]. Another noteworthy example is the various forms of viral infection and associated inflammation.

Induction of iNOS is necessary and sufficient for a substantial antiviral effect of IFN-γ and for the ability of the body to fight viruses. Thus, inhibition of iNOS enhances viral replication and leads to an increase in lethality in systemic viral infection models [77]. In other words, the potential beneficial anti-inflammatory effects of iNOS inhibition are canceled out by independent, deleterious actions of iNOS inhibition. However, in other types of viral infection, where clearance of the virus is independent of interferon-γ, the anti-inflammatory effect of iNOS inhibition becomes prominent. For example, viral pneumonitis induced by influenza A virus is abolished in iNOS deficient mice [78].

4. Conclusions-future directions

Figure 1 provides a framework for the events in zymosan-induced shock related to NO. Dominant in this scheme is the role of iNOS, which is up-regulated during shock by

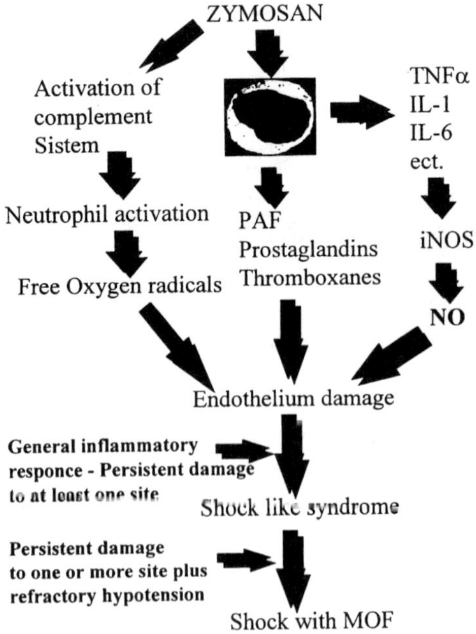

Figure 1. The suggested framework for the events in zymosan-induced shock

mechanisms that are currently not understood. After zymosan administration superoxide is produced, leading to the formation of peroxynitrite. Peroxynitrite alone or with NO then can mediate direct toxicity, as well as activate inflammatory cascades via NFkB. Combined, these events lead to organ injury and contribute to organ failure. Many gaps exist in this paradigm. Nonetheless, the recent observations leading to this proposed scheme form the basis for novel therapeutic interventions.

Neither NO nor superoxide alone act as strong oxidants towards most types of organic compounds. The evidence presented above favors the view that the reaction of NO and superoxide yields peroxynitrite, which, under many conditions, enhances the cytotoxic potential of its "precursors". The finding that peroxynitrite can be formed by the combination of superoxide with NO produced by ecNOS in pathophysiological conditions, such as the early phases of shock and reperfusion injury, has important practical applications, because it challenges the conventional wisdom, which proclaims that small amounts of NO are beneficial, while large amounts of NO are toxic. Based on these recent findings, one can propose that: (a) selective inhibition of iNOS does not always or necessarily prevent all NO-related cytotoxicity, and (b) in some circumstances (e.g., early phase of ischemia-reperfusion), inhibition of constitutively produced NO can be beneficial. Nevertheless, because of the numerous side effects of constitutive NO synthesis inhibition, it is difficult to envision how such intervention may result in a clinically useful therapeutic approach. Selective pharmacological inhibition of iNOS in shock and inflammation is expected to exert its therapeutic benefit, without disturbing the physiological functions of ecNOS (such as inhibition of platelet and white cell adhesion, maintenance of vasodilator tone, etc.). This approach would be beneficial because it would inhibit self-amplifying circles of inflammation in the early phase of shock, and it would inhibit one important final common mediator of shock. In view of our current knowledge, however, such an approach would not completely eliminate the formation of peroxynitrite. Another therapeutic option would be to use scavengers that remove excess NO, superoxide, and/or peroxynitrite. This could preserve the essential basal levels of NO and suppress NO-mediated toxicity. Additional approaches may target delayed effectors of NO or/and peroxynitrite cytotoxicity such as PARS or caspase-3. Because of massive synergism and redundancy in the pathways of inflammation associated with zymosan shock, it is likely that combined or appropriate sequential targeting of the above listed pathways will be more effective than targeting a single pathway.

5. References

[1] M.A. Marletta, Nitric oxide synthase structure and mechanism, *J Biol Chem* 268 (1993) 12231-12234

[2] D.J. Stuehr, Structure function aspects in the nitric oxide synthases, *Ann Rev Pharmacol Toxicol* 37 (1997) 339–359.

[3] D.A. Geller and T.R. Billiar, Molecular biology of nitric oxide synthases, *Cancer Metastasis Rev* 17 (1998) 7–23.

[4] L.J. Ignarro, Signal transduction mechanisms involving nitric oxide, *Biochem Pharmacol* 41 (1991) 485–490.

[5] S. Moncada, R.M.J. Palmer and E.A. Higgs, Nitric oxide: physiology, pathophysiology, and pharmacology, *Pharmacol Rev* 43 (1991) 109–141.

[6] G.J. Dusting, Nitric oxide in cardiovascular disorders, *J Vasc Res* 32 (1995) 14361.

[7] C. Nathan, Nitric oxide as a secretory product of mammalian cells, *FASEB J* 6 (1992) 3051–3064.

[8] D. Salvemini and J.L. Masferrer, Interactions of nitric oxide with cyclooxygenase: *in vitro*, ex vivo, and *in vivo* studies, *Methods Enzymol* 269 (1996) 1225.

[9] O. Khatsenko, S.S. Gross, A.B. Rifkind, J.R. Vane, Nitric oxide is a mediator of the decrease in cytochrome P450-dependent metabolism caused by immunostimulants, *Proc Natl Acad Sci USA* 90 (1993) 11147–11151.

[10] E. Felley-Bosco, S. Ambs, C.J. Lowenstein, L.K. Keefer and C.C. Harris, Costitutive expression of inducible nitric oxide synthase in human bronchial epithelial cells induces c-fos and stimulates the cGMP pathway, *Am J Respir Cell Mol Biol* 11 (1994) 159–164.
[11] N. Peunova and G. Enikolopov, Amplification of calcium-induced gene transcription by nitric oxide in neuronal cells. *Nature* 364 (1993) 450–453.
[12] W.C. Sessa, The nitric oxide family of proteins, *J Vasc Res* 31 (1994) 131–143.
[13] B.G. Harbrecht, B. Wu, S.C. Watkins, H.P. Marshall Jr, A.B. Peitzman and T.R. Billiar: Inhibition of nitric oxide synthase during hemorrhagic shock increases hepatic injury, *Shock* 4 (1995) 332–327.
[14] T. Adachi, S. Hori, K. Miyazaki, M. Nakagawa, S. Inoue, Y. Ohnishi, H. Nakazawa, N. Aikawa and S. Ogawa, Inhibition of nitric oxide synthesis aggravates myocardial ischemia in hemorrhagic shock in constant pressure model, *Shock* 9 (1998) 204–209.
[15] Z. Todorovic, M.S. Prostran, V. Varagic, G. Zunic, J. Savic and S. Vujnov, The cardiovascular effects of the administration of L-NAME during the early posthemorrhagic period, *Gen Pharmacol* 30 (1998) 763–769.
[16] B.G. Harbrecht, B. Wu, S.C. Watkins, T.R. Billiar and A.B. Peitzman: Inhibition of nitric oxide synthesis during severe shock but not after resuscitation increases hepatic injury and neutrophil accumulation in hemorrhaged rats, *Shock* 8 (1997) 415–421.
[17] C. Szabò, M. Faragò, I. Horvàth, Z. Lohinai and A.G.B. Kovàch, Hemorrhagic hypotension impairs endothelium-dependent relaxations in the renal artery of the cat, *Circ Shock* 36 (1992) 238–241.
[18] P.Wang, Z. Ba and I.H. Chaudry, Endothelial cell dysfunction occurs very early following trauma-hemorrhage and persists despite fluid resuscitation. *Am J Physiol* 265 (1993) H973–979.
[19] C. Szabò and C. Thiemermann, Role of nitric oxide in haemorrhagic, traumatic and anaphylactic shock and in thermal injury. *Shock* 2 (1994) 145–155.
[20] Q. Nathan and Q. Xie: Nitric oxide synthases: roles, tolls, and controls. *Cell* 78 (1994) 915–918.
[21] M.E. De Vera, R.A. Shapiro, A.K. Nussler, I.S. Mudgett, R.L. Simmons, S.M. Morris and T.R. Billiar, Transcriptional regulation of human inducible nitric oxide synthase gene by cytokines: initial analysis of human NOS2 promoter, *Proc Nad Acad Sci USA* 93 (1996) 1054–1059.
[22] C. Linn, P.J. Morelli, I. Edry, S.E. Cottongim, C. Szabò and A.L. Salzman, Promoter analysis of the human inducible nitric oxide synthase gene suggests both transcriptional and post-transcriptional regulation, *Am J Physiol* 272 (1997) GI499–508.
[23] S.S. Gross and R. Levi, Tetrahydrobiopterin synthesis: An absolute requirement for cytokine-induced nitric oxide generation by vascular smooth muscle, *J Biol Chem* 267 (1992) 25722–25729.
[24] A.K. Nussler, T.R. Billiar, Z.Z Liu and S.M. Morris, Coinduction of nitric oxide synthase and arginosuccinate synthetase in a murine macrophage cell line, *J Biol Chem* 269 (1994) 1257–1261.
[25] R.M.J. Palmer, L. Bridge, N.A. Foxwell and Moncada, The role of nitric oxide in endothelial cell damage and its inhibition by glucocorticoids, *Br J Pharmacol* 105 (1992) 11–12.
[26] M.A. Moro, V.M. Darley-Usmar, I. Lizasoain, Y. Su, R.G. Knowles, M.W. Radomski and S. Moncada, The formation of nitric oxide donors from peroxynitrite, *Br J Pharmacol* 116 (1995) 1999–2004,
[27] M.M. Tarpey, J.S. Beckman, H. Ischiropoulos, J.Z. Gore and T.A. Brock, Peroxynitrite stimulates vascular smooth muscle cell cyclic GMP synthesis, *FEBS Lett* 364 (1995) 314–318.
[28] B. Mayer, A. Schrammel, P. Klatt, D. Koesling and K. Schmidt, Peroxynitrite induced accumulation of cyclic GMP in endothelial cells and stimulation of purified soluble guanylyl cyclase: Dependence on glutathione and possible role of S-nitrosation, *J Biol Chem* 270 (1995) 17355–173560.
[29] C.A. Davidson, P.M. Kaminski, M.S. and Wolin, NO elicits prolonged relaxation of bovine pulmonary arteries via endogenous peroxynitrite generation, *Am J Physiol* 273 (1997) L437–444.
[30] J.S. Beckman and W.H. Koppenol Nitric oxide, superoxide, and peroxynitrite: the good, the bad, and ugly, *Am J Physiol* 271 (1996) CI424–437.
[31] C. Szabò, The role of peroxynitrite in the pathophysiology of shock, inflammation and ischemia-reperfusion injury, *Shock* 6 (1996) 79–88.
[32] N. Kooy, J. Royall, H. Ischiropoulos and J. Beckman, Peroxynitrite-mediated oxidation of dihydrorhodamine 123, *Free Rad Biol Med* 16 (1994) 149–155.
[33] C. Szabò, A.L. Salzman and H. Ischiropoulos, Endotoxin triggers the expression of an inducible isoform, of nitric oxide synthase and the formation of peroxynitrite in the rat aorta *in vivo*, *FEBS Lett* 363 (1995) 235–238.
[34] S. Cuzzocrea, B. Zingarelli, G. Costantino, A. Szabò, A.L. Salzman, A.P. Caputi and C. Szabò, Beneficial effects of 3-aminobenzamide, an inhibiltoor of poly (ADP-ribose) synthetase in a rat model of splanchnic artery occlusion and reperfusion. *Br J Pharmacol* 121 (1997) 1065–1074.
[35] Cuzzocrea S, de Sarro G, Costantino G, Mazzon E, Laura R, Ciriaco E, de Sarro A, Caputi AP: Role of interleukin-6 in a non-septic shock model induced by zymosan. *Eur Cytokine Netw.* (1999) 10:191–203.

[36] H. Ischiropoulos, L. Zhu, J. Chen, M. Tsai, J.C. Martin, C.D. Smith and J.S. Beckman, Peroxynitrite-mediated tyrosine nitration catalyzed by superoxide dismutase. *Arch Biochem Biophys* 298 (1992) 431–437.
[37] J.P. Eiserich, C.E. Cross, A.D. Jones, B. Halliwell and A. Van der Vliet, Formation of nitrating and chlorinating species by reaction of nitrite with hypochlorous acid. A novel mechanism for nitric oxide-mediated protein modification. *J Biol Chem* 271 (1996) 19199–191208.
[38] J.P. Eiserich, M. Hristova, C.E. Cross, A.D. Jones, B.A. Freeman, B. Halliwell and A. Van der Vliet, Formation of nitric oxide derivatives catalysed by myeloperoxidase in neutrophils, *Nature* 391 (1997) 393–397.
[39] B. Halliwell, What nitrates tyrosine? Is nitrotyrosine specific as a biomarker of peroxynitrite formation *in vivo*?, *FEBS Lett* 411 (1997) 157–160.
[40] S. Cuzzocrea, G. Costantino, E. Mazzon, A. De Sarro and A.P. Caputi, Beneficial effects of Mn(III)tetrakis (4-benzoic acid) porphyrin (MnTBAP), a superoxide dismutase mimetic, in zymosan-induced shock, *Br J Pharmacol* 128 (1999) 1241–1251.
[41] C. Szabò and A.L. Salzman, Endogenous peroxyrutrite is involved in the inhibition of cellular respiration in immuno-stimulated J774.2 macrophages, *Biochem Biophys Res Comm* 209 (1995) 739–743.
[42] C. Szabò, B. Zingarelli and A.L. Salzman, Role of poly-ADP ribosyltransferase, activation in the nitric oxide- and peroxynitrite-induced vascular failure, *Circ Res* 78 (1996) 1051–1063.
[43] D.C. Hooper, O. Bagasra, J.C. Marini, A. Zborek, S.T. Ohnishi, R. Kean, J.M. Whampion, A.B. Sarker, L. Bobroski, J.L. Farber, T. Akaike and H. Maeda, Prevention of experimental allergic encephalomyelitis by targeting nitric oxide and peroxynitrite: implications for the treatment of multiple sclerosis, *Proc Natl Acad Sci USA* 94 (1997) 2528–2533.
[44] B.N. Ames, R. Cathcart, E. Schwiers, P. Hochstein, Uric acid provides an antioxidant defense in humans against oxidant- and radical-caused aging and cancer: a hypothesis. *Proc Natl Acad Sci USA* 78 (1981) 6858–6862.
[45] B. Frei, L. England and B.N. Ames, Ascorbate is an outstanding antioxidant in human blood plasma. *Proc Natl Acad Sci USA* 86 (1989) 6377–6381.
[46] H. Rubbo, R. Radi, M. Trujillo, R. Telleri, B. Kalyanaraman, S. Barries, M. Kirk and B.A. Freeman, Nitric oxide regulation of superoxide and peroxynitrite dependent lipid peroxidation. Formation of novel nitrogen-containing oxidized lipid derivatives. *J Biol Chem* 269 (1994) 26066–26075.
[47] A.M. Miles, D.S. Bohle, P.A. Glassbrenner, B. Hansert, D.A. Wink and M.B. Grisham, Modulation of superoxide-dependent oxidation and hydroxylation reactions by nitric oxide, *J Biol Chem* 271 (1996) 40–47.
[48] C. Szabò and H. Ohshima, DNA injury induced by peroxynitrite, *Nitric Oxide Biol Chem* 1 (1997) 323–385.
[49] C. Szabò, Role of poly (ADP-ribose) synthetase in inflammation, *Eur J Pharmacol* 350 (1998) 1–19.
[50] C, Szabò and V.L. Dawson, Role of poly (ADP-ribose) synthetase activation in inflammation and reperfusion injury. *Trends Pharmacol Sci* 19 (1998) 287–298.
[51] L. Viràg, G.S. Scott, A.L. Salzman and C. Szabò, Peroxyrutrite-induced thymocyte apoptosis: the role of caspases and poly-(ADP-ribose) synthetase (PARS) activation. *Immunology* 94 (1998) 345–355.
[52] L. Viràg, D.J. Marmer and C. Szabò, The crucial role of apopain/caspase-3 in the peroxynitrite-induced apoptotic DNA fragmentation, *Free Rad Biol Med* 21 (1998) 1075–1082.
[53] L. Viràg, A.L. Salzman and C. Szabò, Poly (ADP-ribose) synthetase activation mediates mitochondrial injury during oxidant-induced cell death, *J Immunol* 161 (1998) 3753–3759.
[54] M. O'Connor, A.L. Salzman and C. Szabò, Role of peroxynitrite in the protein oxidation and apoptotic DNA fragmentation in immunostimulated vascular smooth muscle cells, *Shock* 8 (1997) 439–443.
[55] M. Leist, D. Single, G. Kunstle, C. Volbracht, H. Hentze and P. Nicotera, Apoptosis in the absence of poly-(ADP-ribose) polymerase, *Biochem Biol Res Comm* 233 (1997) 518–522.
[56] C. Szabò, S. Cuzzocrea, B. Zingarelli, M. O'Connor and A.L. Salzman, Endothelial dysfunction in endotoxic shock: importance of the activation of poly (ADP ribose) synthetase (PARS) by peroxynitrite, *J Clin Invest* 100 (1997) 723–735.
[57] S. Cuzzocrea, B. Zingarelli, G. Costantino, A. Sottile, D. Teti and A.P. Caputi, Protective effect of poly(ADP-ribose) synthetase inhibition on multiple organ failure after zymosan induced peritonitis in the rat, *Crit Care Med* 27 (1999) 1517–1523.
[58] S. Cuzzocrea, A. Filippelli, B. Zingarelli, M. Falciani, A.P. Caputi and F. Rossi, Role of nitric oxide in a nonseptic shock model induced by zymosan in the rat, *Shock* 7 (1997) 351–357.
[59] K.T. Lin, J.Y. Xue, M.C. Lin, E.G. Spokas, F.F. Sun and P.Y. Wong, Peroxynitrite induces apoptosis of HL-60 cells by activation of a caspase-3 family protease, *Am J Physiol* 274 (1998) C855–860.
[60] H. Yaoita, K. Ogawa, K. Maehara and Y. Maruyama, Attenuation of ischemia/ reperfusion injury in rats by a caspase inhibitor, *Circulation* 97 (1998) 276–281.

[61] M. Endres, S. Namura, M. Shimizu-Sasamata, C. Waeber, L. Zhang, T. Gomez-Isla, B.T. Hyman and M.A. Moskowitz, Attenuation of delayed neutonal death after mild focal ischernia in mice by inhibition of the caspase family, *J Cereb Blood Flow Metab* 18 (1998) 238–247.
[62] Y.M. Kim, R.V. Talanian and T.R. Billiar, Nitric oxide inhibits apoptosis by preventing increases in caspase-3-like acivity via two distinct mechanisms, *J Biol Chem* 272 (1997) 31138–31148.
[63] C. Szabo, L.H. Lim, S. Cuzzocrea, S.J. Getting, B. Zingarelli, R.J. Flower, A.L. Salzman, and M. Perretti, Inhibition of poly (ADP-ribose) synthetase attenuates neutrophil recruitment and exerts antiinflammatory effects. *J Exp Med* 186 (1997) 1041–1049.
[64] S. Cuzzocrea, E. Mazzon, L. Dugo, V. Lepore, M.L. Terranova, A. Ciccolo, M.T. Fonti and A.P. Caputi, Inducible nitric oxide synthase knock out mice exhibit a resistance to multiple organ failure induced by zymosan, *Shock* (2001) in press,.
[65] C. Hierholzer, B. Harbrecht, J.M. Menezes, J. Kane, J. MacMicking, C.F. Nathan, A.B. Peitzman, T.R. Billiar and D.J. Tweardy, Essential role of induced nitric oxide in the initiation of the inflammatory response after hemorrhagic shock, *J Exp Med* 187 (1998) 917–928.
[66] S. Cuzzocrea, E. Mazzon, A. De Sarro and A.P. Caputi, Role of free radicals and poly(ADP-ribose) synthetase in intestinal tight junction permeability, *Mol Med* 6 (2000) 766–778.
[67] S. Cuzzocrea, G. Costantino, Mazzon E and A.P. Caputi, Protective effect of N-acetylcysteine on multiple organ failure induced by zymosan in the rat, *Crit Care Med* 27 (1999) 1524–1532.
[68] T. Brenner, S. Brocke, F. Szafer, R.A. Sobel, J.F. Parkinson, D.H. Perez and L. Steinman, Inhibition of nitric oxide synthase for treatment of experimental autoimmune encephalomyelitis, *J Immunol* 158 (1997) 2940–2946.
[69] J.P. Pelletier, D. Jovanovic, J.C. Fernandes, P. Manning, J.R. Connor, M.G. Currie, J.A. Di Battista and J. Martel-Pelletier, Reduced progression of experimental osteoarthritis *in vivo* by selective inhibition of inducible nitric oxide synthase, *Arthritis Rheum* 41 (1991) 1275–1286.
[70] E. Bralm, M.L. Banquerigo, G.S. Firestein, D.L. Boyle, A.L. Salzman and C. Szabò, Collagen-induced arthritis: reversal by mercaptoethylguanidine, a novel anti-inflammatory agent with a combined mechanism of action, *J Rheumat* 25 (1998) 1785–1793.
[71] Diefenbach, H. Schindler, N. Donhauser, E. Lorenz, T. Laskay, J. Mac Micking, M. Rollinghoff, I. Gresser and C. Bogdan, Type 1 interferon (IFNalpha/beta) and type 2 nitric oxide synthase regulate the innate immune response to a protozoan parasite. *Immunity* 8 (1998) 77–87.
[72] M.N. Ajuebor, L. Virag, R.J. Flower, M. Perretti and C. Szabò, Role of inducible nitric oxide synthase in the regulation of neutrophil migration in zymosan-induced inflammation: relationship to IL-10 and chemokine production, *Immunology* 95 (1998) 625–630.
[73] J. Zielasek, S. Jung, R. Gold, F.Y. Liew, K.V. Toyka and H.P. Hartung, Administration of nitric oxide synthase inhibitors in experimental autoinumme neuritis and experimental autoimmune encephalomyelitis, *J Neuroimmunol* 58 (1995) 81–88.
[74] T. Brenner, S. Brocke, F. Szafer, R.A. Sobel, J.F. Parkinson, D.H. Perez and L. Steinman, Inhibition of nitric oxide synthase for treatment of experimental autoimmune encephalomyelitis, *J Immunol* 58 (1997) 2940–2946.
[75] Y. Okuda, S. Sakoda, H. Fujimura and T. Yanagihara, Aminoguanidine, a selective inhibitor of the inducible nitric oxide synthase, has different effects on experimental allergic encephalomyelitis in the induction and progression phase, *J Neuroimmunol* 81 (1998) 201–210.
[76] J.E. Fenyk-Melody, A.E. Garrison, S.R. Brunnert, J.R. Weidner, F. Shen, B.A. Shelton, J.S. Mudgett, Experimental autoimmune encephalomyelitis is exacerbated in mice lacking the NOS2 gene, *J Immunol* 160 (1998) 2940–2946.
[77] G. Karupiah, Q.W. Xie, R.M. Buller C., Nathan, C. Duarte and J.D. MacMicking, Inhibition of viral replication by interferon-gamma-induced nitric oxide synthase, *Science* 261 (1993) 1445–1448.
[78] G. Kaupiah, J.H. Chen, S. Mahahngarn, C.F. Nathan and J.D. MacMicking, Rapid interferon gamma-dependent clearance of influenza A virus and protection from consolidating pneumonitis in nitric oxide synthase 2-deficient mice, *J Exp Med* 188 (1998) 1541–1546.

Nitric Oxide
R.J. Gryglewski and P. Minuz (Eds.)
IOS Press, 2001

Nitric Oxide – a Safeguard of Pulmonary Microcirculation in Early Endotoxaemia

Stefan Chłopicki*, Joanna B. Bartuś, Michał Walski[1]
Paweł Wołkow and Ryszard J. Gryglewski
Department of Pharmacology, Jagiellonian University Medical College, Grzegórzecka 16 31-531 Kraków, Poland
[1]Medical Research Center PAN, Pawińskiego 5, 02-106 Warszawa, Poland

Abstract: It is generally accepted that hypotension, vasoplegia and metabolic disorders are mediated in the course of endotoxic shock by NO produced by NOS-2. On the other hand, NO generated by NOS-3 seems to play a cytoprotective role in endotoxaemia. Our present data point to the protective role of NO in early endotoxaemia and to its particular importance in the pulmonary microcirculation. LPS-induced, complement-mediated activation of platelets and neutrophils leading to elimination of bacterial endotoxin from pulmonary circulation, if not tempered by endogenous NO, is associated with acute microvascular lung injury and fatal pulmonary oedema.

1. Introduction

Numerous reports show that inhibition of NOS-2 in endotoxaemia or septic shock reverses hypotension, and restores vascular responsiveness to vasoconstrictor agents [1;2]. However, in many of these studies, in order to inhibit inducible isoform of NO synthase (NOS-2) nonselective inhibitors were used, such as L-NAME or L-NMMA [1;2] which at the same time limited consitutive isoenzyme NOS-3 in endothelium. This is why, after L-NAME -treatment of animals with endotoxic shock, excessive vasoconstriction appeared which led to subnormal organ perfusion [3], as well as depression of cardiac output [4], aggravated liver [5], intestine [6] and lung [7;8] injury. Moreover, non-selective NOS inhibition could be associated with an increase of mortality rate in septic shock [9]. Despite several attempts at elucidation, mechanisms determining the protective role of NO generated by NOS-3 in host response to LPS remain unclear. It is proposed that the protective role of endothelial NO in this setting results from its vasodilator [3], antiplatelet [10] or antileukocyte [11] activities. Interestingly, it was freshly proposed that increased resistance to LPS-induced mortality of transgenic mice with overexpression of NOS-3 resulted from cytoprotective action of NO in the lung [12]. The aim of the present article is to summarize our data on the cytoprotective role of endogenous NO in the immediate phase of lung response to endotoxin.

* Corresponding author: Stefan Chłopicki E-mail: mfschlop@cyf-kr.edu.pl

2. Methods

2.1 Rat model of endotoxaemia in vivo

Under thiopentone anesthesia (120 mg/kg) Wistar rats (Lod:WIST BR) were intubated intratracheally and ventilated with room air (tidal volume 10 ml kg^{-1}, 50 breaths min^{-1}). Polyurethane catheters were inserted into the left femoral vein and carotid artery for drug administration and for recording of arterial blood pressure, respectively. LPS (10 mg kg^{-1}) was infused at a rate of 2 mg^{-1} kg^{-1} min^{-1} for 10 min (total dose 20 mg kg^{-1}). L-NNA was given as an intravenous bolus injection (10 mg kg^{-1}) 45 minutes before or after infusion of LPS (L-NNA+LPS or LPS+L-NNA, respectively).

2.2 Isolated blood-perfused endotoxaemic lung

Lungs were isolated from Wistar rats weighing 200–250g (Lod:WIST BR) and mounted in an isolated rat lung set-up (Hugo Sachs Elektronik) as described previously [13]. Briefly, anaesthetised rats (thiopentone 120 mg kg^{-1}, i.p.), after laparatomy, were exsanguinated by incision of left renal artery and lungs were exposed via medial sternotomy. Pulmonary artery and left atrium were cannulated via right and left atrium, respectively. Immediately after cannulation lungs were dissected from the thorax and mounted in the water-jacketed (37°C), air-tight glass chamber (HSE), and ventilated with negative pressures. The end-expiratory pressure in the chamber was set to be -2 cm H_2O and inspiratory pressure was adjusted between -6 to -10 cm H_2O to yield the initial tidal volume (TV) of about 2.0 ml. Breathing frequency was set to be 80 breaths/minute and a duration of inspiration versus expiration was 1:1 in each breath. Every 5 min throughout the experiments a deep breath of end-inspiratory pressure of -18 cm H_2O was automatically initiated by VCM module (HSE) to avoid atelectasis.

Lungs were perfused with various fluids using a peristaltic pump (ISM 834, HSE) at constant flow of about 16 ml/min. The venous pressure was set at a level of 2-5 cm H_2O.

Following parameters were measured: arterial and venous pulmonary pressures (PAP, PVP), tidal volume (TV) and the weight of lungs. Scheme of the isolated lung set-up (HSE) is shown in fig. 1.

The following fluids were used to perfuse the lung preparation: rat whole blood, rat platelet rich plasma (PRP), rat platelet poor plasma (PPP) or Krebs-Henseleit buffer, obtained as described previously [13].

All lung preparations were allowed to equilibrate for the first 15 min until baseline PAP, PVP, TV and weight became stable. LPS was injected 45 minutes after the beginning of the experiment. L-NAME or L-NIL were administered 15 min prior to LPS (L-NAME+LPS and L-NIL+LPS, respectively) or 30 min after LPS (LPS+L-NAME). Camonagrel (300 μM), WEB 2170 (100 μM), MK 571 (100 μM), anti-CD62P (1 μg/ml) sCR1 (100 μg/ml) were added 30 min prior to injection of LPS.

Data were expressed as means ± SEM of changes in TV (ΔTV), PAP (ΔPAP), PVP (ΔPVP), or weight (Δweight) as compared to respective values 30 min before LPS injection.

2.3 Light and electron microscopy of the isolated lung

For light microscopy whole lungs were fixed immediately after termination of respective experiments in 10% buffered formalin (pH 7.4) for about 24 hours at room temperature, and then washed for 2 hours, dehydrated through a graded ethanol series, placed in xylen (3 times for 15 minutes), embedded in parafin (58°C, 3 hours) and cast into blocks. Tissue

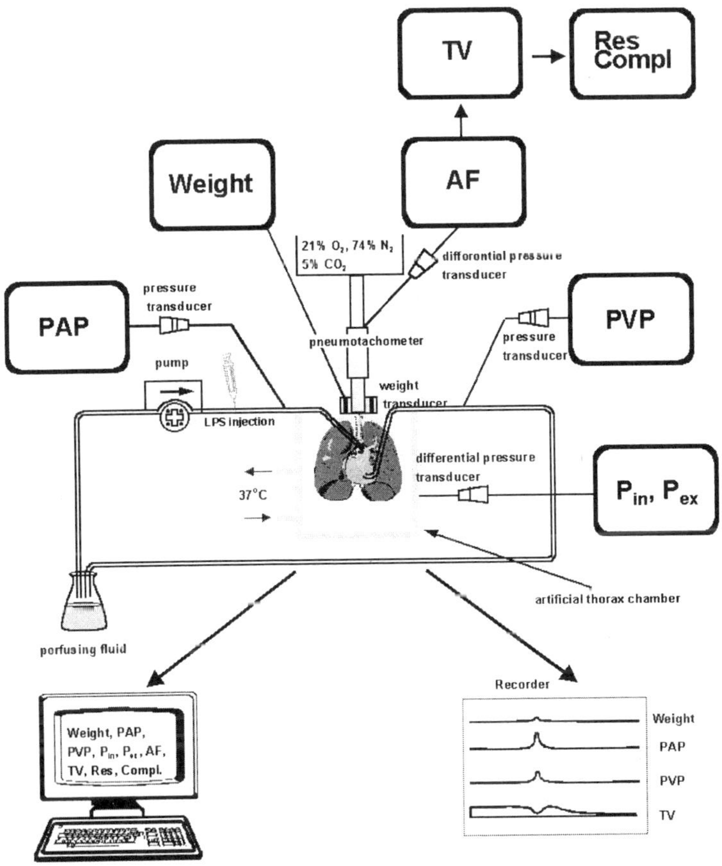

Fig. 1. Isolated lung set-up. See methods for details.

samples were cut into approximately 5 μm - thick sections by microtome, placed on glass slides and stained by hematoxylin and eosin. Light microscopic examination and photographic documentation were performed using a Zeiss microscope.

For electron microscopy lungs were fixed immediately after termination of the respective experiment using a mixture of 2.5 % glutaraldehyde, 2% freshly prepared paraformaldehyde in 0.1 mol/l cacodylate buffer at a pH 7.4. Ventilated lung was fixed by immersion for 5–10 minutes, gently dismantled from the apparatus and fixed for 4–6 h at room temperature. Subsequently 5-10 pieces of left and right lung tissue were gently dissected and fixed at room temperature for further 24 hours. After rinsing with 0.1 mol/L cacodylate buffer (12 hours) lung tissue blocks (approximately 3 x 3 x 3 mm^3) were post-fixed, dehydrated and embedded in Spur resin by routine procedure. Lung ultrastructure was viewed and photographed using Transmission EM Jem 1200 Ex.

2.4 Reagents

Lipopolysaccharide from *Escherichia coli* (LPS, B:0127), non-selective NOS inhibitors, N^G-nitro-L-arginine methyl ester (L-NAME) and N^G-nitro-L-arginine (L-NNA) were purchased from Sigma Chemicals International. Selective NOS-2 inhibitor L-N^6-(1-

iminoethyl)lysine (L-NIL) was obtained from Caymann Biochemicals, USA, anti P-selectin blocking antibody (antiCD62P) from Pharmingen (Germany), cysteinyl leukotriene antagonists (MK 571 - (3-(3-(2-(7-chloro-2-quinolinyl)ethenyl)phenyl)((3-dimethylamino-3-oxopropyl)thio)methyl)thio propanoic acid)) from Biomol Research Lab. Inc., USA. PAF receptor antagonist (WEB 2170 - (5-(2-chlorophenyl)-3,4-dihydro-10-methyl-3-[(4-morpholinyl)carbonyl]-2H,7H-cyclopenta[4,5]thieno[3,2-f][1,2,4]triazolo[4,3-a][1,4]diazepine)) was a gift from Boehringer Ingelheim, Germany; TXA_2 synthase inhibitor (Camonagrel - ([+]-5-[2-imidazole-1-ethyloxy]-1-indan-carboxylic acid hydrochloride)) was a gift from Ferrer inc. Spain. sCR1, an alternative and classic pathway complement inhibitor (C3 and C5 convertase inhibitor) was a gift from T Cell Science Boston.

2.5 Statistical analysis

Data are shown as means ± SEM. Differences between groups were by one-way ANOVA followed by *post hoc* Fisher test. $P < 0.05$ was considered as statistically significant.

3. Results

3.1 Rat model of endotoxaemia in vivo

In the rat model of endotoxaemia, infusion of LPS (i.v. 20 mg kg $^{-1}$) induced a lethal shock with a survival time of 278 ± 49 min (fig. 2). Infusion of L-NNA (10 mg kg $^{-1}$) followed by LPS (i.v. 20 mg kg $^{-1}$) (L-NNA+LPS) induced a rapid fall in arterial blood pressure accompanied by symptoms of acute respiratory failure which shortened the survival time of animals down to 22 ± 5 min. In contrast, infusion of L-NNA after LPS (LPS+L-NNA) did not affected survival time of endotoxaemic rat (314 ± 77 min) (fig. 2).

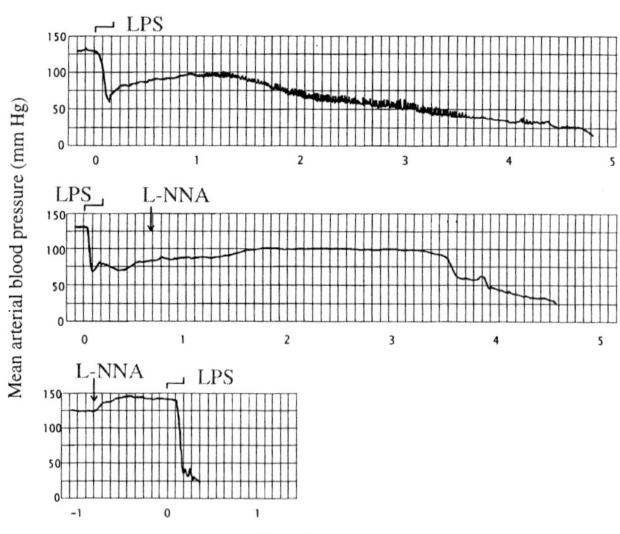

Fig.2. Original tracing of mean arterial blood pressure in Wistar rats after intravenous infusion of A/ lipopolysaccharide (LPS, 20 mg kg^{-1}, B/ LPS (20 mg kg$^{-1)}$ followed by non-selective NOS inhibitor N^G-nitro-L-arginine (L-NNA, 10 mg kg^{-1}) and C/ LPS (20 mg Kg$^{-1)}$ preceded by L-NNA (10 mg kg^{-1}).

3.2 Isolated rat lung preparation

Injection of LPS (300 μg/ml) into isolated blood-perfused rat lungs induced a biphasic response. An immediate transient phase of this response consisted of an increase in pulmonary arterial (PAP) and venous (PVP) pressures (by 10.7 ± 1.7 and 6.5 ± 1.9 cm H_2O, respectively) with a simultaneous decrease in tidal volume (TV) (by 0.6 ± 0.1 ml). This response was not accompanied by a significant change in weight of the lung. A delayed phase of lung response to LPS consisted of a gradual decline in TV without accompanying changes in PAP, PVP and weight (fig. 3).

After pretreatment with L-NAME (300 μM), injection of LPS (300 μg/ml) led to an immediate arrest of lung function. L-NAME + LPS - induced immediate increase in PAP and PVP (by 44 ± 1 and 45 ± 1 cm H_2O, respectively) and a fall of TV (by 2.01 ± 0.3) were accompanied by a sudden increase in lung weight (by 3.5 ± 0.4 g). In contrast with the abrupt response induced by L-NAME+LPS, isolated lung response to LPS+L-NAME or to L-NIL+LPS did not differ from that evoked by LPS alone (fig. 3).

At the macroscopic level, either LPS alone or LPS+L-NAME or L-NIL+LPS did not change the uniform bright-red appearance of the isolated blood perfused lung surface. However, in the L-NAME pretreated lung within a few minutes after injection of LPS deep-red heamorrhagic patches appeared on the lung surface. Approximately at the time point when the increase in weight achieved 3–4 grams, foamy and bloody exudate emerged in trachea of the lung and when dismounted lungs were heavy, firm and boggy.

At the microscopic level the immediate transient lung response induced by LPS did not leave any changes in lung histology when examined by light microscopy (fig. 4). In the ultrastructure of capillary-alveolar barrier only an increased number of pinocytic vessels was noted (fig. 5). On the other hand, irreversible lung response induced by L-NAME+LPS was associated with alveolar haemorrhagic oedema (fig. 4) and a profound injury of capillary-alveolar barrier (fig. 5). Importantly, electron microscopy revealed that the immediate phase of lung response to LPS and to L-NAME+LPS were associated with sequestration of neutrophils and platelets in pulmonary microcirculation and with internalisation of LPS by neutrophils (fig. 5).

Immediate responses to LPS and L-NAME + LPS involved blood cells since they were not seen in lungs perfused with either plasma or Krebs-Henseleit buffer. Moreover, there was a pool of leukocytes (mainly neutrophils) and platelets sequestered within pulmonary circulation during immediate response to LPS and L-NAME+LPS. Interestingly, number of sequestrated leukocytes and platelets during immediate response to LPS and L-NAME+LPS was comparable (data not shown).

Immediate lung responses to LPS or L-NAME+LPS were both abrogated by a complement inhibitor sCR1 (100 μg/ml) but they were affected differently by inhibitors of lipid mediators. Immediate transient lung response to LPS was inhibited by PAF receptor antagonist, WEB 2170 (100 μM), and a TXA_2 synthase inhibitor, Camonagrel (300 μM), but not by a leukotriene receptor antagonist MK 571 (300 μM), whereas immediate irreversible lung response to L-NAME+LPS was inhibited by WEB 2170 (100 μM), Camonagrel (300 μM) as well as by MK 571 (300 μM). Interestingly, response induced by L-NAME+LPS but not that induced by LPS alone was inhibited by antibody against selectin P (anti-CD62P, 1μg/ml) (data not shown).

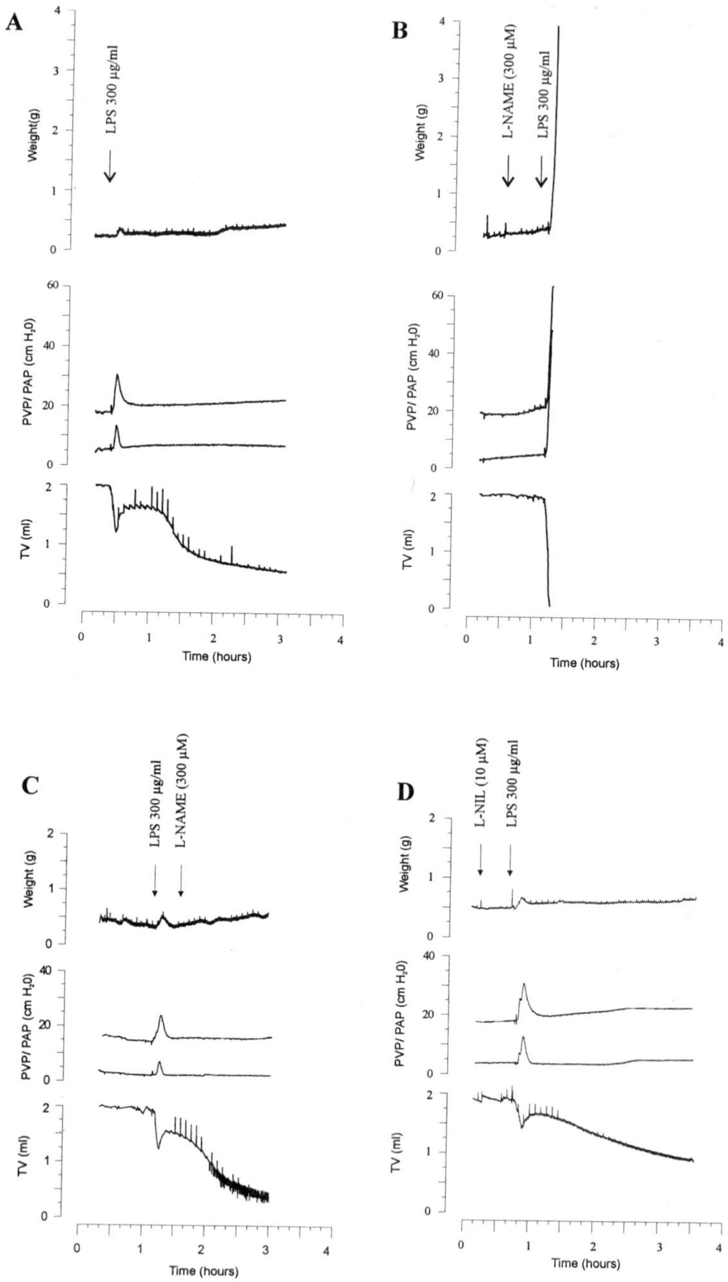

Fig. 3. Tracings from the isolated blood perfused rat lung preparation showing responses evoked by injection into pulmonary circulation of A/ lipopolysaccharide (LPS, 300 µg ml -1), B/ non-selective NOS inhibitor NG-nitro-L-arginine methyl ester (L-NAME, 300 µM) followed by LPS (300 µg ml -1) (L-NAME+LPS), C/ LPS followed by L-NAME (300 µM) (LPS+L-NAME) and D/ selective NOS-2 inhibitor L-NIL (10 µM) followed by LPS (L-NIL+LPS). Only lung response to L-NAME+LPS was associated with an abrubt and irreversible rise in pulmonary arterial and venous pressure (PAP and PVP) and a fall in tidal volume (TV) to zero, accompanied by pulmonary oedema.

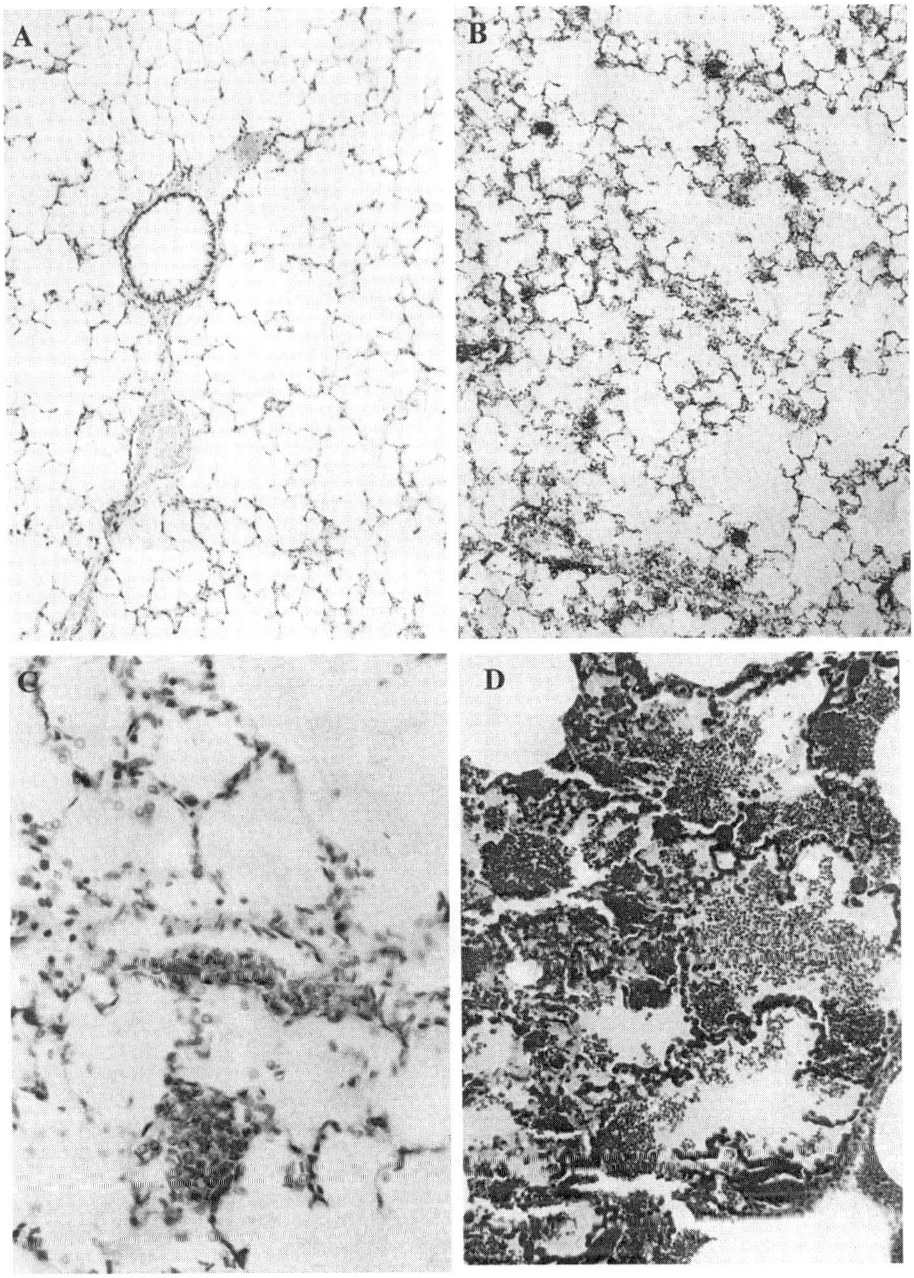

Fig. 4. Light micrograph of isolated blood perfused rat lung shortly after a challenge with LPS (A) and L-NAME+LPS (B–D). Immediate transient phase of lung response to LPS was not associated with changes in lung hitology (A) whereas irreversible lung response induced by L-NAME+LPS was associated with massive alveolar oedema and alveolar haemorrhages of various intensity (B,C,D). Hematoxylin-eosin stain. Objective magnification 20x (A,B) and 40x (C,D)

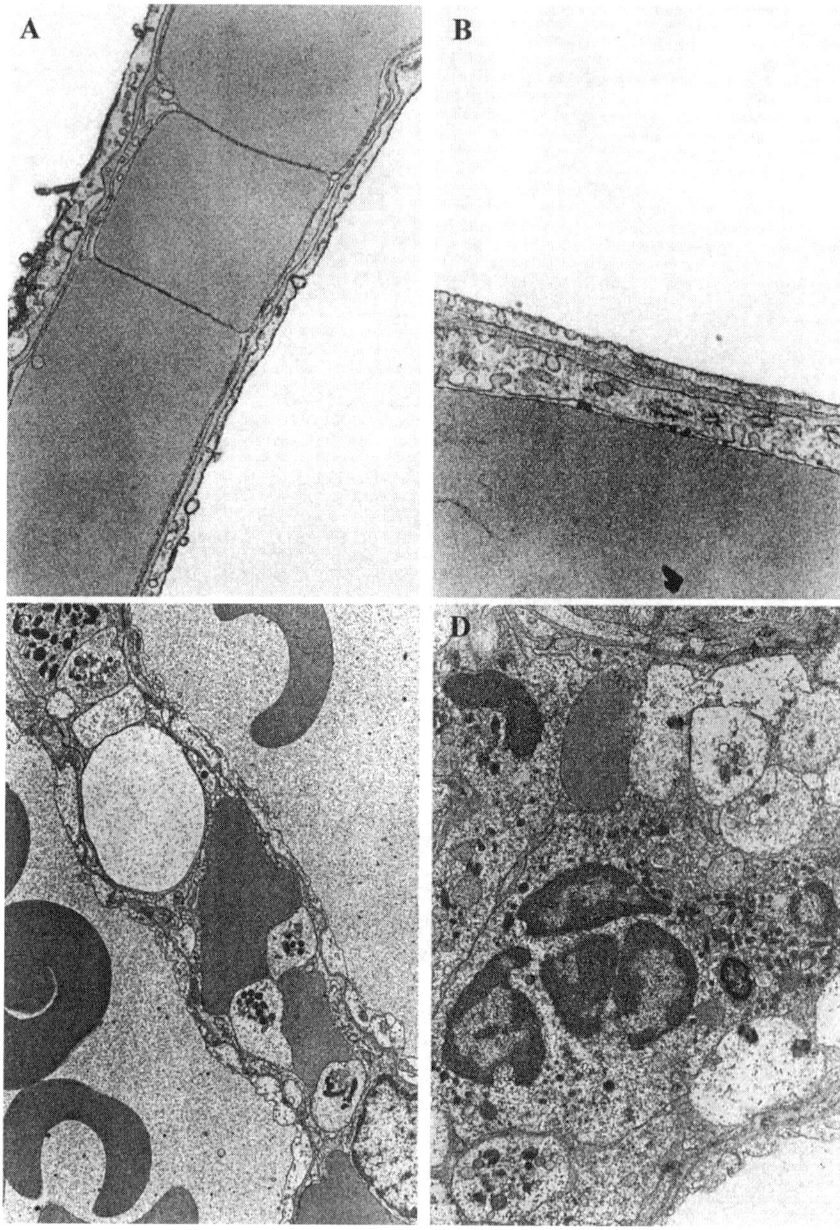

Fig. 5. Electron micrograph of pulmonary capillaries from non-treated isolated blood-perfused rat lung (A) shortly after a challenge with LPS (B) and after L-NAME+LPS (C,D). Immediate transient phase of LPS response was associated with subtle changes in ultrastructure of capillary-alveolar barrier such as increase number of pinocytic vessels in endothelial and epithelial cells (B). Immediate irreversible response evoked by L-NAME-LPS was associated with capillary endothelial cells and epithelial cell type I damage as evidenced by decrease in electron density and oedema of these cells (C). Moreover, inside capillary vessel sequestrated platelets and neutrophils whereas in alveolar space protein rich fluid and erythrocytes were seen (C,D). Inside a neutrophil internalised LPS with electron dense lipid A, and less osmophilic polisccharide chains of the LPS structure were visible (D). Magnification A/20000x, B/50000x, C/6000x, D/10000x.

4. Discussion

Non-selective NOS inhibition prior to LPS injection dramatically shortened the survival time of endotoxaemic rats *in vivo* due to acute lung injury. This injury was reproduced in the isolated blood-perfused rat lung by its treatment with L-NAME+LPS. The unique feature of isolated blood-perfused rat lung is that is permits simultaneous on-line measurement of circulatory and respiratory lung functions. At the same time, one may record the weight of the lung that would not be feasible *in vivo* [14]. Taking advantage of this preparation we showed that in the isolated blood-perfused rat lung after non-selective inhibition of NOS by L-NAME, LPS had caused an immediate microvascular lung injury and massive haemorrhagic oedema leading to the arrest of all lung functions. On the other hand, selective inhibition of NOS-2 by L-NIL [15] had no effect on the lung response to LPS. Indeed, even though as early as 30 min after LPS treatment NOS-2 mRNA was already detected [16], by that time immediate response to LPS had been completed (fig. 3). These data suggest that pneumoprotective NO is generated by endothelial NOS-3.

Although in many experimental studies [3–7] it was reported that NO from NOS-3 played a cytoprotective role in host response to LPS, here we show that NO produced by pulmonary endothelium is a life-saving mediator of endotoxaemic animals by its capability to safeguard pulmonary microcirculation when exposed to a high dose of LPS.

In the isolated blood-perfused rat lung an immediate bronchoconstrictor and vasoconstrictor responses to bacterial lipopolysaccharide (LPS) were mediated by platelet activating factor (PAF) and thromboxane A_2 (TXA_2) released from activated neutrophils and platelets, respectively, as shown previously *in vivo* in other species [17;18]. These responses were transient, since LPS-induced, complement-dependent activation of platelets and neutrophils as well as their intercellular interactions were tempered by endogenous NO, released from endothelial NOS-3. Indeed, increase in NO release in the immediate phase of lung response to LPS was previously demonstrated [8]. We may hypothesise that mechanisms involved in NO release in an immediate phase of lung response to LPS may involve several pathways; mechanical stimulation of endothelium evoked by pulmonary vasoconstriction [19], receptor mediated stimulation of endothelium evoked by LPS [20], by bradykinin cleaved off by activated complement [21], or by lipid mediators released from activated blood cells [22]

Whatever were the mechanisms of pulmonary NO release, inhibition of its generation in early endotoxaemia had been fatal as it allowed LPS to induce acute microvascular lung injury and haemorrhagic lung oedema. In absence of endogenous NO the immediate response to LPS was no longer transient and involved not only PAF and TXA_2 but also cysteinyl leukotrienes (cys-LTs) and selectin P-dependent intercellular interactions. We believe that these findings can be explained as follows. Uninhibited by endogenous NO, LPS induced complement-dependent activation of platelets and neutrophils was potentiated, leading to selectin P expression [23;24] and to unrestricted adhesion of platelets and neutrophils to each other and to endothelial cells. These intercellular interactions draw a scenario favourable for an increased generation of PAF and TXA_2 [25;26] as well as for transcellular synthesis of cys-LTs [27]. Moreover, in absence of endogenous NO, outburst of TXA_2, PAF and cyst-LTs may further amplify activation of blood cells and their intercellular interactions [28;29].

Although PAF and cys-LTs increase endothelial permeability and activate endothelial cells to express selectin P [28], their direct action on vascular wall could not confer such a profound microvascular injury encompassing extravasation of erytrocytes through an alveolar-capillary barrier of the lung. Proteolytic enzymes e.g. elastase [30] or metalloproteinases [31], released by activated neutrophils are likely to be involved.

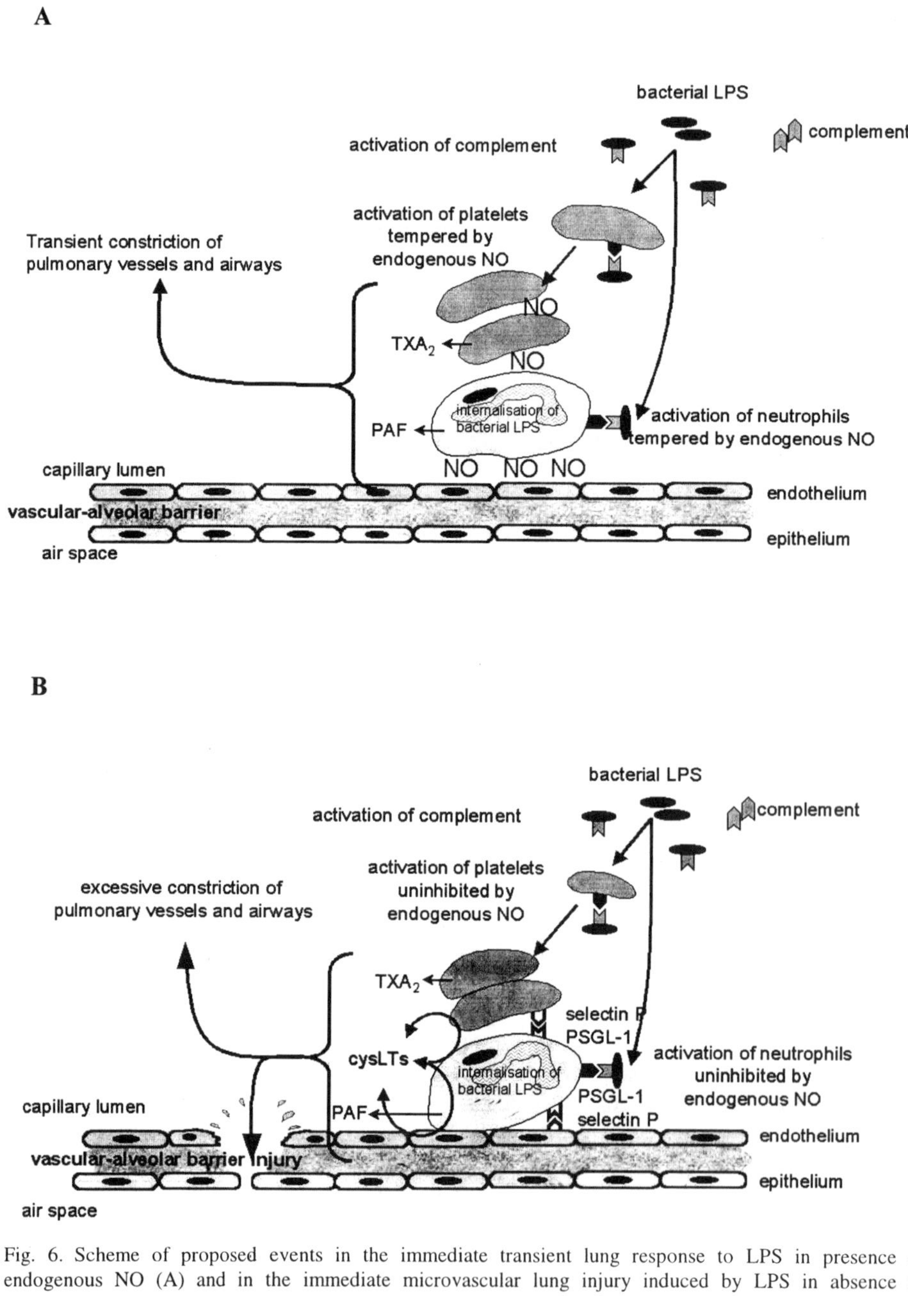

Fig. 6. Scheme of proposed events in the immediate transient lung response to LPS in presence of endogenous NO (A) and in the immediate microvascular lung injury induced by LPS in absence of endogenous NO (B).

Receptor mediated stimulation of endothelial cells to release NO is associated with a coupled release of PGI_2 [32]. However, as shown previously, it is NO but not PGI_2 that protects the lung against LPS-induced damage [8;32]. It looks as if the safeguard role of NO in pulmonary endothelium in early endotoxaemia cannot be replaced by PGI_2. Indeed, it is NO but not PGI_2 that inhibits intercellular adhesion [33]. PGI_2 being a potent inhibitor of platelet aggregation, hardly influences platelet adhesion [34] and it has virtually no effect on neutrophil function [35].

The immediate phase of lung response to bacterial LPS was associated with internalisation of LPS by neutrophils. Indeed, the main biological role of complement system is to accelerate phagocytosis by immunoadherence. Neutrophils are main phagocytosing cells in blood [36] and their ability to phagocytise bacteria is markedly enhanced in the presence of platelets [34;37]. If so, it is tempting to speculate that the immediate vascular and bronchial response to bacterial LPS is an epiphenomenon associated with the internalisation of LPS in which complement-activated neutrophils and platelets and lipid mediators play a role. Indeed, lipid mediators released by blood cells were shown to be involved in the removal of bacteria from blood [38].

Interestingly, it was shown that removal of LPS from circulation occurs within minutes [39] (fig. 5). Needless to say, the quick elimination of bacterial LPS from blood limits its spreading out to other parts of circulation. It could well be that huge surface of pulmonary microcirculation operates as a sieve whereby sequestrated neutrophils and platelets in their cooperative gesture remove bacterial LPS from blood (fig. 6). If so, pulmonary NO appears to be an endogenous mediator that safeguards pulmonary microcirculation keeping under control dangerous multicellular events which are triggered to defend the host and to rapidly remove bacterial LPS from pulmonary circulation.

5. Acknowledgements

This study was supported by the State Committee for Scientific Research (grant No 4 PO5A 118 19 and 4 PO5A 044 19).

6. References

[1] Thiemermann C: Nitric oxide and septic shock. Gen.Pharmacol. 29:159–166, 1997.

[2] Kirkeboen KA, Strand OA: The role of nitric oxide in sepsis--an overview. Acta Anaesthesiol.Scand. 43:275–288, 1999.

[3] Spain DA, Wilson MA, Garrison RN: Nitric oxide synthase inhibition exacerbates sepsis-induced renal hypoperfusion. Surgery 116:322–330, 1994.

[4] Offner PJ, Robertson FM, Pruitt BA, Jr.: Effects of nitric oxide synthase inhibition on regional blood flow in a porcine model of endotoxic shock. J.Trauma 39:338–343, 1995.

[5] Harbrecht BG, Billiar TR, Stadler J, Demetris AJ, Ochoa JB, Curran RD, Simmons RL: Nitric oxide synthesis serves to reduce hepatic damage during acute murine endotoxemia. Crit Care Med. 20:1568–1574, 1992.

[6] Laszlo F, Whittle BJ, Moncada S: Time-dependent enhancement or inhibition of endotoxin-induced vascular injury in rat intestine by nitric oxide synthase inhibitors. Br.J.Pharmacol. 111:1309–1315, 1994.

[7] Park JH, Chang SH, Lee KM, Shin SH: Protective effect of nitric oxide in an endotoxin-induced septic shock. Am.J.Surg. 171:340–345, 1996.

[8] Gryglewski RJ, Wolkow PP, Uracz W, Janowska E, Bartus JB, Balbatun O, Patton S, Brovkovych V, Malinski T: Protective role of pulmonary nitric oxide in the acute phase of endotoxemia in rats. Circ.Res. 82.819–827, 1998.

[9] Cobb JP: Use of nitric oxide synthase inhibitors to treat septic shock: the light has changed from yellow to red. Crit Care Med. 27:855–856, 1999.

[10] Shultz PJ, Raij L: Endogenously synthesized nitric oxide prevents endotoxin-induced glomerular thrombosis. J.Clin.Invest 90:1718–1725, 1992.

[11] Laszlo F, Whittle BJ, Moncada S: Attenuation by nitrosothiol NO donors of acute intestinal microvascular dysfunction in the rat. Br.J.Pharmacol. 115:498–502, 1995.

[12] Yamashita T, Kawashima S, Ohashi Y, Ozaki M, Ueyama T, Ishida T, Inoue N, Hirata K, Akita H, Yokoyama M: Resistance to endotoxin shock in transgenic mice overexpressing endothelial nitric oxide synthase. Circulation 101:931–937, 2000.

[13] Chlopicki S, Bartus JB, Gryglewski RJ: Biphasic response to lipopolysaccharide from E. coli in the isolated ventilated blood-perfused rat lung. J.Physiol Pharmacol. 50:551–565, 1999.

[14] Uhlig S, Wollin L: An improved setup for the isolated perfused rat lung. Journal of Pharmacological & Toxicological Methods 31:85–94, 1994.

[15] Moore WM, Webber RK, Jerome GM, Tjoeng FS, Misko TP, Currie MG: L-N6-(1-iminoethyl)lysine: a selective inhibitor of inducible nitric oxide synthase. J.Med.Chem. 37:3886–3888, 1994.

[16] Uracz W, Gebska A, Ziemianin B, Chlopicki S, Gryglewski RJ: Role of nitric oxide synthase types II and III in early protection against endotoxin-induced lung injury. Experimental and Clinical Cardiology 3:78–86, 1998.

[17] Olson NC, Hellyer PW, Dodam JR: Mediators and vascular effects in response to endotoxin. British Veterinary Journal 151:489–522, 1995.

[18] Hellewell PG, Henson PM, Downey GP, Worthen GS: Control of local blood flow in pulmonary inflammation: role for neutrophils, PAF, and thromboxane. J.Appl.Physiol 70:1184–1193, 1991.

[19] Ballermann BJ, Dardik A, Eng E, Liu A: Shear stress and the endothelium. Kidney Int.Suppl 67:S100-S108, 1998.

[20] Salvemini D, Korbut R, Anggard E, Vane J: Immediate release of a nitric oxide-like factor from bovine aortic endothelial cells by Escherichia coli lipopolysaccharide. Proc.Natl.Acad.Sci.U.S.A 87:2593–2597, 1990.

[21] Morrison DC, Ulevitch RJ: The effects of bacterial endotoxins on host mediation systems. A review. Am.J.Pathol. 93:526–617, 1978.

[22] Noguchi K, Matsuzaki T, Shiroma N, Ojiri Y, Sakanashi M: Involvement of nitric oxide and eicosanoids in platelet-activating factor-induced haemodynamic and haematological effects in dogs. Br.J.Pharmacol. 118:941–950, 1996.

[23] Armstead VE, Minchenko AG, Schuhl RA, Hayward R, Nossuli TO, Lefer AM: Regulation of P-selectin expression in human endothelial cells by nitric oxide. Am.J.Physiol 273:H740–H746, 1997.

[24] Michelson AD, Benoit SE, Furman MI, Breckwoldt WL, Rohrer MJ, Barnard MR, Loscalzo J: Effects of nitric oxide/EDRF on platelet surface glycoproteins. Am.J.Physiol 270:H1640–H1648, 1996.

[25] Coeffier E, Delautier D, Le Couedic JP, Chignard M, Denizot Y, Benveniste J: Cooperation between platelets and neutrophils for paf-acether (platelet- activating factor) formation. J.Leukoc.Biol. 47:234–243, 1990.

[26] Maugeri N, Evangelista V, Piccardoni P, Dell'Elba G, Celardo A, de Gaetano G, Cerletti C: Transcellular metabolism of arachidonic acid: increased platelet thromboxane generation in the presence of activated polymorphonuclear leukocytes. Blood 80:447–451, 1992.

[27] Sala A, Rossoni G, Berti F, Muller-Peddinghaus R, Folco G: Transcellular synthesis of Cys-LT: from isolated cells to complex organ system. Adv.Exp.Med.Biol. 433:95–98, 1997.

[28] Pedersen KE, Bochner BS, Undem BJ: Cysteinyl leukotrienes induce P-selectin expression in human endothelial cells via a non-CysLT1 receptor-mediated mechanism. J.Pharmacol.Exp.Ther. 281:655–662, 1997.

[29] Wiles ME, Welbourn R, Goldman G, Hechtman HB, Shepro D: Thromboxane-induced neutrophil adhesion to pulmonary microvascular and aortic endothelium is regulated by CD18. Inflammation 15:181–199, 1991.

[30] Carden D, Xiao F, Moak C, Willis BH, Robinson-Jackson S, Alexander S: Neutrophil elastase promotes lung microvascular injury and proteolysis of endothelial cadherins. Am.J.Physiol 275:H385-H392, 1998.

[31] Carney DE, Lutz CJ, Picone AL, Gatto LA, Ramamurthy NS, Golub LM, Simon SR, Searles B, Paskanik A, Snyder K, Finck C, Schiller HJ, Nieman GF: Matrix metalloproteinase inhibitor prevents acute lung injury after cardiopulmonary bypass. Circulation 100:400–406, 1999.

[32] de Nucci G, Gryglewski RJ, Warner TD, Vane JR: Receptor-mediated release of endothelium-derived relaxing factor and prostacyclin from bovine aortic endothelial cells is coupled. Proc.Natl.Acad.Sci. U.S.A. 85:2334–2338, 1988.

[33] Kanwar S, Kubes P: Nitric oxide is an antiadhesive molecule for leukocytes. New Horiz. 3:93–104, 1995

[34] Radomski MW: Vascular endothelium in the processes of hemostasis and thrombosis. The role of prostacyclin and EDRF. Acta Physiol Pol. 40 Suppl 33:97–109, 1989.

[35] Gryglewski RJ, Szczeklik A, Wandzilak M: The effect of six prostaglandins, prostacyclin and iloprost on generation of superoxide anions by human polymorphonuclear leukocytes stimulated by zymosan or formyl-methionyl-leucyl- phenylalanine. Biochem.Pharmacol. 36:4209–4213, 1987.

[36] Hampton MB, Kettle AJ, Winterbourn CC: Inside the neutrophil phagosome: oxidants, myeloperoxidase, and bacterial killing. Blood 92:3007–3017, 1998.

[37] Peters MJ, Dixon G, Kotowicz KT, Hatch DJ, Heyderman RS, Klein NJ: Circulating platelet-neutrophil complexes represent a subpopulation of activated neutrophils primed for adhesion, phagocytosis and intracellular killing. Br.J.Haematol. 106:391–399, 1999.

[38] Stanley DW: Eicosanoids mediate insect cellular immune reactions to bacterial infections. Adv.Exp.Med.Biol. 433:359–362, 1997.

[39] Herring WB, Herion JC, Walker RI, Palmer JG: Distribution and clearance of circulating endotoxin. J.Clin.Invest 42:79–87, 1963.

Nitric Oxide
R.J. Gryglewski and P. Minuz (Eds.)
IOS Press, 2001

Therapeutic delivery of the scaffolding domain inhibits nitric oxide synthesis and reduces inflammation.

Mariarosaria Bucci[1,2], Giuseppe Cirino[2] and William C. Sessa[1]

[1]*Department of Pharmacology, Boyer Center for Molecular Medicine, Yale university School of Medicine, New Haven, CT USA,* [2]*Dipartimento di Farmacologia Sperimentale, Università di Napoli "Federico II", via D. Montesano Napoli, Italy*

ABSTRACT

Caveolin-1, the primary coat protein of caveolae, has been implicated as a regulator of signal trasduction through binding of its "scaffolding domain" to key signaling molecules. However, the physiological significance of caveolin-1 as a regulator of signaling has been difficult to separate from its traditional roles in caveolae assembly, transcytosis, and cholesterol transport. To directly address the importance of the caveolin-1 scaffolding domain in vivo, we generated a chimeric peptide with a cellular internalization sequence fused to caveolin-1 scaffolding domain (aminoacids 82-101). The chimeric peptide was efficiently taken up into blood vessels and endothelial cells resulting in selective inhibition of acetylcholine-induced vasodilation and nitric oxide production, respectively. More importantly, systemic administration of the peptide to mice blunted acute inflammation and vascular leak to the same extent as a glucocorticoid or a nitric oxide synthase inhibitor. These data imply that the caveolin-1 scaffolding domain can selectively regulate signal trasduction to eNOS in endothelial cells and that small molecule mimicry of this domain may afford a novel therapeutic approach.

Introduction

Caveolins are cholesterol binding proteins that can regulate a variety of signal trasduction pathways (Smart et al., 1999, Kurzchalia et al, 1999). For example, numerous laboratories have demonstrated localisation of proteins in caveolae, interaction of these proteins with caveolins, and the ability of over-expressed caveolins or peptide derived from caveolins to suppress or stimulate signaling functions in vitro or cultured cells (Li et al., 1996; Razani et al., 1999; Garcia-Cardena et al., 1997). However, the importance of caveolins as modulators of signal trasduction in vivo is controversial since caveolin-1 and 3, per se, are cholesterol binding proteins that deliver cholesterol from the endoplasmic reticulum to the plasmalemma thereby regulating signal tranduction indirectly by modulating the cholesterol content of lipid raft domains and/or caveolae (Roy et al., 1999; Sternberg et al., 1999).

Our recent work has focused on the subcellular trafficking and regulation of endothelial nitric oxide synthase (eNOS). eNOS derived NO is necessary for the maintenance of systemic blood pressure, vascular remodeling, angiogenesis and wound healing (Huang et al., 1995; Murohara et al., 1998; Rudic et al., 1998; Lee et al., 1999). eNOS is a dually acylated, peripheral membrane protein that targets to lipid raft domains and/or caveolae (Garcia-Cardena et al.,1996; Liu et al., 1997). In caveolae, eNOS can physically interact with caveolins-1 and –3 by binding to their putative scaffolding domain located between amino acids 82-101 (Li et al., 1996) and that this interaction renders eNOS in its "less active" state (Garcia-Cardena et al., 1997; Ju et al., 1997; Michel et al., 1997). The data for this model was largely elucidated in vitro either using over-expression system, fusion proteins or yeast-two hybrid screening to map the interacting domains. In support of caveolin as a negative regulator of eNOS are studies showing that peptides derived from the scaffolding domain of caveolin-1 will disrupt the binding of eNOS to caveolin and dose-dependently inhibit NOS activity in vitro (IC_{50}=1-3 μM) by slowing electron flux from the reductase to the oxygenase domain of NOS (Garcia-Cardena et al., 1997; Ju et al., 1997; Ghosh et al., 1998). Therefore, the purpose of this study was to develop technology that would permit investigation of the importance of this protein-protein interaction in vivo and to directly examine if the caveolin scaffolding domain can regulate vascular function via eNOS in vivo.

1. Actions of caveolin fusion peptide on intact blood vessels

1.1 Inhibition endothelium-dependent relaxations by a caveolin fusion peptide

To examine the importance of caveolin as a negative regulator of eNOS in intact blood vessels, we synthesized the primary eNOS binding domain of caveolin-1 (the scaffolding domain; amino acids 82-101) as a carboxy terminal fusion protein with the homeodomain of Antennapedia (AP), a *Drosophila* transcription factor (AP-CAV). The AP protein facilitates the homogeneous uptake of peptides or oligonucleotides into cultured mammalian cells through a non-endocytic, non-degradative pathway (Derossi et al., 1998; Derossi et al., 1996; Derossi et al., 1994). As seen in Fig. 1B, incubation of mouse aortic rings with AP-CAV (10μM) attenuated the ability of acetylcholine (Ach), an endothelium-dependent vasodilator, to elicit relaxation of the blood vessel (trace in right panel): However, incubation of aortic rings with a scrambled version (AP-CAVx; 10μM; trace in left panel) or AP alone (30μM) did not influence Ach-induced relaxation of the blood vessel. These bioassay data are summarised in 1C. The inhibitory effects of AP-CAV on Ach-induced relaxations were dose-dependent and time-dependent; in fact in a first set of experiments isolated blood vessels were incubated for 20 hours in presence of AP-CAV 1,

3, 10μM or AP-CAVx 10μM; successively blood vessels were incubated for 1, 6, 12 and 20 hours with AP-CAV or AP-CAVx 10μM and following a phenylephrine (PE) pre-contraction the vessels were then subjected to increasing concentration of Ach (30 nM to 30μM expressed as Log of molar concentration). Consistent with the ability of AP-CAV to inhibit eNOS derived NO, pre-incubation of AP-CAV, but not AP-CAVx, resulted in potentiation of the vasoconstrictor actions of PE, an alpha-1 adrenergic receptor agonist (Fig.1D).

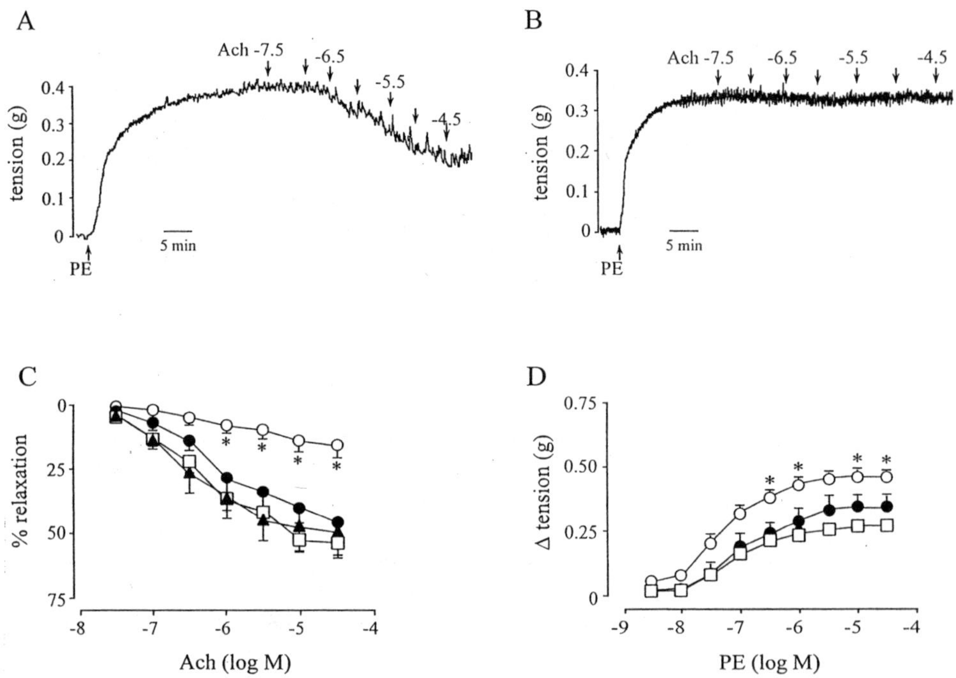

Figure 1: Caveolin scaffolding domain peptide inhibits acetylcholine (Ach)-induced relaxation and potentiates phenylephrine-induced contraction of the isolated mouse aorta. Mouse aortic rings were incubated for 20h with either AP-CAVx (10 μM; **a**) or AP-CAV (10 μM; **b**) and mounted in organ bath chambers. Following phenylephrine (PE) pre-contraction, the vessels were subjected to increasing concentration of Ach (30 nM to 30μM expressed as Log of molar concentration). Panel **c** shows summary of the data presented in **a** and **b** (□ vehicle; •: AP-CAVx, 10 μM; o AP-CAV, 10 μM). Note that AP alone (30

μM) does not affect Ach-induced relaxations. Panel **d** shows the effect of vehicle (□), AP-CAVx (•10 μM), or AP-CAV (o10 μM) peptides on PE concentration-response curves in isolated aortic rings. *$p<0.05$ compared to vehicle alone; n=6-8 rings per treatment.

1.2 ENOS is a primary biological target of AP-CAV in intact blood vessels

To test the hypothesis that inhibitory actions of AP-CAV were occurring through inhibition of eNOS, we examined the vasoconstrictor actions of nitro-L-arginine methyl ester (L-NAME) in mouse aortic rings previously pre-incubated with AP-CAV or AP-CAVx (Fig.2). In vessels pre-treated with AP-CAVx (Fig.2a), Ach induced relaxation of the vessel segments and the addiction of L-NAME, an inhibitor of NOS, further increased isometric tension development due a inhibition of basal NO synthesis. This is in contrast to vessels treated with AP-CAV (Fig.2b). Pre-incubation of the vessel with AP-CAV inhibited Ach-induced relaxation of the vessel (consistent with NOS inhibition) and also diminished the ability of L-NAME to further increase vascular tension. The inability of L-NAME to further increase isometric tension suggest eNOS is the primary molecular target of AP-CAV in the endothelium of an intact blood vessel (summary data in Fig. 2c). Furthermore, AP-CAV, AP-CAVx and AP had no effect of the direct vasorelaxant properties of the NO donor drug, sodium nitroprusside (SNP). SNP (10μM) induced relaxation of PE pre-contracted blood vessel was assessed in either control isolated mouse aortas or following a 20 hours incubation of the vessels with vehicle (DMSO). To further test the specificity of AP-CAV, we measured Ach-induced prostacyclin production. Ach binds to the muscarinic receptors (M1 and M3) on endothelium and activates a c-src, G-protein dependent increase cytoplasmatic calcium. The increase in calcium will activate calmodulin to stimulate phospholipases C and A_2, arachidonic acid release and its subsequent metabolism by cyclooxygenase and prostacyclin synthase to generate prostacyclin (Smith et al., 1997). Neither AP-CAVx nor AP-CAV influenced basal or Ach stimulated prostacyclin production (measured as 6-keto-PGF$_{1\alpha}$) from mouse aortic ring suggesting that AP-CAV was not interfering with signal transduction mechanisms leading to prostacyclin production whereas ibuprofen, a cyclooxygenase inhibitor, abolished Ach-induced prostacyclin production (Fig.2d).

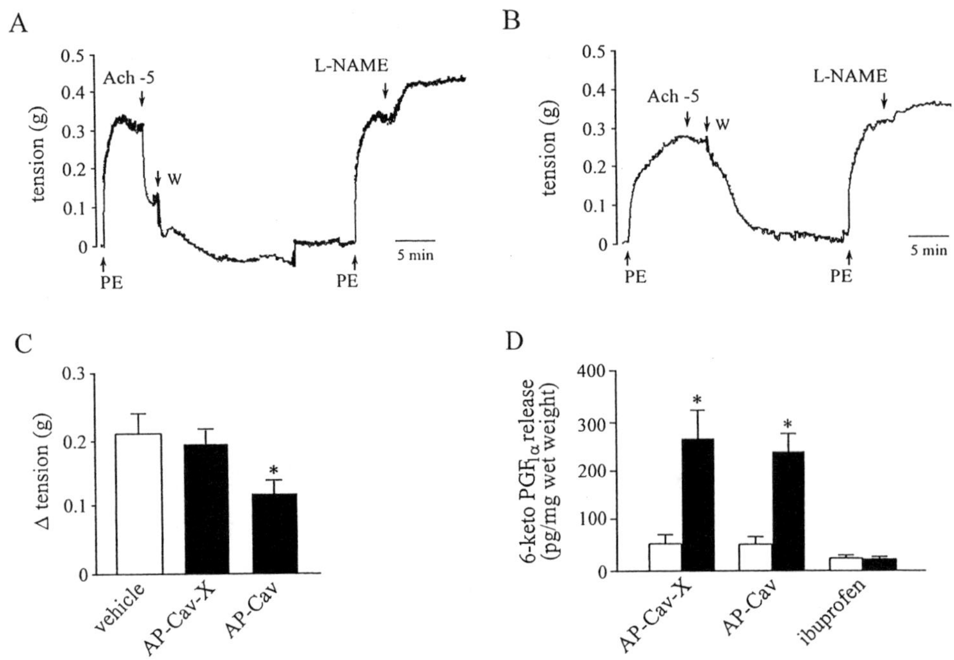

Figure 2: Evidence that eNOS is primary biological target of the caveolin scaffolding domain peptide in intact blood vessels. Upper panels show typical recording of L-NAME (100 μM) induced contraction of AP CAVx (**a**, 10 μM) and AP-CAV (**b**, 10 μM) treated vessels. Vessels were pre-contracted with PE to yield similar levels of tone, followed by Ach. Vessels were washed (w), PE pre-constricted again, followed by the addition of L-NAME. AP CAV pre-treatment attenuates L-NAME induced contraction of mouse aorta. Panel **c** shows the summary of the further increase in tension elicited by L-NAME in **a** and **b**. Panel **d** shows AP-CAVx and AP-CAV do not affect basal (open bars) or Ach-induced (10 μM; closed bars) prostacyclin release from mouse aortic rings. Prostacyclin release was

assessed through measurement of the levels of the stable metabolite of prostacyclin, 6-keto-$PGF_{1\alpha}$ by EIA in rings treated with AP-CAV, AP-CAVx or ibuprofen, before and after Ach stimulation (30 minutes) of the vessels. *$p<0.05$ vs basal prostacyclin levels, * $p<0.05$ compared to vehicle alone, n=5-7 per treatment.

Next we wanted to examine the uptake of AP-CAV into the vessel wall, therefore, we synthesized a biotinylated version of AP-CAV and AP-CAVx. Biotinylated AP-CAV (Biot-AP-CAV) but not Biot-AP-CAVx, inhibited Ach induced relaxation of mouse aorta demostrating that biotin conjugate did not interfere with the biological activity of the peptide (Fig.3a). Incubation of mouse aorta with Biot-AP peptides resulted in the uptake of both AP-CAV and AP-CAVx into the endothelium and adventitial lining of the vessel (Fig.3b). The uptake of both AP-CAV and AP-CAVx were identical suggesting that the inhibitory effect of AP-CAV on eNOS was indeed specific. Because the endothelium, but not the cells of the adventitia contain eNOS (Pollock et al., 1993) the inhibitor action of AP-CAV is most likely due to inhibition of eNOS in the endothelium. Collectively these data (Fig.2 and 3) demonstrate that AP-CAV is selectively inhibiting Ach-induced eNOS dependent vasorelaxation in intact blood vessels.

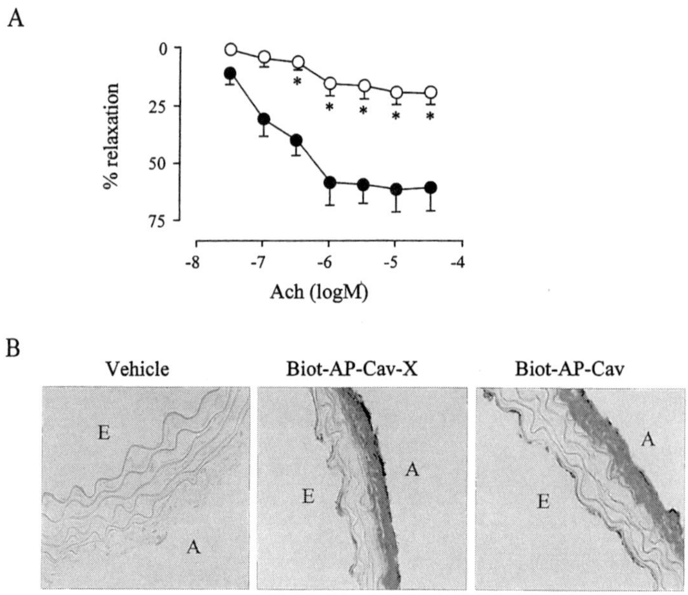

Figure 3: Biotinylated AP-caveolin peptide inhibits vascular relaxation and localises to endothelium and adventitia of mouse aorta. **a**, mouse aortas were incubated with either biotinylated (biot) AP-CAV (o100 μM) or biot-AP-CAVx (•100 μM) for 20h and subjected to Ach concentration-response curves. *$p<0.05$ vs biot-AP-CAVx, n=5 rings per treatment. **b**, biot-AP-CAV and biot-AP-CAVx treated cross sections of mouse aortas stained using HRP-streptavidin. E denotes the endothelium and A denotes adventitia.

2. Actions of caveolin fusion peptide on endothelial cells

2.1 AP-CAV attenuates eNOS dependent NO production cells

Ach induced relaxation of the mouse aorta is primarily eNOS mediated since this response is eliminated in eNOS knockout mice (Huang et al., 1995). However, to verify that AP-CAV indeed blocked NO release, we examined the uptake and effect of AP-CAVx and AP-CAV on NO release from cultured aortic endothelial cells. As seen in Fig.4a, incubation of bovine aortic endothelial cells with rhodamine labeled AP-peptides resulted in efficient uptake of the peptide into the cytosol and nucleus. Under these conditions, AP-CAV, but not AP-CAVx blocked calcium ionophore stimulated NO release from endothelial cells as quantified by NO specific chemiluminescence (Fig.4b). To examine if the AP-CAV can also influence iNOS derived NO production, we transfected COS cells with either the iNOS or eNOS cDNAs and examined NO production. As seen in Fig. 4c, transfection of iNOS resulted in marked accumulation of NO_2^-, an effect blocked by L-NAME (1 mM), but not by AP-CAV or AP-CAVx. In contrast, in eNOS transfected cells calcium ionophore stimulated NO_2^- release, an effect blocked by both L-NAME and AP-CAV, but not by AP-CAVx. These data in conjunction with results in isolated blood vessels indicates that AP-CAV is an inhibitor of eNOS, but not iNOS in intact cells.

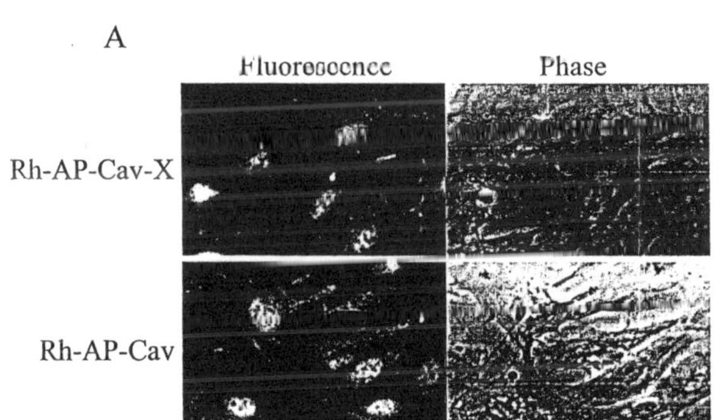

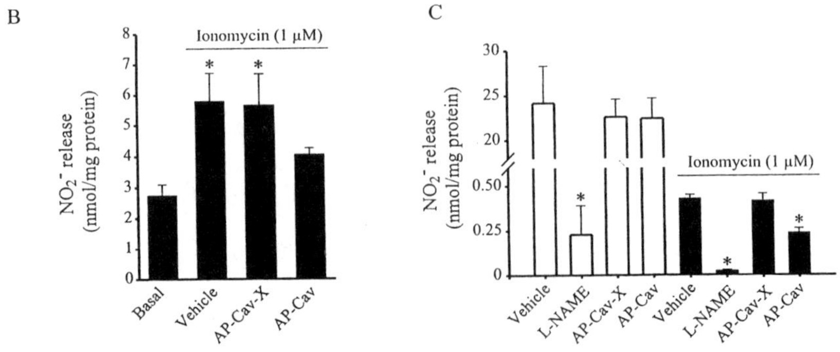

Figure 4: Cellular uptake and inhibition of nitric oxide release by AP-caveolin scaffolding domain peptide in cultured endothelial cells and eNOS trasfected cells. **a**, confocal microscopic imaging of bovine aortic endothelial cells (BAEC) incubated for 6h with either rhodamine labeled AP-CAVx or AP-CAV (1 μM). Following incubation with the labeled peptides the cells were washed and fixed for fluorescence (left panels) and phase-contrast imaging (right panels). **b**, nitric oxide (NO) release from cultured BAEC (measure as levels of NO_2- released in media) pre-incubated for 6h with AP-CAVx or AP-CAV (1 μM of each). Cells were washed, stimulated with ionomycin (1 μM for 30 minutes) and NO_2^- release was measured using a NO specific chemiluminescence analyser (Sievers). *$p<0.05$ when compared to basal NO levels, n=5. **c**, nitric oxide release from COS-7 cells trasfected with iNOS (open bars) or eNOS (closed bars) cDNAs. Cells were trasfected then treated with either vehicle, AP-CAV, AP-CAVx for 6h or the arginine analogue, L-NAME (1 mM) for 30 minutes and basal NO_2^- accumulation measured for iNOS trasfected cells or ionophore induced NO_2^-, COS cells were transfected with cDNA for β-galactosidase and the level of NO_2^- in the media subtracted from levels measured in NOS transfected cells. Data are representative of a single experiment in triplicate that was repeated an additional time. *p<0.05 when compared to NO levels from vehicle treated cells.

3. Actions of caveolin fusion peptide *in vivo*

3.1 AP-CAV blocks inflammation in mice

In various models of acute and chronic inflammation, the production of NO has been linked to increase in vascular permeability, oedema formation and disease progression. Therefore, we treated mice with subplantar injection of carageenan to induce inflammation followed by intraperitoneal administration of vehicle, AP-CAV (0.3 and 1 mg/kg) or AP-CAVx (1 mg/kg) at time 0, 24 and 48h. in this model, the first 24h post-carrageenan are characterised by vascular leakage, oedema formation followed by cellular infiltration into exudate (after 24h). As seen in Fig. 5a at 24 and 48h post carrageenan injection, administration of AP-CAV, but not AP-CAVx, dose-dependently suppressed carrageenan induced oedema formation (n=5 mice per group; *$p<0.05$). Interestingly, the antiinflammatory efficacy of AP-CAV was comparable to that of the potent antiinflammatory steroid, dexamethasone (0.1 mg/kg administered at 0, 24 and 48h subcutaneously). Histological analysis of carrageenan injected paws demonstrated diffuse inflammation characterised by interstitial oedema formation and the presence of inflammatory cells (see control and vehicle treated sections in left and right upper panels). Mice treated with AP-CAV or AP-CAVx did not show marked differences in the cellular infiltration (bottom left and right panels) into the sites of inflammation. However, AP-CAV treated mice exhibited less interstitial oedema (left panel, arrow) consistent with a reduction in paw volume suggesting that the active peptide reduced vascular permeability during inflammation. Under these conditions when AP-CAV reduced oedema formation, AP-CAV did not influence systemic arterial blood pressure or heart rate (89±8, 88±8 and 80±7 mmHg and 340±27, 280±32 and 320±29 bpm for saline, vehicle or AP-CAV treated mice respectively; n=4 mice per group).

To directly assess if the AP-peptides can influence vascular leakage in an additional model, we treated mice with vehicle, AP-CAV, AP-CAVx (2 mg/kg, i.p. for each; 45 minutes pretreatment) or L-NAME (30 mg/kg, i.p., 35 minutes pretreatment) and examined the ability of mustard oil to induce the extravasation of Evans blue, as an index of vascular permeability. Both AP-CAV and L-NAME significantly attenuated vascular leak and overall interstitial oedema (Fig.5c). As illustrated in Fig.5d, AP-CAV (top panel) reduced the amount of Evans blue in the ear skin and surrounding tissue compared to AP-CAVx (bottom panel). Thus, in two distinct models, AP-CAV reduces inflammation likely through attenuating eNOS-dependent blood flow and/or permeability changes evoked by inflammatory stimuli.

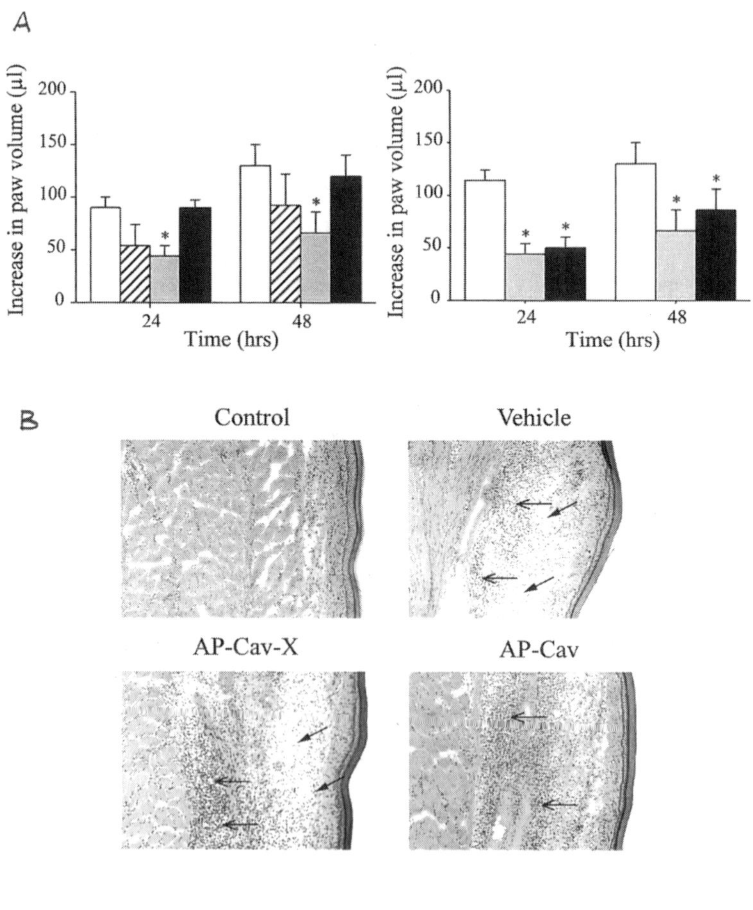

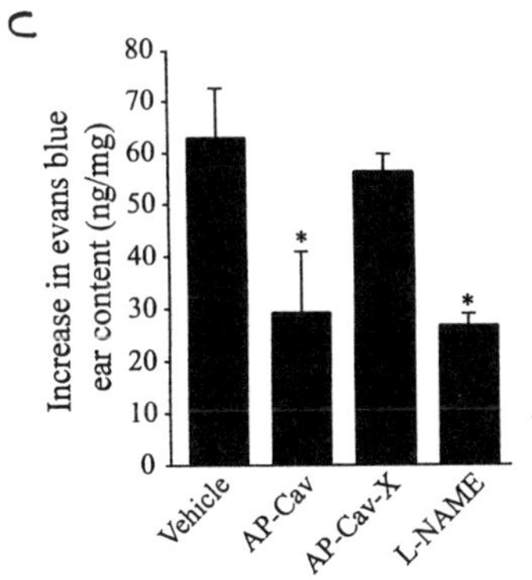

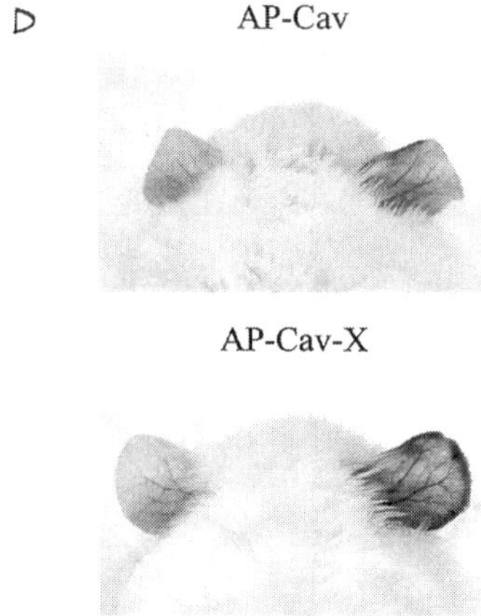

Figure 5: AP-caveolin peptide inhibits edema formation and vascular leakage. **a**, subplantar injection of 1% λ carrageenan (50 μl) was injected in both hind paws and the increase in paw volume was monitored at 24 and 48h. Vehicle (open bar); AP-CAV, 0.3 mg/kg (dotted bar), or 1 mg/kg (grey bar) and AP-CAVx, 1 mg/kg (closed bar) were administered intraperitonealy (i.p.) at time 0,24 and 48h following carrageenan injection (left panel). In a separate experiment, the antiinflammatory effects of AP-CAV 1 mg/kg (grey bar) were compared to dexamethasone 0.1 mg/kg (closed bar right panel). *$p<0.05$ vs vehicle, n=6 mice per group. **b**, Representative ematoxylin/eosin stained cross sections from control (upper left panel) or carrageenan injected mouse paws 48h post-injection. Carrageenan injected mice were treated with either vehicle (upper right panel), AP-CAVx (1 mg/kg; lower left panel) or AP-CAV (1 mg/kg; lower right panel) as described above. The cross section is oriented with the plantar region of the paw on the right side of each panel. Open arrows show increased cellular infiltration and filled arrows indicate interstitial edema. **c**, mice were pre-treated with AP-CAV, AP-CAVx (2 mg/kg of each, i.p.) or L-NAME (30 mg/kg, i.p.) for 45 minutes followed by Evans blue administration (30 mg/kg, i.v.) Mustard oil was applied to ventral and dorsal side of the right ear 1 minute after Evans blue injection. The extent of Evans blue extravasation was assessed after 30 minutes (n=4 mice per group for vehicle and L-NAME treated mice and n=5 for AP-CAV and AP-CAVx

treated mice; *p<0.05). **d**, Representative pictures of mice subjected to mustard oil induced inflammation. Note the appearance of dilated vessels and diffuse blue dye in the periphery of the right ear in lower panel (AP-CAVx) compared to AP-CAV treated mice (upper panel).

Discussion

Our findings underscore the importance of the caveolin scaffolding domain as a regulator of eNOS and perhaps other signaling molecules in vivo. Prior to our work, the scaffolding domain peptide inhibited the functional activity of broad classes of proteins in vitro (including eNOS, G-protein alpha subunits, Ras, Src) (Okamoto et al., 1998) suggesting that this domain either was fundamentally important for many aspects of signal transduction or that may exert non-selective, inhibitory actions on disparate classes of proteins in vitro. A major finding of this study is that the AP-linked scaffolding domain of caveolin-1 is sufficient to block NO release from intact blood vessels and cultured endothelial cells, while not blocking prostacyclin release. This demonstrates that transduction of the endothelium with caveolin-1 scaffolding domain does not broadly inhibit all signaling pathways (i.e. c-src dependent calcium fluxes, phospholipases) but selectively targets NOS.

The inhibitory actions of AP-CAV on Ach-stimulated relaxations of mouse aorta and on NO release from endothelial cells is consistent the notion that caveolin-1, by virtue of its scaffolding domain, tonically inhibits eNOS activity in resting endothelial cells (Garcia-Cardena et al., 1997). Whether this mechanism can solely account for the antiinflammatory actions of AP-CAV in vivo is not known. However, it has been shown that L-arginine based NOS inhibitors block increases in vascular permeability and oedema formation in rats in an acute inflammation setting and after vascular endothelial growth factor treatment (Salvemini et al., 1996; Ialenti et al., 1992; He et al., 1997; Hobbs et al., 1999; Murohara et al., 1998). Our data support a role for eNOS in the hyperemia and permeability changes associated with acute inflammation since L-NAME and AP-CAV both suppress the extravasation of Evans blue while AP-CAVx does not. Interestingly, using the dosage regimen that blunted oedema formation, the AP-CAV peptide did not influence systemic arterial blood pressure suggesting that the peptide may preferentially inhibit eNOS in the inflamed tissue. The ability of AP-CAV to inhibit oedema and vascular leak is unlikely attributed to inducible NOS (iNOS) since iNOS dependent NO production was not blocked by AP-CAV. Furthermore, in the vascular leak model of inflammation, Evans blue extravasation was assessed after 30 minutes of application of the irritant and cytokine induction of iNOS protein expression typically requires a much longer time frame (several hours).

Another salient feature of our paper is the usage of AP fusions to deliver bioactive peptides to cells of the vasculature in vitro and in vivo. Previous work utilising this this method of delivery has focused on the fusion of AP with oligonucleotides and small peptides for treatment of cells in culture. The uptake of AP bound cargo into cells is rapid, independent of membrane fluidity and is not affected by extremes in temperature (Derossi et al., 1996). In the present experiments, AP-CAV either in its biotinylated or rhodamine labeled form was internalised by the endothelium and adventitia. Moreover, based on the antiinflammatory actions of AP-CAV in vivo, the fusion peptide must be stable enough to survive first pass metabolism in the liver and pulmonary circuit to deliver the active peptide to the site of inflammation. Thus, AP may be a useful approach to deliver anti-sense oligonucleotides or as a part of viral delivery system, for therapeutic cardiovascular gene targeting in vivo.

In conclusion, the potent biological activity of the caveolin scaffolding domain peptide in vivo supports the idea that caveolin binding to ENO via the scaffolding domain can indeed regulate NO production and modulate vascular function. Considering the plenitude of potential targets that caveolin may interact with in vivo, further elucidation of AP-CAV regulated pathways may reveal insights into the role of caveolin as a signal-transduction molecule in health and disease.

References

1) Smart E.J. *et al.*, Caveolins, liquid-ordered domains, and signal trasduction. *Mol.Cell Biol* **19**, 7289-7304 (1999).

2) Kurzchalia T.V. & Parton R.G. Membrane microdomains and caveolae. *Curr.Opin.Cell Biol.* **11**, 424-431 (1999).

3) Li S., Couet J. & Lisanti M.P. Src tyrosine kinases, G-alpha subunits, and H-Ras share a common membrane-anchored scaffolding protein, caveolin. Caveolin binding negatively regulates the auto-activation of src tyrosine kinases. *J.Biol.Chem.* **271**, 29182-29190 (1996).

4) Razani B., Rubin C.S. & Lisanti M.P. Regulation of cAMP-mediated signal trasduction via interaction of caveolins with the catalytic subunit of proteine kinase A. *J.Biol.Chem.* **274**, 26353-26360 (1999).

5) Nasu Y., *et al.*, suppression of caveolin expression induces androgen sensitivity in metastatic androgen-insensitive mouse prostate cancer cells. *Nat.Med.* **4**, 1062-1064 (1998).

6) Garcia-Cardena G. *et al.*, Dissecting the interaction between nitric oxide synthase (NOS) and caveolin. Functional significance of the nos caveolin binding domain in vivo. *J.Biol.Chem.* **272**, 25437-25440 (1997).

7) Roy S. *et al.*, Dominant-negative caveolin inhibits H-Ras funciotn by disrupting cholesterol-rich plasmamembrane domain. *Nat.Cell Biol.* **1**, 98-105 (1999).

8) Sternberg P.W. & Schmid S.L. Caveolin, cholesterol and Ras signalling. *Nat.Cell Biol.* **1**, E35-37 (1999).

9) Huang P.L. *et al.*, Hypertension in mice lacking the gene for endothelial nitric oxide synthase. *Nature* **377**, 239-242 (1995).

10) Murohara T. et al., Nitric oxide synthase modulates angiogenesis in response to tissue ischemia. *J.Clin.Inv.* **101**, 2567-2578 (1998).

11) Rudic R.D. et al., Direct evidence for the importance of endothelium-derived nitric oxide in vascular remodeling. *J.Clin.Inv* **101**, 731-736 (1998).

12) Lee P.C. et al., Impaired wound healing and angiogenesis in ENO-deficient mice. Am.J.Physiol. 277, H1600-1608 (1999).

13) Garcia-Cardena G., Oh P., Liu J., Schnitzer J.E. & Sessa W.C. targeting of nitric oxide synthase to endothelial cell caveolae via palmitoylation: implications for nitric oxide signaling. *Proc.Nat.Acad.Sci. USA* **93**, 6448-6453 (1996).

14) Liu J., Huges T.E. & Sessa W.C. The first 35 amino acids and fatty acylation sites determine the molecular targeting of endothelial nitric oxide synthase into the Golgi region of cells: a green fluorescent protein study. *J.Cell Biol.***137**, 1525-1535 (1997).

15) Ju H., Zou R., Venema V.J. & Venema R.C. Direct interaction of endothelial nitric-oxide synthase and caveolin-1 inhibits synthase activity. *J.Biol.Chem*. 272, 18522-18525 (1997).

16) Michel J.B., Feron O., Sase K., Prabhakar P. & Michel T. Caveolin versus calmodulin. Counterbalancing allosteric modulators of endothelial nitric oxide synthase. *J.Biol.Chem* **272**, 25907-25912 (1997).

17) Ghosh S. et al., Interaction between caveolin-1 and reductase domain of endothelial nitric-oxide synthase. Consequences for catalysis. *J.Biol.Chem* **273**, 22267-22271 (1998).

18) Derossi D., Chassaing G. & Prochiantz A. Trojan peptides: the penetratin system for intracellular delivery. *Trends Cell Biol.* **8**, 84-87 (1998).

19) Derossi D. et al., Cell internalization of the third helix is receptor-independent. *J.Biol.Chem* **271**, 18188-18193 (1996).

20) Derossi D., Joliot A.H., Chassaing G. & Prochiantz A. The third helix of the Antennapaedia homeodomain traslocates through biological membranes. *J.Biol.Chem* **269**, 10444-10450 (1994).

21) Smith W., Molecular biology of prostanoid biosynthetic enzymes and receptors. *Adv.Exp.Med.Biol.*, 989-1011, (1997).

22) Pollock J.S. et al., Characterization and localization of endothelial nitric oxide synthase using specific monoclonal antibodies. *Am.J.Physiol.* **265**, C1369-1387 (1993).

23) Okamoto T., Schlegel A., Scherer P.E. & Lisanti M.P. Caveolins, a family of scaffolding proteins for organising & quot; preassembled signaling complexes & quot

24) Salvemini D. et al., Nitric oxide: a key mediator in the early and late phase of carrageenan-induced rat paw inflammation. *Br.J.Pharmacol.* **118**, 829-838 (1996).

25) Ialenti A., Ianaro A., Moncada S. & Di Rosa M. Modulation of acute inflammation by endogenous nitric oxide. *Eur.J.Pharmacol.* **211**, 177-182 (1992).

26) He P., Zeng M. & Curry F.E. Effect of nitric oxide synthase inhibitors on basal microvessel permeability and endothelial cell $[Ca^{2+}]i$. *Am.J.Physiol.* **273**, H747-755 (1997).

27) Hobbs A.J., Higgs A. & Moncada S. Inhibition of nitric oxide synthase as a potential therapeutic target. *Annu.Rev.Pharmacol.Toxicol.* **39**, 191-220 (1999).

28) Murohara T. et al., Vascular endothelial growth factor/vascular permeability factor enhances vascular permeability via nitric oxide and prostacyclin. *Circulation* **97**, 99-107 (1998).

29) Garcia-Cardena G. et al., Dynamic activation of endothelial nitric oxide synthase by Hsp90. *Nature* **392**, 821-824 (1998)

Nitric Oxide
R.J. Gryglewski and P. Minuz (Eds.)
IOS Press, 2001

DETECTION OF THE ANTIOXIDANT ACTIVITY OF FLAVONOIDS BY USING A SUPEROXIDE SENSOR

S. IGNATOV
State Research Center for Applied Microbiology,
Obolensk, Moscow Region, 142279, Russia

B. GE, F.W. SCHELLER, F. LISDAT
University of Potsdam, Institute of Biochemistry and Molecular Physiology,
Karl-Liebknecht Straße 24 – 25, Building 25, 14476 Golm, Germany

Abstract

A method for the electrochemical detection of antioxidant activity of the flavonoids has been developed, which is based on a radical measurement with a cytocrome *c* modified electrode. A controlled enzymatic production system for superoxide radicals based on xanthine oxidase was used. The addition of antioxidants facilitated the decomposition of the radical in addition to the spontaneous dismutation. The steady-state of the superoxide generation and its decomposition has shifted to a new situation due to the higher decomposition rate after antioxidant addition. This resulted in a decreased current level at the electrode. Antioxidant activity can be quantified from the response of the sensor electrode by the percentage of the signal decrease. The 50% inhibition value (IC_{50}) for different antioxidants was calculated and a hierarchy of superoxide radical scavenging abilities of flavonoids was established.

Introduction

Flavonoids are a chemical compounds from plants which play a great role in human diet. Flavonoids are formed in plants from he aromatic aminoacids. The chemical structure of flavonoids is based on C15 skeleton with two aromatic rings (A and B) and the one heterocycle ring C (fig 1.)

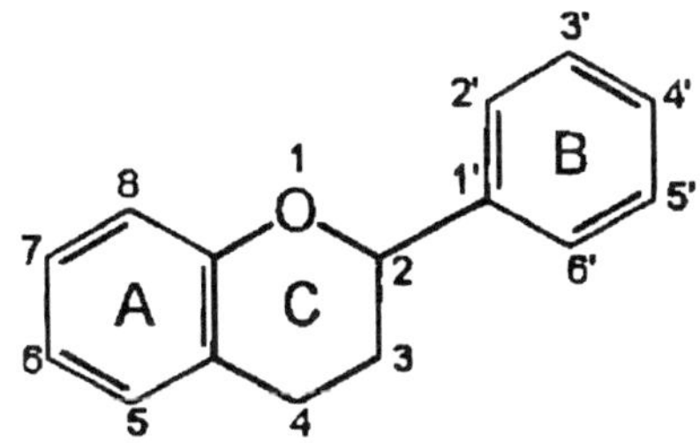

Figure 1. Basic flavonoid structure.

The various classes of flavonoids differ in the level of oxidation and pattern of substitution of the A and B rings. Among the many classes of flavonoids, those of particulat interest to this work are flavanols, flavonols, flavones, flavonones and isoflavonones. It has been shown that flavonoids have a variety of physiological and pharmacological functions [1]. However, most interest has been devoted to the antioxidant activity of flavonoids which is due to their ability to scavenge reactive oxygen species (ROS). The main sources of reactive oxygen species (ROS) are electron transport chains, active phagocytosis and the activity of some enzymes [2, 3, 4]. ROS have been linked to several toxicity processes including damage of proteins, membranes, DNA and enzymes [5] and various pathological situations including cancerogenesis, heart disease, reperfusion injuries, rheumatoid artritis, inflammation and aging [6]. Overproduction of ROS can be harmful. Thus the antioxidant potential of flavonoids is has been actively investigated due to their importance in medicine and food industry [7].

Several methods have been proposed for detection of antioxidants up to now. Methods for total antioxidant detection of biological samples *in vivo* were summarized in a recent review [8]. For *in vitro* characterization photometric, fluorimetric, chromotographic, nuclear magnetic resonance, electron spin resonance, flow cytometry and electrochemical methods have been developed [9-18].

Recently electrode approaches are intensively used for antioxidant's detection. The investigation of the voltammetric behavior of antioxidants was performed by using cylindrical carbon fiber microelectrodes applying differential pulse voltammetry and square wave voltammetry as electroanalytical techniques [19]. A poly-(vinyl chloride)-composite electrode is described for the determination of antioxidants in processed food [20].Cyclic voltammetry as an instrumental tool for the evaluation of the total antioxidant capacity has been also developed [21].

Inherent limitations of the analysis differ from method to method (low sensitivity, interference by reducing substances, slow detection and high price equipment) have stimulated the study of a simple, fast and sensitive method for the antioxidant characterization. Recently superoxide dismutase activity has been measured by using cytochrome *c*-modified electrode in vitro[22, 23]. In the prolongation of these studies a new method of antioxidant detection is described which is based on a sensorial approach. The method uses the redox transformation of cytochrome *c* as an oxidant of superoxide at a modified gold electrode. Therefore, the superoxide level in solution can be detected as an oxidation current at the sensor electrode due to electron transfer from the radical via cytochrome *c* to the electrode.

The objectives of the present study are to evaluate the ability of superoxide-specific amperometric biosensor for antioxidant detection of flavonoids.

Materials and methods

Materials

Gold wire electrodes (diameter 0,5 mm) were purchased from Good Fellow (Bad Nauheim, Germany). 11-mercaptoundecanoic acid (MUA), 1-ethyl-3(3-dimethyl aminopropyl) carbodiimide (EDC), cytochrome C from horse heart (Cyt C), hypoxantine (HX) was supplied by Sigma, (Deisenhofen, Germany), xanthine oxidase from cow milk (XOD, EC 1.1.3.22), superoxide dismutase from bovine erythrocytes (SOD, EC 1.15.1.1) by Boehringer Mannheim and hydrogen peroxide (30% H_2O_2) by Fluka. The other chemicals were of analytical grade and were used without further purification.

Ascorbic acid, rutin, pelargonidine chloride, luteolin, (-) epigallocatechin gallate, hesperetin, tannic acids, genistein, daidzein, apigenin, (+/-) catechin were used from Sigma (Deisenhofen, Germany). Naringenin, quercetin dihydrate, biochanin, myricetin, kaempferol were from ICN Biomedicals Inc.. Trolox was from Fluka.

0.1 M Na-phosphate buffer (pH 7.5) was used for the antioxidant measurements. EDTA was added to result in a concentration of 100 µmol/L. During electrode modification, a potassium dihydrogen phosphate/ dipotassium hydrogen phosphate-based buffer (pH 7.0) without EDTA was used (5-10 mmol/L).

Apparatus

Investigations were performed in a stirred electrochemical cell with a three-electrode configuration. The cell with a volume of 1 mL comprised a Ag/AgCl/ 1 M KCl reference electrode and a Pt counter electrode. Cyclic voltammetry was performed using the Autolab system (Eco chemie, Netherlands). For the antioxidant measurements the modified gold electrode was polarized at + 150 mV vs Ag/AgCl. The potentiostat was a "Bioanalyzer" from Kreijci Engineering (Brno, Czech Republic). The measurement medium was a phosphate buffer (pH 7.5, 100 mM) with 0.1 mmol/L EDTA.

Electrode Preparation

Gold wire electrodes (diameter 0.5 mm) were cleaned by boiling in 2 mol/L KOH, incubating in concentrated H_2SO_4 (8 h), concentrated HNO_3 (10 min), and carefully rinsing with water in between. The electrodes were stored in concentrated H_2SO_4. After being rinsed with water and ethanol, the electrodes were incubated in 5 mM solution MUA in ethanol for 24 - 48 h. The modified electrodes were rinsed with ethanol and 10 mM potassium phosphate buffer (pH

7) and then mounted into an electrochemical cell with the same buffer. A cytochrome c solution was added for a final concentration of 20-30 µmol/L. During cytochrome c adsorption, the electrodes were cycled in the range –0.4 to +0.4 V with a scan rate of 50 mV/s for ~10 min. Then EDC was added to give a final concentration of 5 mM, and the electrodes were incubated with the solution for about 20-30 min. After rinsing with 10 mM buffer, the electrochemical behavior of the immobilized cytochrome c was analyzed using cyclic voltammetry.

Antioxidant Measurements

The principle of antioxidant detection is shown in Figure 2. For superoxide generation the xanthine oxidase (XOD) catalysed conversion of hypoxanthine (HX) to uric acid was used. The main path of electron flux from hypoxanthine to oxygen occurs via the two-electron pathway in the flavin – oxygen complex producing hydrogen peroxide. However, a second pathway allows one-electron transfer from the reduced enzyme. Under air saturation and physiological pH conditions ~20 % of the electron flux occurs via a one-electron step and thus leads to the generation of superoxide radicals. Activity of XOD and concentration of HX were optimised to ensure a steady-state superoxide level for several minutes in solution.

Stock solutions of 20 mM were prepared for hypoxanthine in 0.1 mol/L sodium phosphate buffer (pH 7.5). Hypoxanthine was added to the cell (total volume 1 ml) to result in a final concentration of 100 µmol/L. The reaction was started by adding XOD in the concentration of 50 mU/mL. After reaching the steady-state current aliqoutes of the samples with antioxidant activity to be measured were added and the current decrease was recorded (Fig. 2). The current decrease after samples addition compared to the measurement without antioxidants was evaluated as the sensor signal. All antioxidant solutions were freshly prepared in buffer, ethanol (rutin) or in DMSO (biochainin A, daidzein, genistein) before experiments and were used immediately. The volume of antioxidants added was 10-40 μl. Total time of antioxidant assay was not more 5 minutes (Fig. 3)

To test the action of potential interfering substances 5-100 μL of pure ethanol or DMSO was added to the measuring cell. No current change was observed.

Results and discussion

We have studied the antioxidative activity of flavonoids based on their scavenging effect for superoxide radicals. For the radical determination, a cytochrome *c* modified electrode was operated in the amperometric mode. A general scheme of reactions occurring in the measuring system is given in Fig. 2. Superoxide was enzymatically produced and dismutated spontaneously to oxygen and hydrogen peroxide. Under controlled conditions of superoxide generation such as air saturation of the buffer, optimal hypoxanthine concentration (100 μM) and XOD activity (50 mU/ml) a steady-state superoxide level can be obtained for several minutes.

Since these steady-state superoxide concentrations can be detected by the modified gold electrode conditions were provided to apply the electrode for determination of the antioxidant activity of some substances as scavengers of this radical. In Figure 3 the scheme of electrode response is shown when different concentrations of an antioxidant were injected into the superoxide-generating solution. A decrease of the current level after antioxidant injection was observed. The addition of antioxidant substances facilitated the decomposition of the radical

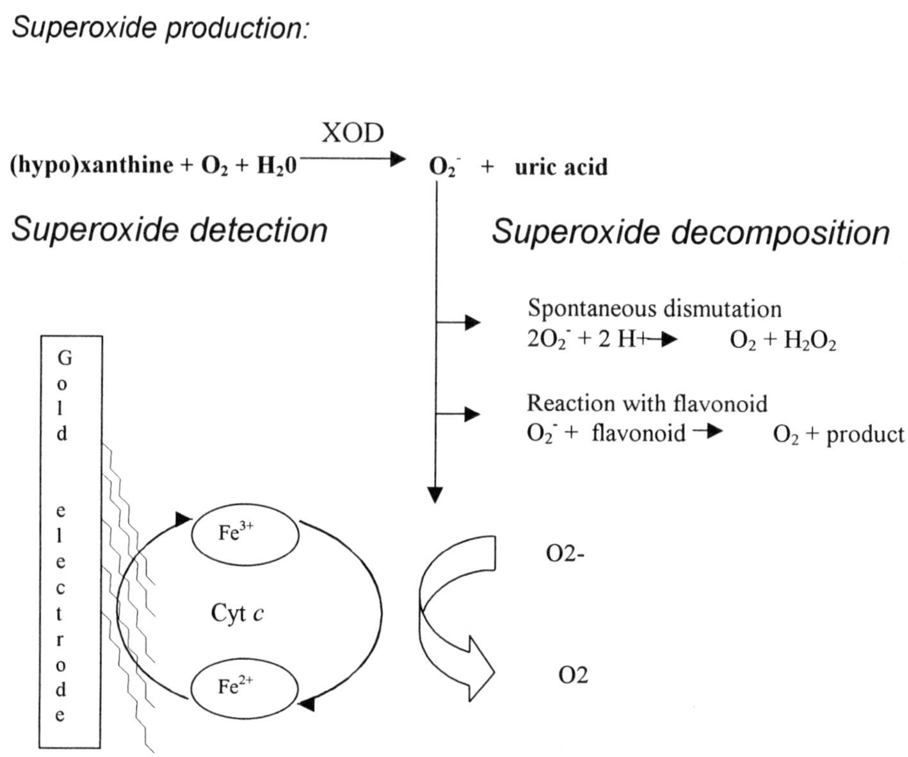

Figure 2. Scheme of the detection principle of superoxide antioxidant scavenging activity of substances. AO is the antioxidant under investigation.

in addition to the spontaneous dismutation. The steady-state situation due to generation and decomposition of superoxide has shifted to a new situation accounting for the higher decomposition rate after antioxidant addition. This can be detected with the electrode by the diminished current levels as shown in Fig. 3. Antioxidant activity can be quantified from the response of the sensor electrode by the percentage of this current decrease (see Fig.3).

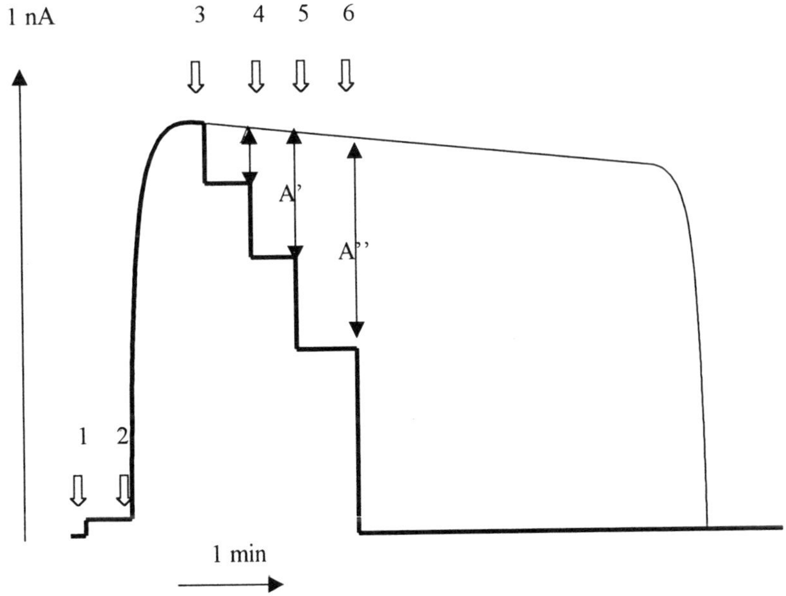

Figure 3. The scheme of response of sensor electrode to superoxide generation, antioxidant and SOD additions. 1 – 100 mM hypoxanthine addition; 2 –XOD addition, 50 mU/ml; 3,4,5 – antioxidant additions; 6 - SOD addition, 2 units/ml. A, A' and A'' are signal depressions to different antioxidant concentrations in the medium. Measuring medium is 0,1 M phosphate buffer pH 7.5 with 100 mM EDTA.

The antioxidant detection system responded rather fast to the new situation after antioxidant injection. Such a behaviour corresponds to the empirical Hill equation which can describe a variety of sensor responses from different biosensors [24] Thus we deduced a hyperbolic function $A = B*X / (C + X)$, where A corresponds to 1-[O_2^-], B - to the maximum current supression (1 or 100%), C - to the antioxidant concentration giving a 50% depression of the radical concentration (IC_{50}) and X corresponds to the antioxidant concentration in solution. With this function it was possible to describe the behaviour of the different antioxidants. The

IC_{50} was used for the estimation of the antioxidant activity of these substances - the smaller the IC_{50} the better the antioxidant.

The current level stabilised within 10 s. The whole measuring cycle with radical generation, antioxidant injections and washing steps can be completed within 300 s. The procedure repeated at least three times. The time scale used here in the experiments seems to be suited for a screening procedure.

For the experiments a cell volume of 1 ml was used allowing sample volumes of about 10-40μl to be analysed. Since the arrangement is a simple three electrode system operated under constant potential the volumes can be easily reduced by one order of magnitude. For the antioxidant analysis with the method proposed rather simple and inexpensive equipment can to be applied. Besides the electrochemical cell there is only need for a potentiostat and an instrument for data representation. This seems to be an advantage in the flexibility of measurement.

In order to compare the efficiency of radical scavenging of various flavonoids ascorbic acid was used as a standard tool substance. A comparison of antioxidant activities of substances based on the ascorbic acid equivalent (AAE) is presented in Table 1.

This comparison shows that group of compounds such as (-) epigallocatechin gallate, (-) epigallocatechin and pyrogallol had significantly higher scavenging activity than ascorbic acid. On the other hand the activities of such compounds as daidzein and genistein were orders of magnitude less compared with ascorbic acid. The low antioxidant activity of genistein against superoxide radicals, which we have found, corresponds to data from the literature [25].

Compound	IC_{50}	AAE
Flavanol		
(-) Epigallocatechin gallate	78 nM	162
(-) Epigallocatechin	94 nM	118
(+) Catechin	538 nM	23
(+/-) Catechin	525 nM	23
Flavonol		
Myricetin	1,8 µM	7
Rutin	2,4 µM	5
Quercetin	7,5 µM	2
Kaempferol	n/d	n/d
Flavone		
Luteolin	5 µM	12
Apigenin	5,5 µM	2
Flavanone		
Hesperetin	57 µM	0,2
Naringenin	0,5 mM	0,03
Hesperidin	no	no
Isoflavone		
Biochainin A	0,27 mM	0,03
Daidzein	0,72 mM	0,01
Genistein	0,37 mM	0,01

Table 1. Hierarchy of the antioxidant capacity of flavonoids. IC_{50} is the concentration which generates a 50% signal depression of sensor. AAE (ascorbic acid equivalent) is ratio of IC_{50} of ascorbic acid to IC_{50} of antioxidant of interest.

The hierarchy of antioxidant activity of flavonoids determined by the amperometric method proposed here was summarized in Table 1. The antioxidant activities can be arranged in the order: flavanols > flavonols > flavones> flavonones > isoflavonones. These results are in agreement with those obtained by Lien et al. for the hierarchy of antioxidant capacity (TEAC) of flavonoids based on Trolox equivalents [26]. This method determines the radical scavenging activity of flavonoids by measuring of their ability to scavenge 2,2'-azinobis-(3-ethylbenzothiazoline-6-sulfonic acid) radical cation ($ABST^{*+}$) optically. But in our case the difference between the antioxidant capacities for various classes of flavonoids was much higher compared with TEAC data of Lien et al.. This can be understood since the scavenging reactions of the substances differ for various radicals and the antioxidant activity of flavonoids against superoxide radical has a wider concentration spectrum compared with those against the artificially generated radical $ABST^{*+}$.

By using our system we could not detect antioxidant activity of some known flavonoids (Table 1.). Two main reasons can be seen. First, some of the compounds (pelargonidine chloride and kaempferol) can be directly oxidized at the electrode. This can interfere with the superoxide measurement. The effect can be also considered with the pro-oxidant ability of these substances since in dependence of concentrations and matrix conditions several antioxidants also show pro-oxidants properties [27, 28].

Flavonoids, for example, can induce infant leukemia at high concentration (50 - 200 μM) [29]. Pro- and antioxidant action originates from their reducing activity (depends on their structure) [28], concentration [30] and available reaction partners (e.g. metal ions).

A second group of antioxidants (hesperidin) was limited in its scavenging activity to other reactive species and had no scavenging activity against superoxide.

Conclusion

A sensitive and inexpensive biosensorial method for the quantification of the scavenging activity of flavonoids has been developed. It is based on a cytochrome *c*-modified gold electrode and on the measurement of steady-state superoxide levels. The concentration which induces 50% inhibition of the superoxide level has been determined for different substances. The applicability of this method was shown by analyzing the hierarchy of antioxidant capacity for flavonoids. The new method can be considered as a real alternative to the numerous complicated and expensive methods for antioxidant activity detection. The electrochemical method described here is fast and allows to use small sample volumes to be analyzed.

Acknowledgement

The authors thank the ISTC (project #890) and DFG for the financial support. We also thank Frau K. Kürschner and Frau E. Micheel for editorial help.

References

1. Das, A., Wang, J. H. And Lein E. J. 1994. Carcinogenicity, mutagenicity and cancer preventing activities of flavonoids: A structure-system-activity relationship (SSAR) analysis. *Prog. Drug Res.* **42** 1, pp. 13-166.
2. Cadenas, E., 1989. Biochemistry of oxygen toxicity. *Annu. Rev. Biochem.* **58**, pp. 79-110.
3. Krause K.-H., 2000. Professional phagocytes: Predators and prey of microorganisms. Schweizerische Medizinische Wochenschrift, **130** 4, pp. 97-100
4. Fridovich, I. 1997. Superoxide Anion Radical (O2), Superoxide Dismutases, and Related Matters. *J.B.C.* **272** 30, pp. 18515-18517.
5. Kowaltowski, A. J. and Vercesi, A.E. 1999. Mitochondrial damage induced by conditions of oxidative stress. *Free Rad. Biol. Med.*, **26** 3-4, pp. 463-471
6. Halliwel, B. 1996. Antioxidants in human health and disease. *Annu. Rev. Nutr.* **16**, pp 33-50.
7. Pieta, P. G. 2000. Flavonoids as antioxidants. *J. Nat. Prod.* **63**, pp.1035-1042.
8. Prior, L. P. and Cao, G. 1999. In vivo total antioxidant capacity: comparison of different analytical methods. *Free Radic. Biol. Med.* **27** 11/12, pp. 1173-1181.
9. Brand-Williams, W., Cuvelier, M. E. and Berset, C. 1995. Use of a free radical method to evaluate antioxidant activity. *Lebensm. Wiss. Technol*, **28** 1, pp. 25-30.
10. Pulido, R., Bravo, L. and Saura-Calixto, F. 2000. Antioxidant activity of dietary polyphenols as determined by a modified ferric reduction/antioxidant power assay, J. Agric. Food Chem., **48** 8, pp. 3396-3402.
11. Fukimoto, L. R. and Mazza G. 2000. Assesing antioxidant and prooxidant acticities of phenolic compounds. J. Agric. Food Chem., **48** 8, pp. 3597-3604.
12. Magnani, L., Gaydou, E.M.and Huband, J.C. 2000. Spectrophotometric measurement of antioxidant properties of flavones and flavanols against superoxide anion. *Anal. Chim. Acta* **411** 2, pp. 209-216.
13. Popov, I.N.and Lewin G. 1994. Photochemiluminescent detection of antiradical activity: II: testing of nonenzymic water0soluble antioxidants. *Free Radic. Biol. Med* **17** 3, pp. 267-271.
14. Van Dyke, K., McConnell, P and Marquart L. 2000. Green tee extract and its polyphenols markedly inhibit luminol-dependent chemiluminescence activated by peroxynitrite or SIN-1. *Luminescence* **15** 1, pp. 37-43.
15. Naguib, Y. M. A. 2000. A fluorometric method for measurement of oxygen radical-scavenging activity of water soluble antioxidants. *Anal. Biochem.* **284** 1, pp.93-96.
16. Sawai, Y, and Sakata, K, 1998. NMR analytical approach to clarify the antioxidative molecular mechanism of catechins using 1.1-Diphenil-2-pickrylhydrazyl. *J. Agricul. Chem.* **46** 1, pp. 111-114.

17. Jorgensen, L.V., Madsen, H.L., Thomsen ,M.K., Dragsted, L.O., Skibsted, L.H.and Sies P.H. 1998. Regeneration of phenolic antioxidants from phenoxyl radicals: an ESR and electrochemical study of antioxidant hierarchy. *Free Radic. Res.* **30** 1, pp. 207-220.
18. Ostrovidov, S. Frank, P., Josef, D., Martarello, L, Kirsch G., Belleville, F., Nabet, P. and Dousset B. 2000. Screening of new antioxidant molecules using flow cytometry. *J. Med. Chem.* **43** 9, pp. 1762-1769.
19. Agui, M. L., Reviejo, A.J., Yanez-Sedeno, P. and Pingarron, J.M. 1995. Analytical applications of cylindrical carbon fiber microelectrodes. Simultaneous voltammetric determination of phenolic antioxidants in food. *Anal. Chem.* , **67**, pp. 2195-2200.
20. Luque, M., Rios, A. And Valcarcel, M. 1999. A poly(vinyl chloride) graphite composite electrode for flow-injection amperometric detection of antioxidants. *Anal. Chim. Acta* **395**, 1/2, pp. 217-223.
21. Chevion, S., Roberts, M. A. and Chevion, M. 2000. The use of cyclic voltammetry for the evaluation of antioxidant capacity. *Free Radic. Biol. Med.* **28** 6, pp. 860-870.
22. Listad, F., Ge, B., Ehrentreich-Forster, E., Reszka R. and Scheller F.W. 1999a. Superoxide dismutase activity measurement using cytochrome *c*-modofied electrode. *Anal. Chem.* **71** 7, pp. 1359-1365.
23. Lisdat, F., B. Ge, R. Reszka and E. Kozniewska 1999 b. An electrochemical method for quantification of the radical scavenging activity of SOD. *Fresenius J Anal. Chem*, 365, 4, p394.
24. Borisov, I. A., Lobanov, A. V., Reshetilov, A. N. and Kurganov B. I. 2000. Quantitative analysis of the calibration dependences for biosensors. *Appl. Biochem. Microbiol. (Russian)* **36**, 3 pp.254-260.
25. Qiuyin, C., Rahn, R.O. and Ruiwen, Z. 1997. Dietary flavonoids, quercetin, luteolin and genistein, reduce oxidative DNA damage and lipid peroxidation and quench free radicals. *Cancer Lett.* **119** 1, pp. 99-107.
26. Lien, J.E., Ren, S., Bui, H.-H. And Wano, R. 1999. Quantitative structure-activity relationship analysis of phenolic antioxidants. *Free Radic. Biol. Med.* **26** 3/4, pp. 285-294.
27. Sugihara, N., Arakawa, T., Ohnishi, M. and Furino, K. 1999. Anti- and pro-oxidative effects of flavonoids on metal-induced lipid hydroperoxide-depended lipid peroxidation in cultured hepatocytes loaded with alfa-linolenic acid. . *Free Radic. Biol. Med.* **27** 11/12, pp. 1313-1323.
28. Cai, Z., Li, X. And Katsumura, Y. 1999. Interaction of hydrated electron with dietary flavonoids and phenolic acids: rate constants and transient spectra studied by pulse radiolisis. . *Free Radic. Biol. Med.* **27** 7/8, pp. 822-829.
29. Strick, R., Strissel, P.L., Borgers, S., Smith, S.L. and Rowlry, J.D. 2000. Dietary bioflavanoids induce cleavage in the *MLL* gene and may contribute to infant leukemia. *Proc. Natl. Acad. Sci.* **97** 9, pp.4790-4795.
30. Eberhardt, M.V., Lee, C.Y. and Liu R.H. 2000. Nutrition: Antioxidant activity of fresh apples. *Nature* **405** 6789, pp. 903-904.

Nitric Oxide
R.J. Gryglewski and P. Minuz (Eds.)
IOS Press, 2001

Nitric Oxide Reverses the Inhibition of Endocytosis Induced by Tumour Necrosis Factor-α in Dendritic Cells

Clara PAOLUCCI*§, Patrizia ROVERE, Samuele BURASTERO*, Angelo A. MANFREDI*, and Emilio CLEMENTI*#

** DIBIT H-San Raffaele Scientific Institute, 20132 Milano, Italy. §Consiglio Nazionale delle Ricerche Centre for Cellular and Molecular Pharmacology, 20129 Milano, Italy. #Department of Pharmaco-Biology, University of Calabria, 87036 Rende, Italy.*

Abstract. Tumour Necrosis Factor-α (TNFα) triggers maturation of dendritic cells (DC), a process which results, among other events, in downregulation of the ability of these cells to phagocytose antigens. We have studied the effects and mechanisms of action of nitric oxide (NO) on endocytosis, investigated with fluorescein-iso-thiocyanate (FITC)-labelled dextran, using human, monocyte-derived DC, both immature and after treatment with TNFα. NO, released by the NO donor DETA-NO, reversed the inhibition of endocytosis observed in DC exposed to TNFα. This effect of NO was mimicked by the membrane-permeant cyclic GMP analogue, 8-Br cyclic GMP, and prevented by inhibition of the soluble guanylate cyclase. The mechanism of action by NO on endocytosis is due to regulation of the generation of ceramide, a lipid messenger involved in the maturation process of DC initiated by TNFα. In particular, NO was found to inhibit, in a cyclic GMP (cGMP)-dependent way, ceramide accumulation triggered by TNFα. NO, however, was found to exert an inhibitory effect also downstream of this event, as NO donors reversed the inhibition of endocytosis induced by the cell-permeant C_2-ceramide. These results indicate that NO, working through cGMP, increases the ability of human DC to internalise antigens and thus modulate the initial steps leading to antigen-specific immune-responses.

1. Introduction

Dendritic cells are professional antigen-presenting cells involved in the initiation of immune responses (1). Immature DC capture antigens at the site of inflammation, process and present them to T cells in secondary lymphoid organs where DC prime the immune response (1). These events are accompanied by a process of DC maturation orchestrated by pro-inflammatory signals generated at the site of infection, among which TNFα plays a pivotal role (1). TNFα triggers DC maturation through the activation of its p55 kDa receptor, whose signal transduction pathway has been recently shown to be under the control of NO (2). This short-lived messenger is known to modulate specific functions of cell populations involved in the immune responses (3). In particular, NO is crucial in enhancing the microbicidal and cytotoxic activities of macrophages, as well as the function of by-stander T lymphocytes (3, 4).

Evidence obtained in the murine system indicates that NO may also modulate the antigen presenting function of DC (5, 6). So far, however, the mechanism of action by NO in the maturation process of DC has not been studied, nor have the intracellular targets of NO been identified.

We have investigated the regulation by NO of the ability of human DC to endocytose antigens, using monocyte-derived DC exposed to TNFα. Our results show that NO prevents in a cGMP-dependent way the downregulation of endocytosis induced in DC by exposure to TNFα. This effect of NO is due to inhibition of TNFα-induced accumulation of the lipid messenger ceramide, as well as to additional effect(s) exerted downstream of this signal transduction event.

Table 1. Effects of DETA-NO, 8-Br cGMP and ODQ on the endocytic activity of DC treated with TNFα.

treatment	endocytosis (% of that observed in immature DC)
TNFα	53 ± 3.6
TNFα + DETA-NO	85 ± 4.6
TNFα + 8-Br cGMP	74 ± 3.8
TNFα + DETA-NO + ODQ	57 ± 3.9
TNFα + decomposed DETA-NO	54 ± 4.1

Immature DC were incubated for 48 h with or without TNFα (100 ng/ml), in the presence or absence of DETA-NO (100 μM), decomposed DETA-NO, the membrane permeant cGMP analogue 8-Br cGMP (3 mM), or the inhibitor of soluble guanylate cyclase ODQ (3 μM), combined as indicated. Endocytosis was measured after 30 min of incubation with or without FITC-dextran (1 mg/ml). The % of cells positive to FITC-dextran vs. cells treated in the same way but not exposed to the fluorescent dye was measured for all preparations. Values obtained (mean ± standard error of the mean; n = 5) are expressed as % of those measured in DC incubated without TNFα (100 %).

2. NO modulates endocytosis of DC undergoing TNFα-induced maturation in a cGMP-dependent way

We have investigated the role of NO in the modulation of the endocytic function of DC using DETA-NO as the NO donor, and the fluorochrome-labelled FITC-dextran to quantify endocytosis. Human monocyte-derived DC were generated by a 7 days treatment with granulocyte-macrophage-colony stimulating factor and interleukin-4 as described (7). Immature DC were incubated with or without TNFα (100 ng/ml, 48 h), in the presence or absence of DETA-NO (100 μM) or 8-Br cGMP (3 mM). In some experiments DETA-NO was combined with the guanylate cyclase inhibitor H-[1,2,4] oxadiazolo [4,3-a] quinoxalin-1-one (ODQ; 3 μM). DC were then resuspended in cytokine-free medium and incubated for 30 min at 37°C with FITC-dextran (1 mg/ml), and the internalisation of the latter measured by flow cytometry as described (7). Treatment of DC with TNFα resulted in a statistically significant inhibition of endocytosis of FITC-dextran (see Table 1). This effect of TNFα was not due to cytotoxicity since cell viability was 96 ± 3.3 % after 48 h of treatment with the cytokine (n = 5). DETA-NO and 8-Br cGMP were able to reverse the inhibition of endocytosis induced by TNFα. When DETA-NO was administered together with ODQ its ability to restore endocytosis was lost. Decomposed DETA-NO, which was unable to release NO as measured with a NO electrode, had no effects on endocytosis. DETA-NO, 8-Br cGMP and ODQ did not induce cytotoxicity (viability vs. untreated controls was 95 ± 5.0 %, 96 ± 2.1 and 93 ± 4.3 %, respectively, after the 48 h incubation, n = 3).

These results indicate that NO controls the endocytosis regulated by TNFα through the activation of cGMP-dependent pathway(s).

3. NO inhibits accumulation of ceramide induced by TNFα in a cGMP-dependent way

Ceramide, a lipid messenger generated by the activation of the p55 receptor for TNFα, has been shown to be involved in inhibition of endocytosis elicited by the cytokine (8). Basal ceramide levels, measured by the diacylglycerol kinase method (2), was 152 ± 21 pmol/10^6 cells, n = 4. TNFα (100 ng/ml) induced a progressive accumulation of ceramide, which reached a plateau after about 6 h (ceramide level at 6 h was 380 ± 32 pmol/10^6 cells, n = 4).

When DC were incubated with the cytokine in the presence of DETA-NO (100 μM), inhibition of ceramide generation was observed (see Table 2). Incubation with 8-Br cGMP (3 mM) gave effects similar to those of NO, indicating that NO action on ceramide accumulation was mediated through cGMP. This was further confirmed by the lack of effects observed when DETA-NO was added together with ODQ (3 μM).

Table 2. Effects of DETA-NO, 8-Br cGMP and ODQ on the accumulation of ceramide induced by TNFα in DC.

treatment	ceramide accumulation (% of that observed in TNFα-treated DC)
TNFα + DETA-NO	32 ± 3.9
TNFα + 8-Br cGMP	40 ± 2.7
TNFα + DETA-NO + ODQ	82 ± 5.2

Immature DC were incubated for 6 h with or without TNFα (100 ng/ml), in the presence or absence of DETA-NO (100 μM), 8-Br cGMP (3 mM), or the inhibitor of soluble guanylate cyclase ODQ (3 μM), combined as indicated. Ceramide levels were measured on lipid extracts, treated with diacylglycerol kinase in the presence of [γ-^{32}P] ATP. Ceramide phosphate was separated by thin layer chromatography (2). Values obtained (mean ± standard error of the mean; n = 5) are expressed as % of those measured in DC incubated with TNFα (100 %).

4. Regulation by NO of endocytosis in DC treated with TNFα involves events downstream of the TNFα-induced accumulation of ceramide

The membrane permeant, short-chain ceramide, C_2 ceramide, has been previously reported to mimick the effect on endocytosis exerted by TNFα (8). To assess whether the effect of NO on endocytosis was due solely to inhibition of ceramide accumulation, or also to effects exerted downstream of it, DC were incubated with C_2 ceramide (40 μM; 48 h), either alone or together with DETA-NO (100 μM) or 8-Br-cGMP (3 mM). C_2 ceramide inhibited FITC-dextran endocytosis with respect to that observed in untreated, control cells (see Table 3). Incubation of DC with C_2 ceramide in the presence of the NO donor or the cGMP analogue reversed the inhibition of endocytosis caused by the lipid metabolite. The effect of DETA-NO was prevented by incubation with ODQ (3 μM) (Table 3).

These results indicate that the action of NO is exerted not only through the inhibition of ceramide generation by TNFα, but also via mechanism(s) active downstream of it. These other effects of NO also depend on the generation of cGMP.

Table 3. Effects of C_2 ceramide, DETA-NO, 8-Br cGMP and ODQ on the endocytic activity of DC.

treatment	endocytosis (% of that observed in immature DC)
C_2 ceramide	64 ± 4.9
C_2 ceramide + DETA-NO	99 ± 6.6
C_2 ceramide + 8-Br cGMP	91 ± 4.8
C_2 ceramide + DETA-NO + ODQ	67 ± 5.7

Immature DC were incubated for 48 h with or without C_2 ceramide (40 μM), in the presence or absence of DETA-NO (100 μM), 8-Br cGMP (3 mM), or the inhibitor of soluble guanylate cyclase ODQ (3 μM), combined as indicated. Endocytosis was measured after 30 min of incubation with or without FITC-dextran (1 mg/ml). The % of cells positive to FITC dextran vs cells treated in the same way but not exposed to the fluorescent dye was measured for all preparations. Values obtained (mean ± standard error of the mean; n = 5) are expressed as % of those measured in DC incubated without C_2 ceramide (100 %).

5. Conclusions

The results presented above indicate that administration of NO from exogenous sources to DC plays a role in the control of endocytosis during maturation of these cells when exposed to TNFα. In particular, NO reversed the inhibition of endocytosis induced by the cytokine. This effect was mimicked by 8-Br cGMP and abolished by the guanylate cyclase inhibitor ODQ, indicating that the action by NO is mediated through the generation of its second messenger cGMP.

The maturation process of DC is triggered by TNFα via activation of its p55 receptor (9), which induces sphingomyelin breakdown with ensuing accumulation of ceramide, a lipid messenger relevant in mediating downregulation of endocytosis (8, 10). We thus investigated whether the maintenance by NO of the endocytic ability in TNFα-treated DC was due to inhibition of ceramide accumulation. Consistent with this possibility, DETA-NO inhibited the accumulation of ceramide induced by TNFα. We found, however, that NO was regulating endocytosis also at steps downstream of it. Both these effects of NO were mediated through cGMP generation.

Inhibition by NO of the TNFα-induced downregulation of endocytosis is thus exerted at multiple levels along the signal transduction cascade triggered by this cytokine, indicating that NO acts as a wide inhibitor of the action of TNFα on endocytosis rather than as a selective regulator of one single transductional event.

NO generation in peripheral tissues occurs as a consequence of various stimuli (3, 11). The regulation of endocytosis by NO might thus enable DC to increase and prolong their antigen uptake function at the site of inflammation and therefore modify the ensuing immune responses in the lymphoid organs. So far, the role of NO in the maturation process of DC has been investigated in the murine system. These experiments were carried out in DC exposed to cytokines in order to trigger the expression of the inducible isoform of the NO synthase (NOS II) and generate a constant flux of NO in these cells (6). Antigen presentation was impaired under these conditions, suggesting an overall inhibitory role for NO on DC maturation (12). These data appear consistent with the mechanism of action by NO elucidated here for human DC, since maintenance of endocytosis, characteristic of an immature phenotype, is expected to be revealed in vitro as a reduced T cell stimulation.

Regulation by NO of human DC function might, however, be different. NOS II expression and NO generation by these cells have been reported in severe pathological conditions, namely primary biliary cirrhosis and hepatocellular carcinoma (13, 14). By contrast, treatment of human DC with various combinations of TNFα, interferonγ and lipopolysaccharide, to mimic situations presumably found during inflammation in the peripheral tissues, resulted neither in the expression of NOS II nor in generation of NO (7). Regulation of human DC maturation appears therefore to depend on exogenous NO, generated at the site of infection by macrophages as a result of their activation by cytokines produced during the inflammatory response (11). NO might maintain endocytosis of antigens as long as DC are confined to the inflammation site, tuning their response to TNFα. After egress from the inflammation site, DC would no longer be exposed to NO, and could therefore down-regulate their endocytic ability to prevent any interference by irrelevant self-antigens (1). The overall effect of NO on DC function here described might contribute to enhance immune responses.

Acknowledgements
We thank Jacopo Meldolesi and Céline De Nadai for helpful discussion. This work was supported by grants from: Italian Association for Cancer Research, Consiglio Nazionale delle Ricerche, Target Project Biotechnology; Ministero dell'Università e della Ricerca Scientifica e Tecnologica, Cofinanziamento 99.

References

[1] J. Banchereau and R. M. Steinman, Dendritic cells and the control of immunity, *Nature* **392** (1998) 245-252.

[2] C. De Nadai *et al.*, Nitric oxide inhibits tumor necrosis factor-α-induced apoptosis by reducing the generation of ceramide, *Proc. Natl. Acad. Sci. USA* **97** (2000) 5480-5485.

[3] J. MacMicking *et al.*, Nitric oxide and macrophage function, *Annu. Rev. Immunol.* **15** (1997) 323-350.

[4] C. Sciorati *et al.*, Generation of nitric oxide by the inducible nitric oxide synthase protects γδ T cells from Mycobacterium tuberculosis-induced apoptosis, *J. Immunol.* **153** (1999) 1570-1576.

[5] P. G. Holt *et al.*, Downregulation of the antigen presenting cell function(s) of pulmonary dendritic cells in vivo by resident alveolar macrophages, *J. Exp. Med.* **177** (1993) 397-407.

[6] L. Lu *et al.*, Induction of nitric oxide synthase in mouse dendritic cells by IFN-γ, endotoxin, and interaction with allogeneic T cells: nitric oxide production is associated with dendritic cell apoptosis, *J. Immunol.* **157** (1996) 3577-3586.

[7] Paolucci *et al.*, Nitric oxide inhibits the tumor necrosis factor α -regulated endocytosis of human dendritic cells in a cyclic GMP-dependent way, J. Biol. Chem. **275** (2000) 19638-19644.

[8] F. Sallusto *et al.*, Ceramide inhibits antigen uptake and presentation by dendritic cells, *J. Exp. Med.* **184** (1996) 2411-2416.

[9] F. Sallusto *et al.*, Dendritic cells use macropinocytosis and the mannose receptor to concentrate macromolecules in the major histocompatibility complex class II compartment: downregulation by cytokines and bacterial products, *J. Exp. Med.* **182** (1995) 389-400.

[10] R. N. Kolesnick and M.Krönke, Regulation of ceramide production and apoptosis, *Annu. Rev. Physiol.* **60** (1998) 643-665.

[11] S. Moncada and E. A. Higgs, Molecular mechanisms and therapeutic strategies related to nitric oxide, *FASEB J.* **9** (1995) 1319-1330.

[12] C. A. Bonham *et al.*, Nitric oxide production by dendritic cells is associated with impairment of T cell responses, *Transplant. Proc.* **29** (1997) 1116-1117.

[13] K. Yamamoto *et al.*, Increased nitric oxide (NO) production by antigen-presenting dendritic cells is responsible for low allogeneic mixed leucocyte reaction (MLR) in primary biliary cirrhosis (PBC), *Clin. Exp. Immunol.* **114** (1998) 94-101.

[14] T. Ninomiya *et al.*, Dendritic cells with immature phenotype and defective function in the peripheral blood from patients with hepatocellular carcinoma, *J. Hepatol.* **31** (1999) 323 331.

Nitric Oxide
R.J. Gryglewski and P. Minuz (Eds.)
IOS Press, 2001

No Effect of Elevated Glucose on the Generation of Endothelium-Derived Nitric Oxide in Rabbit Aorta

K. Hansen and O.A. Nedergaard
Department of Physiology and Pharmacology, Odense University, Odense, Denmark

Diabetes predisposes to many forms of tissue damage resulting in hypertension, atherosclerosis, ischemic heart disease, blindness and nephropathy. These diabetic complications may possibly be due to dysfunction of vascular endothelium. (Kamata et al. 1992).

When isolated blood vessels derived from healthy (non-diabetic) animals were incubated for a short period in salt solution containing elevated glucose, a functional impairment of endothelium was seen in rat mesenteric arteries (Taylor & Poston 1994); guinea pig aorta (Dorigo et al. 1997); rabbit jugular vein (Davies & Hagen 1996) and aorta (Tesfamariam et al. 1990). Thus, elevated glucose treatment attenuated acetylcholine-evoked relaxation of several precontracted arteries from various species, including rabbit abdominal aorta (Tesfamariam et al. 1990). The in vitro dysfunction resembled the in vivo complications in diabetic patients (Hsueh & Anderson 1992).

We studied the effect of elevated glucose on the acetylcholine-evoked relaxation of rabbit thoracic aorta. The method described by Hansen & Nedergaard (1999) was used. Intact rings of aorta were precontracted with phenylephrine. When a stable tension was obtained, acetylcholine (10-8 – 3 x 10-6 M) was added cumulatively. After a washout period, the rings were incubated in salt solution containing either normal (5.5 mM) or elevated (44 mM) glucose for 6 h. Then, rings were precontracted again with phenylephrine and acetylcholine (10-8 – 3 x 10-6 M) was added cumulatively.

The nitric oxide synthase inhibitor NG-nitro-L-arginine (10-4 M) abolished the relaxing effect of acetylcholine. Short-term (6 h) treatment with elevated glucose (44 mM) attenuated the maximum relaxation (Emax) evoked by acetylcholine by 32% without altering pEC50 (-log M; mean ± S.E.M): 7.1 ± 0.1 (before); 7.0 ± 0.1 (after) (n = 6). The same effect (Emax and pEC50) was seen with normal glucose (5.5 mM) and glucose (5.5 mM) plus mannose (38.5 mM). The antioxidant ascorbic acid (10-4 M) and the cyclooxygenase inhibitor indomethacin (10-5 M) abolished the attenuation of Emax without altering the pEC50.

We conclude (1) that the relaxing effect of acetylcholine is due to nitric oxide; (2) that elevated glucose and hyperosmolarity do not cause an endothelial dysfunction per se; (3) that the attenuation of Emax is due to the formation of reactive oxygen species and vasoconstricting prostaglandins; and (4) that treatment of blood vessels with elevated glucose in vitro is not a suitable model for the endothelial impairment seen in diabetic patients.

References

Davies MG & Hagen P-O (1996). Alterations in venous endothelial cell and smooth muscle cell relaxation induced by high glucose concentrations can be prevented by aminoguanidine. *J Surg Res* 63: 474-479.

Dorigo P, Fraccarollo D, Santostasi G & Maragno I (1997). Impairment of endothelium-dependent but not of endothelium-independent dilatation in guinea-pig aorta rings incubated in the presence of elevated glucose. *Br J Pharmacol* 121: 972-976.

Hansen K & Nedergaard OA (1999). Methodologic aspects of acetylcholine-evoked relaxation of rabbit aorta. *J Pharmacol Toxicol Methods* 41: 153-159.

Hsueh WA & Anderson PW (1992). Hypertension, the endothelial cell, and the vascular complications of diabetes mellitus. *Hypertension* 20: 253-263.

Kamata K, Miyata N, Abiru T & Kasuya Y (1992). Functional changes in vascular smooth muscle and endothelium of arteries during diabetes mellitus. *Life Sci* 50: 1379-1387.

Taylor PD & Poston L (1994). The effect of hyperglycaemia on function of rat isolated mesenteric artery. *Br J Pharmacol* 113: 801-808.

Tesfamariam B, Brown ML, Deykin D & Cohen RA (1990). Elevated glucose promotes generation of endothelium-derived vasoconstrictor prostanoids in rabbit aorta. *J Clin Invest* 85: 929-932.

Nitric Oxide
R.J. Gryglewski and P. Minuz (Eds.)
IOS Press, 2001

APOPTOTIC PROCESS MODULATED BY NITRIC OXIDE IN CHRONIC B-CELL LYMPHOCYTIC LEUKEMIA

Roman, V.[1], Mentz, F.[2], Dugas, N.[3], Dugas, B.[4], Sigaux, F.[5], Kolb, J.P.[6]

1. Center of Immunology, Inst."Stefan S. Nicolau", Bucharest, Romania
2. Hematology Laboratory, Hôpital Pitié Salpêtrière, Paris, France
3. Virus, Neuron, Immunity Laboratory, IFR-Kremlin-Bicetre, Bicêtre, France
4. Oxykine Therapeutics Laboratory, Paris, France
5. U462-INSERM, Saint Louis Hospital, Paris, France
6. U365-INSERM, Curie Institute, Paris, France

B-cell chronic lymphocytic leukemia, or B-CLL, a monoclonal hematopoietic disorder with expansion of small lymphocytes of B cell is the most common leukemia in Western countries. More than 95% of the tumoral cells reside in the G_0 stage of the cell cycle and the accumulation *in vivo* of the malignant clonal B cell does not result from an active proliferation but is a consequence of a failure in apoptosis. The resistance of B-CLL cells to apoptotic cell death contributes to the expansion of the neoplastic cells endowing them with a selective growth advantage over normal lymphocytes.

Although resistant to apoptotic cell death *in vivo*, B-CLL tumoral B cells frequently display more rapid death by apoptosis *in vitro* than their normal counterpart, the small mature B lymphocytes, suggesting that a signal conferring active resistance to apoptosis my be provided *in vivo* by cell-cell interaction or soluble factors. For instance, spontaneous *in vitro* apoptosis of B-CLL is increased via B cell receptor stimulation, but is reverted by some cytokines (IL-4, IL-10, IL-13) and soluble CD23.

Even though various chromosomal abnormalities may explain, in part, resistance to apoptosis, they cannot account for the totality. Other mechanisms of resistance are therefore involved and we have shown previously that NO, released through the endogenous type II NO synthase, is a component of this resistance of B-CLL to apoptosis (1).

Recently, we directly measured the activity of the enzyme *ex vivo* in the leukemic cells by a technique of fluorescence based on the use of the NO specific probe DAF-2DA (diaminofluorescein 2, diacetate). The increase in fluorescence intensity was linear with time and was significantly reduced in the presence of the NOS inhibitor L-NAME (L-nitroarginine methylester).

Inhibition of endogenous NO production by NOS inhibitors was found to increase the apoptotic process in a large proportion of the tumoral cell tested, as demonstrated by the increased percentage of cells displaying phosphatidylserine at the outer leaflet of their plasma membrane after their simultaneous labeling with annexin V-FITC and propidium iodine.

The demonstration of an anti-apoptotic role of NO in B-CLL was further substantiated when the tumoral cells were treated with an anti-Fas mAb together with a NOS inhibitor. The percentage of cells undergoing Fas dependent apoptosis was significant increased, in agreement with several reports of an inhibitory role of NO on Fas-induced apoptosis through an impairment of the downstream caspase cascade via S-nitrosylation of the active site.

We therefore analyzed the possible targets of the NOS inhibitors which could favor apoptosis in B-CLL notably the activation of caspase 3 and the mitochondrial transmembrane potential, $\Delta\Psi$ m.

The effects on these parameters of various NOS inhibitors (L-NAME, 1400W) and chemical NO donors were measured on B-CLL cells. These agents were either tested alone or in combination with known apoptotic drugs, such as etoposide (an inhibitor of topoisomerase II), aminophylline (an inhibitor of cyclic nucleotide phosphodiesterases) and resveratrol (a phytoalexin exhibiting cancer chemopreventive activity).

The NOS inhibitors alone did not elicit an early (1-2 h) activation of caspase 3 in B-CLL. Whether it is now clear that NO interacts directly with caspase leading to their inactivation, a number of reports have described the possibility for NO to indirectly stimulate the process of pro-caspase activation through still poorly recognized pathways (2).

The redox status of the cells is one of the key factors modulating the pro- or anti-apoptotic role of NO, in which glutathione plays a crucial role.

Treatment of the leukemic cells with 50 microM etoposide and with 500 microM aminophylline and with resveratrol 50 microM resulted in a marked activation of caspase 3, as measured by the cleavage of its specific substrate DEVD-pNA (para nitro-aniline).

When tumoral B cells were cultured with the pro-apoptotic drugs in the presence of NO donors, the augmentation in caspase 3 activity observed in leukemic cell lysates was significantly decreased, probably through an inhibition by NO, through S-nitrosation, of the cysteine present in the catalytic site of caspase 3 (3). The specificity of the enzymatic activity was assessed by the fact that total inhibition of the cleavage of the substrate was achieved in the presence of the peptide DEVD-CHO.

A marked decreased in the $\Delta\Psi$ m of these tumoral cells was observed when B-CLL cells were incubated with a chemical NO donor, such as NOC-18 (a slow NO releasing agent with a $t_{1/2}$ of about 5h). A diminution of the $\Delta\Psi$m, measured by flow cytometry with the fluorescent probe DIOC(6)3, was detected as early as 4h after treatment and was more important after 16h of incubation.

Treatment *in vitro* of B-CLL cells with aminophylline resulted, as expected, in an increased percentage of leukemic cells displaying a low $\Delta\Psi$m; this percentage was further increased in the simultaneous presence of the NOS inhibitor L-NAME. Interestingly, resveratrol elicited a marked reduction in iNOS expression and NO release by B-CLL and also an inhibition of bcl-2 expression.

In conclusion, due to the multiplicity of its different effects, the inhibition of NO released was found to modulate the apoptotic process triggered by aminophylline at the level of caspase 3 activation and $\Delta\Psi$m, but not that elicited by etoposide, whereas resveratrol caused a reduction in the anti-apoptotic proteins iNOS and bcl-2. On the other hand, the addition of exogenous NO under the form of chemical NO donors was found to impair the activation of caspase 3 but elicited a decreased $\Delta\Psi$m, suggesting a complex role of NO in B-CLL apoptosis.

References

1. Zhao H, Dugas N, Mathiot C, Delmer R, Dugas B, Sigaux F, Kolb JP. Blood, 92: 1031, 1998.
2. Brune B, von Knethen A, Sandau KB. Eur.J.Pharmacol., 351: 261, 1998.
3. Rossig L, Fichtlschere B, Breitschopf K, Haendeler J, Zeiher AM, Isch A, Dimmeler S. J. Biol.Chem., 274: 6823, 1998.
4. Kolb J.P. Leukemia, 14: 1685, 2000.

Nitric Oxide
R.J. Gryglewski and P. Minuz (Eds.)
IOS Press, 2001

NO-MODULATING SELECTIVE β-BLOCKER NEBIVOLOL: DOES IT IMPROVE ARTERIAL COMPLIANCE IN INSULIN-RESISTANT HYPERTENSIVES ?

Sergey A. Golubev [a], Maxim N. Mily [a], Jeffrey J. Tsai[b], Vjacheslav V. Afanassiev [c]
[a]Basic and Clinical Pharmacology Department, Vitebsk State Medical University and
[c]Vitebsk Regional Cardiology Center, Vitebsk, Republic of Belarus and
[b]Pulse Metric, Incorporated, San Diego, California, USA.

Introduction. Human essential hypertension is associated with a mosaic of metabolic disorders, among which a core value has been attributed to insulin resistance (IR). IR, glucose intolerance, and dyslipidemia appear to jointly contribute to increasing arterial stiffness. Impaired NO action might be the primary defect triggering vascular dysfunction, sympathetic overactivity and IR [1], the relationship between EDRF(s)/NO and arterial compliance being intriguing. Much controversy still exists about vascular actions and metabolic effects of classical β-blockers and their usefulness for hypertensive patients with IR. It would appear that nebivolol, a newer long-acting highly selective β_1-blocker might have metabolic and hemodynamic advantages due to ancillary vasodilating effect, which seems to be NO-mediated [2]. There are a few of clinical studies devoted to the influence of nebivolol on compliance of peripheral arteries in humans. The present study was designed to investigate daily systemic, vascular, and cardiac hemodynamics changes during short-term nebivolol treatment of young overweight insulin-resistant essential hypertensives (EH).

Methods. Ten patients (aged 38.9±8.6 years, body mass index 32.1±5.0 kg/m^2) with IR features were selected using a specially developed algorithm (BMI > 25 kg/m^2, waist-to-hip ratio >0.9 for men and >0.8 for women and any lipid abnormality [3] and fasting insulin>102 pmol/l for men and >90 pmol/l for women or postload insulin >312 pmol/l for men and >354 pmol/l for women [4]) from a group of verified untreated mild-to-moderate EH. Ambulatory blood pressure monitoring (DynaPulse 5000A, Pulse Metric, Inc., USA) as well as 24-hour cardiac, systemic and vascular hemodynamics evaluation by the previously validated pulse dynamics analysis technology [5]; some metabolic tests (serum insulin before and 2 h after standard oral glucose tolerance test; serum total triglycerides, total and HDL cholesterol) were performed before and after 4-week treatment with 5 mg nebivolol o. d.

Results. Average daytime and nighttime systolic blood pressure (BP), diastolic BP and heart rate values were significantly reduced after therapy, without BP variability and BP nightfall changes. The smoothness-indexes of antihypertensive effect were 1.21±0.91 and 0.96±0.63 for systolic and diastolic BP, respectively. Results for the 24-hour hemodynamic variables obtained from the pulse dynamics analyses are shown in table (* P<0.05, ** P<0.01,*** P<0.001, **** P<0.0001 versus baseline).

Variables	Baseline	4 week
Cardiac Left Ventricle (LV) parameters:		
LV ejection time (second)	0.279±0.027	0.318±0.033 ***
LV dP/dT max (mmHg/s)	1329.9±179.5	1165.9±152.7 *
LV contractility (1/s)	14.5±1.7	13.8±1.1
Cardiac index (L/min/m^2)	2.53±0.22	2.30±0.24****
Stroke volume index (mL/ B/ m^2)	34.9±3.4	37.8±2.8**
Systemic Vascular (SV) parameters:		
SV Compliance (mL/mmHg)	1.185±0.118	1.388±0.160*
SV Resistance (mmHg/L/min)	20.7±2.8	21.0±3.8
Brachial Artery (BA) parameters:		
BA compliance (mL/mmHg)	0.087±0.017	0.089±0.022
BA distensibility (%/mmHg)	5.32±0.79	6.37±0.84 *
BA resistance (mmHg/L/min)	2498.9±907.1	3208.0±1388.7**

There were no statistically significant changes in the lipid, glucose and insulin levels after treatment with nebivolol.

Conclusions. Our findings indicate that in the investigated insulin-resistant EH nebivolol 5 mg ensured efficient BP control throughout the 24-hour dosing interval. Under these conditions the apparent cardiac effects of nebivolol as a β-blocker are associated with some favorable vascular changes, which argue in favor of vasodilatation as well as regional and systemic vascular compliance improvement without metabolic deterioration. However, the local vascular effects on the brachial artery level seem to be less pronounced comparing with systemic ones. All of these form a basis for long-term evaluation, which should be performed to establish the persistence and clinical relevance of these findings.

References.

1. Sartori C, Scherrer U. Insulin, nitric oxide and the sympathetic nervous system: at the crossroads of metabolic and cardiovascular regulation. *J Hypertens* 1999; 17:1517-1527.
2. Cockroft JR, Chowienczyk PJ, Brett SE, Chen CP, Dupont AG, Nueten LV, et al. Nebivolol vasodilates human forearm vasculature: evidence for an L-arginine/NO-dependent mechanism. J Pharmacol Exp Ther 1995; 274:1067-1071.
3. Wood D, De Backer G, Faergeman O, Graham I, Mancia G, Pyorala K. Prevention of coronary heart disease in clinical practice. Summary of recommendations of the Second Joint Task Force of European and other Societies on Coronary Prevention. *J Hypertens* 1998; 16:1407-1414.
4. Saint-Remy A, Scheen A, Lefebvre P, Jeanjean M, Rorive G. Are fasting and postload insulin good markers of cardiovascular risk factors? *J Hypertens* 1998; 16 (Suppl.2):S152.
5. Brinton TJ, Cotter B, Kailsam MT, Brown DL, Chio S-S, O'Connor DT, *et al.* Development and validation of a noninvasive method to determine arterial pressure and vascular compliance. *Am J Cardiol* 1997; 80:323-330.

Nitric Oxide
R.J. Gryglewski and P. Minuz (Eds.)
IOS Press, 2001

Vitamin C and lipoic acid lower oxidant production in cultured HAEC: Implications for age-related endothelial dysfunction

Anthony Smith, Francesco Visioli*, Balz Frei, and Tory Hagen

Linus Pauling Institute, Oregon State University, Corvallis, Oregon, USA
**University of Milan, Italy*

Introduction

Endothelial cells exert significant control over vasorelaxation, principally through synthesis and secretion of nitric oxide (NO) [1]. As a consequence, pathophysiological conditions that affect endothelial function could lead to a loss of vasorelaxation and contribute to hypertension, congestive heart failure, and stroke. In particular, reactive oxygen/nitrogen species (ROS/RNS) from a number of endogenous and exogenous sources have been implicated in loss of endothelial-dependent vasorelaxation in a variety of diseases and the aging process [2]. However, the conditions and potential sources of oxidants that lead to loss of endothelial-dependent vessel tone have not been elucidated as yet. The purpose of the present study was to discern the overall appearance of ROS/RNS in cultured HAEC and how addition of important antioxidants necessary for normal cell function, i.e. Vitamin C (AA) and lipoic acid (LA, a potent thiol antioxidant in its reduced form) could modulate ROS/RNS production (Figure 1.). Moreover, the mitochondrial contribution of ROS/RNS was assessed by using specific probes for mitochondrial oxidant production.

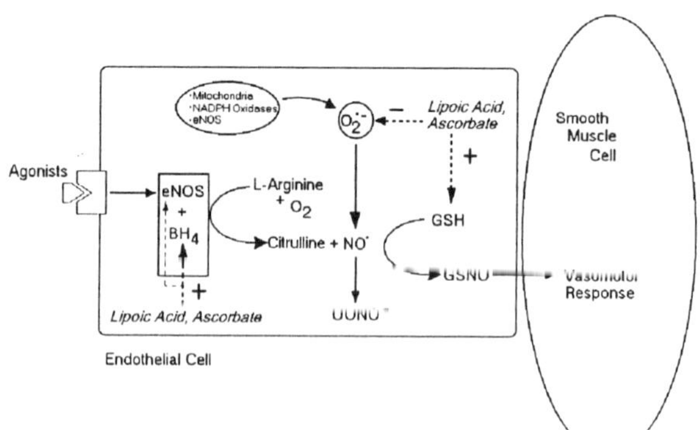

Figure 1. The potential effects of ascorbate and lipoic acid on endothelial nitric oxide.

Materials and methods

HAEC (2.5 x 10^4 cells; Clonetics) were allowed to adhere overnight to gelatin-coated 96-well microtiter plate wells. Cells were washed with HEPES-buffered saline (HBSS) and incubated for 30 min with AA (0.1mM), LA (0.1mM), L-NAME (0.3mM; nitric oxide synthase inhibitor). Oxidant appearance was monitored over time using a Cytofluor fluorescence plate reader after addition of 0.01mM (final concentration) of the following dyes: *Dichlorofluorescin diacetate* (DCFH-DA): a non-fluorescent dye that is taken up rapidly into the cytosol and becomes fluorescent upon contact with ROS or RNS [3]. *Dihydrorhodamine 123* (DHR): a non-fluorescent dye that accumulates rapidly in the mitochondria and becomes fluorescent upon contact with ROS/RNS there [3].

Results and Discussion

Cells grown in normal media are ascorbutic but the addition of ascorbate (0.1mM) rapidly repletes cellular Vitamin C levels (Figure 2, Panel A.). Vitamin C treatment lowers cellular ROS/RNS rates by 62% (Figure 2, Panel B.). Oxygen consumption remained unchanged after treatment showing that the reduction is not a metabolic effect. Lipoic acid lowers cellular ROS/RNS rates by 39% (Figure 2, Panel D). This suggests that other antioxidants can also lower ROS/RNS, but not as extensively as Vitamin C. L-NAME lowers cellular ROS/RNS rates by 56% (Figure 2, Panel C.). A significant percentage of the overall ROS/RNS detected by DCF is thus due to peroxynitrite and other RNS. In cultured cells without Vitamin C, endothelial nitric oxide synthase (eNOS) may contribute significantly to ROS/RNS appearance.

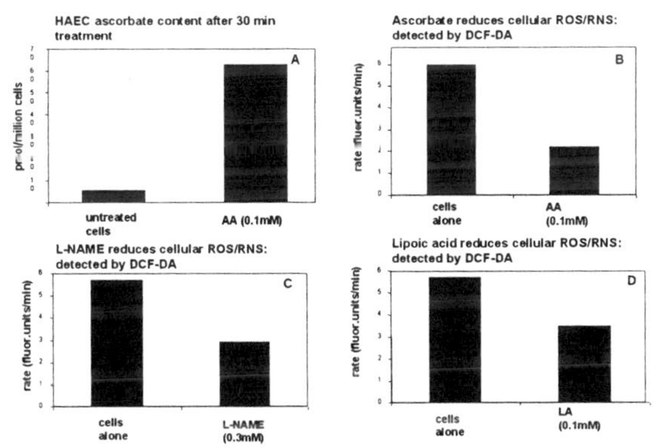

Figure 2. Vitamin C and Lipoic Acid lower cellular ROS/RNS.

Similar results were obtained when the mitochondrial oxidant production was monitored. In particular, mitochondrial oxidant production appears to contribute 30-40% of overall cellular ROS/RNS and the addition of AA, LA, or L-NAME was able to lower ROS/RNS production by 53%, 35%, and 31%, respectively.

Confocal microscopy of HAEC (Figure 3.) confirms that DHR accumulates in the mitochondria, where it becomes fluorescent due to its oxidation by ROS/RNS; the addition of ascorbate remarkably quenches such oxidant-induced fluorescence.

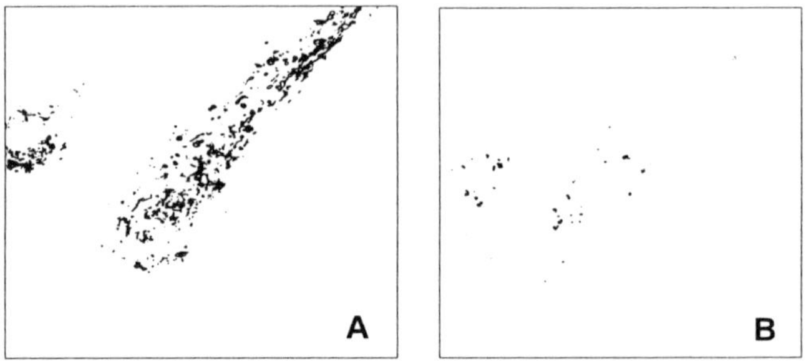

Figure 3. Scanning confocal laser images of DHR stained HAEC untreated (A) and treated with 0.1mM Vitamin C (B). Untreated cells show intense mitochondrial DHR staining, indicating higher ROS/RNS production than in Vitamin C repleted cells. Negative images are presented for clarity.

In conclusion, this investigation shows that HAEC in culture become ascorbutic, thereby increasing ROS/RNS appearance from a variety of sources. Hence, HAEC should be repleted with Vitamin C in order to approximate physiological conditions. Furthermore, eNOS and mitochondria significantly contribute to ROS/RNS appearance in HAEC when Vitamin C is not present. Finally, lipoic acid can also markedly lower ROS/RNS appearance in HAEC, although not to the same extent as Vitamin C.

References

1. R. A. Kelly, J. L. Balligand, and T. W. Smith. Nitric oxide and cardiac function. *Circ Res* (1996) 79: 363-380.
2. J. A. Holland, R. W. O'Donnell, M. M. Chang, D. K. Johnson, and L. M. Ziegler. Endothelial cell oxidant production: effect of NADPH oxidase inhibitors. *Endothelium* (2000) 7: 109-119.
3. J. P. Crow. Dichlorodihydrofluorescein and dihydrorhodamine 123 are sensitive indicators of peroxynitrite in vitro: implications for intracellular measurement of reactive nitrogen and oxygen species. *Nitric Oxide* (1997) 1: 145-157.

Correspondence to Anthony Smith by email: smitha@bcc.orst.edu

The features of NO metabolism under organic nitrates biotransformation in patients with ischemic heart disease

Andriy P. Sisetsky, Nickolaj F. Starodub, Valentyna M. Starodub
Dept. Biochemistry Sensory and Regulatory Systems of A.V.Palladin Institute of Biochemistry of Ukrainian National Academy of Sciences. 9 Leontovicha Str., 01030 Kyiv, Ukraine

1. Introduction

An action of nitrate containing preparations consists of several stages. The initial one is presented by transforming organic nitrates into nitrates and renovating nitrates into NO with creating labile nitrosile iron-sulfuric complexes. Biotransformation at this stage practically has not been studied yet. Relaxing effect of NO or its complexes through guanilatecyclase on a smooth muscle cells is being studied now. Any disorders in organic nitrates metabolism on any stage may decrease its efficacy on cardio-vascular system. Our research is aimed at studying metabolism of organic nitrates at the initial stage of biotransformation in patients with atherosclerotic damage of cardiac arteries.

2. Materials and methods

In 20 patients with ischemic heart disease, verified by data of coronaroventriculography and veloergometry tests, 90 minutes before and after applying 6,4 mg of sustac forte one day after nitropreparates had been stopped, there were taken arterial blood pressure, heart rate and content of nitrites, nitrates, transferine, ceruloplasmine, general Hb, methemoglobine (MetHb) and free radicals in blood samples taken from vein of ulna. The samples of plasma and erythrocyte mass were frozen in liquid nitrogen, further on the spectrum of the electron paramagnetic resonance (EPR) on a radiospectrometer "Varian E-109" was taken at the following parametres: UHF field power 5 mW, frequency modulation 100 kHz, amplitude modulation 8 Hz/s, 3 free radicals were registered at 0,2 mW, 100 kHz, 2 Hz/s respectively. The nitrites were defined with the help of Griss reagent. 200 μl of 1% dilution of sulphanile acid were added to 2 ml of the studied solution. In three minutes 200 μl of 1% dilution of N-1 naphtilethylendiaminchloride and in 10 min there was carried out a study on the "Specord M40" spectrophotometer with the waves about 500-600 nm of length. Nitrates were defined after their renovation into nitrites by classical method of renovation. Especially for this research there was developed micronodification of the method with the use of small disposable cadmium columns devised for samples less than 2 ml, with high renovating ability – 97±8%. The experimental part of the study was carried out at linear rats.

3. Results and discussions

The analyses of blood samples in patients have revealed a big difference between final level in blood nitrates and nitrates and transferine depending on groups they belong to. In the 1st group its content was much lower – 0,22±0,01 mg/l and 0,52±0,04 IU^{-8} respectively than in the 2nd one – 0,83±0,03 mg/l (P<0,01) and 0,84±0,02 IU^{-8} (P<0,01). In patients of the 1st group there was noted higher content of cholesterine, erythrocytes, hemoglobin (Hb), and lower in triglycerides compared with the indices in patients from the second group, however the difference was nor verified. In the patients from the 1st group there was revealed better tolerance towards physical exercise – 100,0±1,5W, than in the patients from the 2nd group – 75,0±1,2W at somewhat higher heart rate (114,0±0,7) and (100,0±0,9) respectively per 1 min ($P>0,05$) and systolic arterial blood pressure – 166,0±1,7 and 143,0±1,3 mm ($P>0,05$). In the patients from the 2nd group there were registered more frequent and intensive attacks of angina pectoris. The patients from both groups developed different reactions to the medicine. In the patients from the 2nd group after administration of organic nitrates there was less expressed decrease of arterial blood pressure and rise of heart rate as compared to the patients from the 1st group. Besides there was noted unexpected decrease of combined concentration of nitrites and nitrates from 0,83±0,04 to 0,34±0,01 mg/l ($P<0,01$), in some patients it was very pronounced as well as MetHb level decrease from 0,22±0,01 to 0,16±0,01 relative units ($P<0,05$) and the amount of oxidised labile iron in blood from 0,84±0,04 to 0,74±0,03 IU^{-8} ($P>0,05$). This was possible due to activation processes of their removal or including in other compounds including the stage of labile NO formation. Excessive amount of NO_2 and NO_3 in blood serum after application of organic nitrates may be caused by significant enzymes activation resulting in renovation of nitrates and nitrites in NO due to hypoxia, so frequent at ishemic heard disease (IHD). Activation of nitritreductase and NO-binding ability of Hb at higher demand of NO as a regulation factor of intracellular concentration of Ca_{2+} ions, insures increase of NO transport into myocardium and other tissues, and promotes the decrease of NO_2 and NO_3 concentration in the blood stream. Resulting from joint function of Ca_{2+} mobilising and nitritereducing (renovating NO_2 and NO_3 in NO) effect in mitochondria and erythrocytes in patients with IHD with separating processes of oxidative phosphorilation (ischemia, hypoxia) at higher NO content there is observed more effective of it and metabolic products (NO_2 and NO_3) in regulation of intracellular CA_{2+}. In the 2nd group patients the joint content of nitrites and nitrates in blood has greatly increased from 0,22±0,01 to 0,435±0,02 mg/l ($P<0,04$). In some patients there was noted significant increase of their concentrations. The increase of nitrate level in blood of patients from that group was followed by the increase of EPR signal from g=6,0 corresponding to MetHb with hemic iron in oxidised high spine form.

Evidently, the increasing of processes of Hb oxidation is caused due to deoxyhemoglobin transformation into oxidized form at the process of nitrites renovation as well as result of oxidation of hemic iron ($Hb0_2$) at interaction with NO and nitrite. Activation of the processes of NO synthesis from exogenic and endogenic sources was followed by the processes of oxidation of plasma labile iron, the EPR signal overlaps with the signal of the threevalent iron, contained by transferrine (g=4,3).

As it was experimentally demonstrated at the overlapping signals from labile and bound oxidised iron the form of the signal was changed. To perform qualitative evaluation of correlation between these forms of iron there was introduced the K index to characterise the prevalence of oxidised nontransferrine iron in blood and is calculated by the formula:

$$K = AC - \frac{BC}{\frac{1}{2}(AC + BC)}$$

where: A,B,C - EPR curve peaks. At K=0 EPR signal with g=4,3 is presented transferrine iron only and at K=1 – nontransferrine threevalent iron.

After the preparation had been administered to the patient of the 1st group K increased from 0,03 to 0,16 relative units ($P<0,01$). Thus the dynamics of the mentioned biochemical parameters after administration of organic nitrites in patients from the 1st group corresponds the general metabolism model of the preparation at the initial stage.

The deviation in NO exchange is defined both by the stage and individual course of disease and its activity which may strongly motivate inefficacy of the nitropreparation. Inadequate assimilation of organic nitrates explains development of tolerance for nitropreparations.

Nitric Oxide
R.J. Gryglewski and P. Minuz (Eds.)
IOS Press, 2001

Genetic Epidemiology of Cardiovascular Diseases in Daghestan Highland Isolates

O. Bulayev, K.Bulayeva

N.I. Vavilov Institute of General Genetics, Russian Academy of Sciences (Moscow, Gubkin street 3, 117809, Russia)

Detection of the susceptibility genes for complex human diseases such as cardiovascular diseases (CVD) is difficult because of phenocopies, reduced penetrance, and genetic heterogeneity. A powerful approach in study of this problem is riding with systematic searches for linkage disequilibrium (LD) in isolated populations [Kruglyak, 1997; Jorde, 1995; Terwilleger, Ott, 1996; Bulayeva et al., 2000]. The effective strategy of these studies is to find human isolates with high aggregation of the single common disease. In human genetic isolates characterized by a small number of founders, it is possible that a fraction of the current disease chromosomes derived from a common ancestor and this excess of the ancestral haplotype can be detected easier [Kruglyak, 1997; Terwilleger and Ott, 1996].

Among the most interesting isolated populations for genetic study are those located in Daghestan (Northern Caucasus, Russia). This region contains approximately two million people and 26 aboriginal ethnic groups. Prolonged reproductive isolation and severe environmental conditions in the highland villages have created genetically diverse 'primary isolates'[Neel 1992]. The rate of the endogamy in these isolates is 85-97%. The average coefficient of the inbreeding F varies from 0.009 to 0.02 that is close to worldwide maximum rate [Bulayeva 1991].

The CVD is a main reason of human deaths in any human populations, incluidng Daghestan. Blood pressure is considered to be elevated when repeated measurement is greater than 140/90 of either the systolic, diastolic, or both measurements. A diagnosis of hypertension is made when a person has had 2 or more elevated readings after the initial assessment. Heredity can be a risk factor if one or more parents have been diagnosed with hypertension. In our genetic epidemiology study were ascertained the Daghestan highland isolates with high aggregation of CVD: affectation of the adult subjects (older 25) by CVD in the isolates is 20- 57% from the totally affected by different complex illnesses. 5 large kindreds with high aggregation of hypertension and IBC ascertained from four ethnic groups (Laks, Dargins and Tindals). Most observed villages members die from heart attack or stroke. Due to high inbreeding of the extended pedigrees (300-700 subjects of 7-11 generations), statistical analysis on these kindreds is quite challenging. Preliminary results of analysis of data collected show the statsitically significant association between inbreeding and affectation by hypertension: the affectation higher among offsprings from consanguinrous marriages in compariosn with offspring from non-consanguineous marriages (χ^2 =3,85, df=1, p=0,050) that may be connecting with the recessive genes involving to the affectation of hypertension. The analysis of the data collected is in progress for study of association measures and fine mapping techniques to narrow regions containing susceptibility genes.

Nitric Oxide
R.J. Gryglewski and P. Minuz (Eds.)
IOS Press, 2001

The role of NO in altered sensitivity of vascular preparations to drugs in diabetic animals

Valéria Kecskeméti, Pál Pacher, Zoltan Ungvári*, Kornélia Tekes, Akos Koller*, Semmelweis University, Faculty of Medicine, Departments of Pharmacology and Pharmacotherapy and Pathophysiology*, Budapest, Hungary

Diabetes mellitus (DM) is known to cause impaired endothelium dependent relaxation of blood vessels in experimental animals as well as in both types of human diabetic patients (1-3). The aim of this study was to characterize the alterations of myogenic response in arterioles of diabetic rats; to compare its responsiveness to various drugs ; to elucidate the responsible mechanisms(nitric oxide, NO, Ca^{2+}) for the altered myogenic responsiveness ; to study the effect of chronic glibenclamide treatment on vasoregulatory function of these diabetic microvessels.

Materials and methods: Diabetes(DM) was induced in male Wistar rats (190-240 g) by a single injection of streptozotocin (65 mg/kg, STZ) into the femoral vein. The blood glucose level rose up to 22.01±2.34 mM by the end of the second week following the STZ injection and remained around this value throughout. Mean blood glucose level of control animals was 6.8±0.48 mM. The experiments were carried out in 10 week of DM. First-order arterioles (i.d.~ 150 μm) were isolated from gracilis muscle, cannulated and pressurized under no-flow conditions (80 mmHg). Changes in internal diameter were evaluated by microvideoangiometry. Endothelium was removed by injection of air into the lumen for 1 min at a perfusion pressure of 20 mmHg (4).

Results: Step increases in perfusion pressure (PP from 10 to 140 mmHg) elicited significantly greater and faster contraction in DM than in C gracilis arterioles both in the presence and absence of endothelium (E). Dilatations to acetylcholine (ACh)(EC_{50}, C: 4.0± 0.9x 10^{-9}M, DM: 4.8 ± 2.0x10^{-8} M) were impaired in diabetic arterioles and unaffected by inhibition of NO synthesis (in the presence of 10^{-4} M N^{ω}-nitro-L-arginine, L-NNA). Constrictions to norepinephrine (NE) were significantly greater in DM compared to those of C rats (EC_{50}, C: 6.2± 0.6x10^{-7}M, DM: 8.0 ± 2.0x10^{-8}M) both in the presence and absence of E. Endothelium-independent dilations to sodium nitroprusside were not different in arterioles from C and DM rats. Dilations to pinacidil (10^{-8}-10^{-5}M) were impaired by glibenclamide (10^{-6}M) both in C and DM arterioles. Chronic treatment of glibenclamide (5 mg/kg for 10 weeks) partially restored DM-induced change in NE responsiveness and normalised dilatory response to ACh of diabetic arterioles. Constrictions to increases in Ca^{2+} (0.25-7.5 mM), or Ca^{2+} channel agonist Bay K 8644 (10^{-11}-10^{-7}M), or protein kinase C activator phorbol 12-myristate 13-acetate (PMA, 10^{-9}-10^{-7} M), were significantly greater in diabetic arterioles than in control.

Conclusion: Our data suggest that experimental diabetes is associated with increased myogenic tone, increased norepinephrine sensitivity and impaired endothelium-dependent dilatation in resistance arterioles. The increased myogenic response is likely to be associated with increased activity of Ca^{2+} signaling mechanisms (Ca^{2+} channels, protein kinase C) and in the smaller part with the impaired endothelial function. It seems that in vitro glibenclamide does not modify, however, chronic glibenclamide treatment may improve by an unclear (probable antioxidant) mechanism the diabetes- induced vascular function.

This work was supported by grants from Hungarian Research Foundation OTKA- T 29624, T 23863.

References

1. Pieper GM, Jordan M, Adams MB, Roza AM. Diabetes Res Clin Practice, 31 Suppl: S143-155, 1996.
2. MacNally PG, et al. Clin Sci, 87: 31-36, 1994.
3. Koltai MZ et al. Cardiovasc Res, 34: 157-163, 1997.
4. Sun D, Kaley G, Koller A. Am J Physiol, 266: H1177-1183, 1994.

Address of correspondance author:
Valeria Kecskeméti, M.D., PhD.
Department of Pharmacology and Pharmacotherapy, Semmelweis University, Faculty of Medicine.
Budapest, Nagyvarad ter 4, P.O.Box 370, 1445 Hungary

Nitric Oxide
R.J. Gryglewski and P. Minuz (Eds.)
IOS Press, 2001

The Effects of a HMG CoA Reductase Inhibitor, Atorvastatin on Human Isolated Bypass Graft Materials

B.S. Uydeş-Doğan[1], S. Takır[1], M. Güden[2], B. Akpınar[2], E. Sağbaş[2], O. Özdemir[1]

[1]Istanbul University Faculty of Pharmacy Department of Pharmacology
[2]Florance Nightingale Hospital Department of Cardiovascular Surgery
34452 Beyazıt, Istanbul, Turkey.

Hypercholesterolemia is a severe risk factor for coronary artery diseases in which a progressive dysfunction develops in the endothelial cells (1). Cholesterol lowering therapy with HMG CoA reductase inhibitors, namely statins, has been shown to improve the vasomotor tone and the impaired endothelium-dependent relaxation response of the coronary arteries as well as peripheral circulation (2-3). They also exert a direct antiatherosclerotic effect on the arterial wall (4). Recent observations on endothelial cell cultures suggest that statins could directly affect endothelial cell function. An increase in endothelial nitric oxide (NO) production was demonstrated by statins which is correlated to direct upregulation of ecNOS expression and activity (5). Thus, statin therapy may have importance in hypercholesterolemic patients who had undergone coronary bypass operations not only by maintaining the patency rate in native coronaries and graft materials due to inhibition of smooth muscle cell proliferation, but also possibly by modulating the vascular tone directly.

In the present study, we investigated direct vascular effects of atorvastatin (a gift from Adilna Sanovel) on human isolated internal mammary artery (IMA) and saphenous vein (SV) which are the vessels used in coronary revascularization. IMA and SV specimens were obtained from coronary bypass operations, prepared as ring preperations and mounted in an organ bath system under optimal resting tensions. Following standardization of the vessel rings by repeated exposure to KCl (40 mM), preperations were precontracted either with noradrenaline (NA, $5x10^{-6}$M and 10^{-7}M, respectively) or KCl (20 and 40 mM, respectively). In IMA, the effects of atorvastatin were also examined in the presence of a nitric oxide synthase inhibitor, N^G-nitro-L-arginine (L-NOARG), a cyclooxygenase inhibitor, indomethacin or a K_{ATP} channel blocker, glibenclamide in order to identify the mechanism of its action.

Atorvastatin (10^{-6}-$5x10^{-4}$M) caused concentration-dependent relaxations in IMA and SV rings. Lower concentrations of atorvastatin were found more effective in SV than in IMA rings. However, the maximal relaxant responses (IMA;79.65±3.40 % n=24 vs SV; 78.79±5.40 % n=13 $p<0.05$) of two grafts were not different. Atorvastatin also caused concentration-dependent relaxant responses in KCl precontracted IMA rings (E_{max}; 80.22±7.90 % n=9). In the presence of L-NOARG, indomethacin or glibenclamide, the relaxant effects of atorvastatin at low concentrations were significantly inhibited however the maximal responses did not change (e.g. in the presence of L-NOARG: EC_{50} values 3.86±0.07, n=6 vs control; 4.05±0.12 n=6, $p<0.05$ and E_{max} values 73.68±7.64 % n=6 vs control; 76.76±7.24 %, n=6, $p>0.05$).

The present study showed that atorvastatin has relaxant effects on isolated IMA and SV rings. Experiments carried out in the presence of selective inhibitors suggested that the release of NO and prostacyclin from the endothelium and also K_{ATP} channel opening may partly mediate the relaxant effects of atorvastatin in IMA rings.

References:

1. Casino P.R., Kilcoyne C.M., Cannon R.O., III et al. Am J Cardiol. 1995,75,40-44.
2. Egashira K., Hirooka Y., Kai H. et al. Circulation 1994, 89, 2519-2524.
3. Muramatsu J., Kobayashi A., Hasegawa N. et al. Atherosclerosis. 1997,130, 179-182.
4. Corsini A., Bernini F., Quarato P. et al. Cardiology. 1996,87,548-468.
5. Laufs U., La Fata V., Plutzky J., et al. Circulaiton. 1998,97,1129-1135.

This work was supported by the Research Fund of the University of Istanbul.
Project number: T-747/251099.

Ischaemic Preconditioning of Coronary Vasculature

M.Lomnicka, S.Chlopicki, R.J.Gryglewski

Chair of Pharmacology, Jagiellonian University,
Grzegórzecka 16, 31-531 Cracow, Poland

Introduction: Murry et al. (1986) found that myocardium exposed to a brief transient ischaemia would delay postischaemic necrosis. The above ischaemic preconditioning (IPC), protects also against ischaemia/reperfusion-induced arrhytmias. However, it is not clear if IPC can (Kaeffer et al. (1996)) or cannot protect (Bauer et al. (1993)) coronary vasculature against such injury. Here we address this question.

Methods: In the isolated guinea pig hearts impairment of endothelium-dependent (Substance P (SP, 10 nM), Acetylcholine (Ach, 100 nM), postocclusive hyperaemia (POH, 10 sec occlusion of coronary flow) and endothelium-independent (Sodium nitroprusside (NaNP, 3 μM), PGE1 (100 pmoles) responses after 30 min of subglobal (reduction of coronary flow to 5%) ischaemia were analysed in hearts which were not preconditioned or preconditioned by various protocols. Preconditioning consisted of single 5 min ischaemia (IP5) or single 10 min ischaemia (IP10) or 5 min ischaemia repeated twice (IP2x5).

Results: Thirty minutes of ischaemia followed by reperfusion reduced both endothelium-dependent and endothelium-independent responses approximately by 30-50 % and slightly suppressed basal coronary flow by 10%. IP5 protected against post-ischaemic impairment of responses to NaNP (92±6% of preischaemic response vs 56±3% control group, p=0.007) or to PGE1 (81±6% vs 52±9%, p=0.04), but not against post-ischaemic impairment of SP (65±9% vs 50±7%, NS), acetylcholine (57±13% vs 42±9%, NS) and POH (61±5% vs 73±8%, NS) responses. The endothelium-dependent responses and post-ischaemic impairment of basal coronary flow were protected by IP10 only (Ach 101±11% p=0.014, SP 165±20%, p<0.001, POH 110±26%, p=0.02). This protocol protects also against impairment of endothelium-independent responses (NaNP 132±11%, p=0.002, PGE_1 86±6%, p=0.04). Interestingly, IP2x5 reversed post-ischaemic suppression of endothelium-independent responses (NaNP 104±16%, p=0.001 and PGE_1 97±9%, p=0.01), but not endothelium-dependent responses (Ach 86±14%, SP 80±9%, POH 63±9%, all NS).

Conclusions: It is concluded that the ischaemia/reperfusion procedure impairs functioning of coronary vascular bed of guinea pig heart. Ischaemic preconditioning protects both endothelial and smooth muscle cells against this impairment. As compared with coronary musculature, the endothelial cells require a longer lasting preconditioning stimulus to trigger protective mechanisms.

Bibliography:

Murry C, Jennings R, Reimer K. (1986) Circulation 74:1124-36

Kaeffer N, Richard V, Francois A, Lallemand F, Henry J, Thuillez C. (1996) American Journal of Physiology 271(3 Pt 2):H842-9

Bauer B, Simkhovich B, Kloner R, Przyklenk K. (1993) Circulation 88:659-72

Nitric Oxide
R.J. Gryglewski and P. Minuz (Eds.)
IOS Press, 2001

Modulatory role of nitric oxide on the PAF-induced cardiopulmonary effects: interaction between nitric oxide synthase and cyclooxygenase pathways

Fulvia Fabi, Roberto Calabrese, Tonino Stati, Livio Argiolas & Paola del Basso

Laboratory of Pharmacology, Istituto Superiore di Sanità, Viale Regina Elena 299, 00161 Rome, Italy

Introduction

It is now well established that nitric oxide (NO) modulates the pulmonary functions contributing to the maintenance of the vascular and bronchial tone. However, although the effects and the mechanisms of the platelet-activating factor (PAF), the phospholipid mediator generated in inflammatory and allergic reactions, have been extensively studied, the functional role of endogenously produced NO on the bronchoconstrictor and pulmonary hypertensive effects of PAF has not been clarified. Therefore, the main purpose of our work was to verify the role of endogenous NO on the pulmonary actions induced by PAF.

Moreover, since secondary mediators from the cyclo-oxygenase (COX) pathway play a pivotal role in these PAF-induced effects, we want to assess *i)* the involvement of the COX-2 isoform and *ii)* the functional effects of a possible interaction between COX and NO pathways on the cardiopulmonary actions of PAF.

Methods

The experiments were carried out in the blood-perfused heart-lung preparation of the guinea-pig (HLP), which makes it possible to assess not only bronchial and pulmonary vascular parameters, but cardiac performance as well. HLPs were performed according to the procedure already described (Argiolas L., Fabi F. & Del Basso P., *Br. J. Pharmacol.* **114,** 203-209, 1995).

PAF (L-α-phosphatidylcholine,β-acetyl-γ–O-hexadecyl; 1-64 ng) was administered by bolus injection into the venous cannula. Inhibitors were added to the perfusing blood.

Results

The presence of the nitric oxide synthase (NOS) inhibitor L-NNA (Nω-nitro-L-arginine; 0.2 mM) significantly potentiated ($P<0.001$) the bronchoconstrictor and pulmonary hypertensive (PAP and PVR) responses induced by PAF (n=7), which were higher than those of control group (n=7). The hyperreactivity to PAF by L-NNA was evident also after subliminal doses of PAF (1-2 ng), i.e. at doses which did not induce by themselves any functional response in control preparations. At the same time, the presence of L-NNA significantly potentiated ($P<0.001$) the decrease of cardiac output and the increase in right atrial pressure produced by PAF when compared to the response of control HLPs. The potentiation by L-NNA of the bronchoconstrictor and haemodynamic effects elicited by PAF was completely prevented by the contemporary presence of L-arginine (20 mM; n=6).
Adding to the perfusing blood the specific COX-2 inhibitor NS 398 (N-(2-Cyclohexyloxy-4-nitrophenyl)methanesulfonamide; 15 μM; n=7) significantly reduced ($P<0.001$) the pulmonary hypertensive, without affecting the bronchoconstriction induced by PAF in comparison to control preparations.
In HLPs perfused with blood containing the COX-2 inhibitor, further addition of L-NNA (n=7) failed to potentiate the bronchial responses elicited by PAF, whereas a slight increase of the pulmonary hypertensive responses induced by the highest doses of PAF was still observed.

Conclusions

Our results indicate that blocking NO production by L-NNA, the bronchoconstrictor as well as the pulmonary hypertensive effects of PAF are potentiated. The hyperreactivity caused by L-NNA on the PAF-induced pulmonary hypertension is accompanied by a strong rise of the right atrial pressure and by a reduction of the cardiac output. Therefore, the data suggest that endogenous production of NO - generated through the L-arginine: NO pathway - plays a modulatory role on the PAF-induced cardiopulmonary effects, acting as a counter-regulatory mechanism to attenuate not only the bronchial and pulmonary vascular constriction, but also the heart dysfunction induced by the phospholipid mediator.
Moreover, our experiments show that *i)* COX-2 isoform is involved in the PAF-induced pulmonary hypertension and *ii)* after the blockade of the COX 2 pathway, L-NNA fails to potentiate the pulmonary responses to PAF. This leads us to hypothesize the existence of a functional interaction between COX-2 and NOS activities.

Nitric Oxide
R.J. Gryglewski and P. Minuz (Eds.)
IOS Press, 2001

CHARACTERIZATION OF NO SPIN TRAPS: EPR PROPERTIES AND TRAPPING EFFICIENCY

SAVIANA NEDEIANU and TIBOR PÁLI
Institute of Biophysics, Biological Research Centre Szeged,
P.O. Box 521, H-6701, Szeged, Hungary,
E-mail: saviana@nucleus.szbk.u-szeged.hu

Nitric oxide (NO) is a small, reactive free radical gas that readily diffuses into cells and cell membranes. At physiological concentrations NO is a key signalling molecule in host-defence, neuronal communication and vascular regulation but excessive amounts of NO can cause or contribute to vascular shock, stroke, diabetes, neurodegeneration, arthritis and chronic inflammation [1]. NO can become even more dangerous by interaction with active oxygen species such as superoxide radical, O2-, producing cytotoxic radicals such as peroxynitrite, ONOO-, and hydroxyl radical, HO, which have been thought to lead to carcinogenesis and mutagenesis [2].

Although NO is paramagnetic, it is not directly detectable by conventional electron paramagnetic resonance (EPR) spectroscopy - the main technique for measuring paramagnetic molecules - due to its short lifetime. Upon reaction with a spin trap molecule, however, the unpaired electron of NO can be transferred onto a molecular orbital where it becomes detectable by EPR. Well known spin traps for NO are nitroxide compounds and haemoglobin [3, 4].

In the present study, we used $Fe^{2+}(DETC)_2$ (iron-diethyldithiocarbamate) as the hydrophobic spin trap and $Fe^{2+}(MGD)_2$ (iron-N-methyl-D-glucamine-dithiocarbamate) as the hydrophilic spin trap, which, with NO, form the paramagnetic ternary complexes, NO-Fe^{2+}-$(DETC)_2$ and NO-Fe^{2+}-$(MGD)_2$, respectively.

These two complexes gave rise to a characteristic three-line EPR spectrum [5]. NO was generated by dissolving an NO donor, sodium nitroprusside (SNP) or sodium nitrite ($NaNO_2$) in a solution with one of the two spin traps. SNP ($Na_2[(CN)_5FeNO]\cdot 2H_2O$), a known hypotensive agent can produce NO by light-catalyzed reaction or in reducing conditions [6]. EPR spectra of trapped NO, together with field position standards, were recorded at room temperature (298K) and at 165K. A detailed characterization of the EPR spectra of the above paramagnetic NO complexes, based on fitting theoretical lineshapes to the experimental ones, concerning g-value, hyperfine splitting and linewidth is presented.

The present study provides spectroscopic data required to develop a quantitative and sensitive detection system for nitric oxide both in hydrophobic and hydrophilic aqueous medium.

REFERENCES

1. Gross, S.S. and Wolin, M.S. (1995): Annu. Rev. Physiol., 57: 737-769.
2. Salgo, M.G., Stone, K., Squadrito, G.L., Pryor, W.A. (1995): Biochem. Biophys. Res. Commun., 210: 1025-1030.
3. Arroyo, C., Kohno, M. (1991): Free Radic. Res. Commun., 14: 145-155.
4. Greenberg, S.S., Wilcox, D.E., Rubanyi, G.M. (1990): Circ. Res., 67: 1446-1452.
5. Komarov, A.M., Mattson, D., Jones, M.M., Singh, P.K., Lai, C.S. (1993): Biochem. Biophys. Res. Commun., 195: 1191-1198.
6. Singh, R.J., Hogg, N., Neese, F., Joseph, J., Kalyanaraman, B. (1995): Photochem. Photobiol., 61(4): 325-330.

Nitric Oxide
R.J. Gryglewski and P. Minuz (Eds.)
IOS Press, 2001

Bradykinin as a strong inducer of COX-2 mRNA and a weak inducer of NOS-2 mRNA in cultured endothelial cells from human umbilical vein

Danuta Uracz, Wojciech Uracz and Ryszard J. Gryglewski

Chair of Pharmacology, Jagiellonian University Medical College, Grzegórzecka 16, 31-531 Cracow, Poland. miuracz@kinga.cyf-kr.edu.pl

Background: Exogenous bradykinin stimulates endothelial release of EDRF(NO) [1], prostacyclin [2] and endothelium derived hyperpolarizing factor (EDHF) [3]. In humans, endogenous bradykinin controls vasomotor function of coronary circulation via stimulation of kinin B_2 receptors [4]. Bradykinin is inactivated by the kininase II function within angiotensin converting enzyme (ACE). Therefore, accumulation of bradykinin may explain some of the endothelium-dependent vascular effects of ACE inhibitors (ACE-I) [5]. We demonstrated elsewhere that bradykinin as compared to ADP, acetylocholine, adrenaline, 5-hydroxytryptamine or even calcium ionophore is the most potent agent in rising free cytosolic calcium level $[Ca^{2+}]i$ and NO release from cultured bovine aortic endothelial cells (BAEC) [6]. Apart from its classical role of a tissue hormone, bradykinin may act as a growth factor, and this is why it may behave as a cytokine [7]. For instance, bradykinin was reported to induce COX-2 in cultured human airway smooth muscles cells [8] and in HUVEC [9]. Because of its low molecular weight bradykinin is not usually classified as a cytokine or a growth factor. In this respect "minicytokine" may be a more appropriate name for bradykinin. Indeed, we describe here an example of the "minicytokine" type action of bradykinin. Four hours incubation of cultured human umbilical vein endothelial cells (HUVEC) with bradykinin (10 nM!) resulted in induction of the mRNAs for haemooxygenase 1 (HO-1), cyclooxygenase 2 (COX-2), prostaglandin E synthase (PGE-S) and, insignificantly, for nitric oxide synthase 2 (NOS-2).

Methods: HUVEC were isolated and cultured according to Jaffe *et al.* [10] with modifications as described previously [11]. Briefly, HUVEC were used for experiments upon reaching 90% confluency in the third passage ($1x10^6$ cells) Then after 24 hours of HUVEC incubation in OPTI-MEM I without foetal bovine serum, HUVEC were stimulated with bradykinin (Bk, 10 nM) for 4 hours and total RNA was isolated by the guanidinum isothiocyanate method as described by Chomczynski and Sacchi [12] using TRIZOL®Reagent (Gibco BRL,UK). Reverse transcription (RT) of total RNA (1 μg) was performed with oligo-$(dT)_{12-18}$ primer and M-MLV reverse transcriptase (Gibco

BRL,UK) for 2 hours at 42^0C. Then, cDNA (1 μl) was amplified with HotStar Taq Polimerase (QIAGEN, USA) for both quantitative and qualitative RT-PCR. Specific mRNAs for HO-1, COX-2, COX-1 and NOS-2 were quantified by Gene Specific Relative RT-PCR Kit (Ambion, USA) using multiplex RT-PCR protocol with endogenous 18S rRNA standard and Competimer® technology. PCR reactions were set up with GeneAmp 9600 machine (Perkin-Elmer, USA) of the following profile: 95°C – 15 min and 35 cycles: 94°C – 30 s, 59°C – 30 s, 72°C – 30 s followed by 72°C – 10 min elongation and then samples were placed on ice. Equal volumes of each sample (5 μL) were run on 3% agarose containing ethidium bromide and images were captured electronically with DC40 digital camera (Kodak, USA) and the bands were quantified using image analysis software (NIH Image, USA). Results were expressed as ratio of pixel density units for specific mRNA to internal standard (18S rRNA). For position and size of observed specific bands, marker M1 (pUC19/MspI, DNA-Gdańsk II, Poland) was run in parallel. For qualitative RT-PCR for PGE-S following specific primers were used: primer 1 (sense) 5'-ATGCCTGCCCACAGCCTG-3' (40nM) and primer 2 (antisense): 5'-TCACAGGTGGCGGGCCGC-3'. Oligonucleotide primers were constructed by Jakobson *et al.*[13], but truncated for cloning restriction sites. Primer sets for NOS-2 and internal standard (18S rRNA) was obtained commercially (Ambion, USA).

Results: To compensate for variations in total RNA quantity, initial quantification errors and random tube-to-tube variations in reverse transcription and PCR, we used multiplex RT-PCR with two primer sets in one tube in a single PCR (multiplex). One set of primers was used to amplify the cDNA of interest: HO-1, COX-2, COX-1, and NOS-2 while a second set of primers was used to amplify an invariant endogenous control 18S rRNA. In HUVEC incubated for 4 hours with Bk at a concentration of 10 nM, a significant increase in mRNAs for HO-1 (4.5-fold) and COX-2 (2.5 fold), but not for COX-1 (1.35 fold) and NOS-2 (0.3 fold) was observed. An assay of the induction of specific PGE-S mRNA was accomplished using classical RT-PCR system. Then, total RNA content was controlled by amplifying invariant control 18S rRNA.

Discussion and Conclusions: In cultured HUVEC incubated with bradykinin at a low concentration of 10 nM for four hours we observed an induction of transcripts for four enzymes: haemooxygenase 1 (HO-1), prostaglandin E synthase (PGE-S), cyclooxygenase 2 (COX-2) and NO synthase 2 (NOS-2). The potency of induction for corresponding mRNAs decreased in the following order: HO-1>PGE-S>COX-2>>NOS-2. Actually, the expression of NOS-2 m-RNA was only slightly touched by bradykinin. Activation of HO-1, apart from the CO generation may also lead to a deficiency in intracellular haeme required as a coenzyme for both COX and NOS. We have not been as yet studying effects of bradykinin on corresponding enzymic proteins or on their activities. Would they go along with their mRNAs, then bradykinin acts as a real "minicytokine" enzymatic inducer. Induction of HO-1 is associated not only with an increased production of CO but also with an increased consumption of haeme required as a prosthetic moiety both in cyclooxygenase and in NO synthase, as well as in a number of other haemoprotein enzymes [14]. The above pattern of induction of endothelial mRNAs may suggest that the delayed microcytokine-like effect of bradykinin on activities of COX-2 and NOS-2 will be tempered by a concomitant rise in HO-1 activity, and accordingly the net "inductive" effect of bradykinin on prostanoid generation might be stronger than its effect on NO generation. This assumption requires an experimental confirmation, especially that an increased generation of CO by inducible HO-1 may assist vasodepressor function of PGI_2 or NO [15].

References

[1] Cherry PD, Furchgott RF, Zawadzki JV, Jothianandan D. Role of endothelial cells in relaxation of isolated arteries by bradykinin. *Proc Natl Acad Sci U S A* (1982) **79**(6):2106-2110.

[2] Gryglewski RJ, Moncada S, Palmer RM. Bioassay of prostacyclin and endothelium-derived relaxing factor (EDRF) from porcine aortic endothelial cells. *Br J Pharmacol* (1986) **87**(4):685-694.

[3] Nakashima M, Mombouli JV, Taylor AA, Vanhoutte PM. Endothelium-dependent hyperpolarization caused by bradykinin in human coronary arteries. *J Clin Invest* (1993) **92**(6):2867-2871.

[4] Goves P, Kurz S, Just H, Drexler H. The role of endogenous bradykinin in human coronary vasomotor control. *Circulation* (1995) **92**:3424-3430.

[5] Mombouli JV, Vanhoutte PM. Heterogeneity of endothelium-dependent vasodilator effects of angiotensin-converting enzyme inhibitors: role of bradykinin generation during ACE inhibition. *J Cardiovasc Pharmacol* (1992) **20 Suppl 9**:S74-S82.

[6] Gryglewski RJ, Chlopicki S, Uracz W, Marcinkiewicz E. Significance of endothelial prostacyclin and nitric oxide in peripheral and pulmonary circulation. *Med Sci Monit* (2001) **7**(1):1-16.

[7] Patel KV, Schrey MP. Inhibition of DNA synthesis and growth in human breast stromal cells by bradykinin: evidence for independent roles of B1 and B2 receptors in the respective control of cell growth and phospholipid hydrolysis. *Cancer Res* (1992) **52**(2):334-340.

[8] Pang L, Knox AJ. PGE2 release by bradykinin in human airway smooth muscle cells: involvement of cyclooxygenase-2 induction. *Am J Physiol* (1997) **273**(6 Pt 1):L1132-L1140.

[9] Yamasaki S, Sawada S, Komatsu S *et al.* Effects of bradykinin on prostaglandin I(2) synthesis in human vascular endothelial cells. *Hypertension* (2000) **36**(2):201-207.

[10] Jaffe EA, Nachman RL, Becker CG, Minick CR. Culture of human endothelial cells derived from umbilical veins. Identification by morphologic and immunologic criteria. *J Clin Invest* (1973) **52**(11):2745-2756.

[11] Ziemian B, Olszanecki R, Uracz W, Marcinkiewicz E, Gryglewski RJ. Thienopyridines: effects on cultured endothelial cells. *J Physiol Pharmacol* (1999) **50**(4):597-604.

[12] Chomczynski P, Sacchi N. Single-step method of RNA isolation by acid guanidinium thiocyanate-phenol-chloroform extraction. *Analytical Biochemistry* (1987) **162**(1):156-159.

[13] Jakobsson PJ, Thoren S, Morgenstern R, Samuelsson B. Identification of human prostaglandin E synthase: a microsomal, glutathione-dependent, inducible enzyme, constituting a potential novel drug target. *Proc Natl Acad Sci U S A* (1999) **96**(13):7220-7225.

[14] Wagener FA, da Silva JL, Farley T *et al.* Differential effects of heme oxygenase isoforms on heme mediation of endothelial intracellular adhesion molecule 1 expression. *J Pharmacol Exp Therap* (1999) **291**(1):416-423.

[15] Johnson RA, Lavesa M, Askari B, Abraham NG, Nasjletti A. A heme oxygenase product, presumably carbon monoxide, mediates a vasodepressor function in rats. *Hypertension* (1995) **25**(2):166-169.

Nitric Oxide
R.J. Gryglewski and P. Minuz (Eds.)
IOS Press, 2001

MODULATION OF NITRIC OXIDE PRODUCTION BY LYSOPHOSPHATIDYLCHOLINE IN HUMAN ENDOTHELIAL CELLS

O. Konopatskaya, J.L. Whatmore, A.C. Shore, J.E. Tooke.
Dept. Vascular Medicine, PGMS, University of Exeter, Exeter, UK

The naturally occurring phospholipid, lysophosphatidylcholine (lyso-PC), is an important mediator and constituent of the atherogenic effects of oxidized low density lipoproteins. (Kume N et al. J. Clin. Invest. 1992;90:1138-1144). Lyso-PC has been implicated in the modulation of vascular cell activation processes, including synthesis of nitric oxide (NO) by endothelial cells. (Zembrowicz A et al. J Biol Chem 1995;28:17006-17010). However, lysoPC at the concentrations previously used (100μM) becomes toxic due to micelle formation (Matsuda Y., et. al, Circ Res, 1993;72:1103-09). This study aimed to investigate the influence of non-toxic (less than 30 μM) concentrations of lyso-PC on the production of biologically active NO.

Human umbilical vein endothelial cells at passage 1 were stimulated with 10 or 20 μM of lyso-PC for 24 hours, washed and incubated for 15 min in Krebs solution containing IBMX and either 20 or 10 μM lyso-PC. NO production was assessed by quantitating intracellular cGMP accumulation, using radioimunoassay.

Lyso-PC significantly increased NO production in a concentration dependent manner. Stimulated levels were 211.30± 64.84% ($p=0.002$) at 10 μM and 487.11±249.89% ($p=0.002$) at 20 μM compared to controls (100%). Preliminary time course data indicate that there is a rapid response occurring within 15 minutes.

These observations indicate that lyso-PC stimulates the formation of NO in human umbilical vein endothelial cell at low, non-toxic concentrations.

Nitric Oxide
R.J. Gryglewski and P. Minuz (Eds.)
IOS Press, 2001

The role of nitric oxide, prostacyclin and melatonin in aspirin-sensitive asthma

Helen V. Evsyukova, M.D., Ph.D.
The I.P.Pavlov Medical University,
Department of Hospital Therapy, St.Petersburg, Russia

Nitric oxide (NO) and prostacyclin (PgI_2) play an important role in the mechanisms of platelet-vessel haemostasis and lung microcirculation, which have been shown by our studies to be disturbed in aspirin-sensitive asthma (ASA). The pineal hormone melatonin (MT) was shown to inhibit several physiological processes in platelets and causes a relaxation of precontracted pulmonary arteries and veins. However the role of melatonin in the pathogenesis of bronchial asthma remains unknown.

Our aim was to study the production of NO, PgI_2 and MT in ASA-patients and to associate these findings with blood platelet function.

In 35 patients with ASA, 25 patients with aspirin-tolerant asthma (ATA) and 24 healthy volunteers the nitrates and nitrites (ΣNO_2^- and NO_3^-) excretion in diurnal (from 09:00 to 21:00) and in nocturnal (from 21:00 to 09:00) portions of urine were determined by modified Griess method (1991). We measured also the excretion of melatonin major metabolite (6-sulphatoxymelatonin-aMT6s) in the same portions of urine, the plasma content of stable PgI_2 and TxA_2 metabolites and ADP-induced platelet aggregation in the morning hours. Patients were examined when they were in stable clinical condition.

The results of the study showed the high level of nitrite and nitrate excretion in urine during the night (88,1±12,7 nmol /ml) in ASA patients compared with that of ATA patients (59.9±9,9 nmol/ml, p=0,01) and controls (65,9±8,9 nmol/ml, p<0,05). The circadian rhythm of NO generation remained unchanged in ASA patients and corresponded to that of MT production.

The ASA patients showed the low daytime excretion aMT6s (12,7±2,7 ng/ml) than that of ATA patients (32,7±9,9 ng/ml, p<0,01) and controls (19,9±5,8 ng/ml, p<0,01). The NO production in normal and ATA humans at night correlated with the MT production in the daytime (R = 0.7, p = 0.03 and R = O.6, p=0.04,respectively). However no such correlation was observed in the ASA patients.

The ASA patients showed more intense ADP-induced platelet aggregation and a higher rate of its first wave.The aggregation parameters in ASA patients didn't correlate with the excretion of NO_2^- and NO_3^- in daytime in contrast to the ATA patients, indicating the lack of NO regulation of the platelet functional activity in these patients.

Conclusion

The higher level of NO produced during the night in ASA patients may have been due to the lack of melatonin control inhibiting the activity of i-NOS. The increased production of NO at night in ASA patients is hypothesized to lead to the generation of peroxynitrite. This in turn would stimulate platelet aggregation, overcome the beneficial effects of endothelial NO and along with low PgI_2 production lead to lung microcirculation disturbances which may be responsible for the protracted course of aspirin-sensitive asthma.

Nitric Oxide
R.J. Gryglewski and P. Minuz (Eds.)
IOS Press, 2001

MECHANISMS OF REGULATION OF NEUROTRANSMITTER RELEASE BY NITRIC OXIDE

Sónia M. Sequeira, Arsélio P. Carvalho, Caetana A. M. Carvalho and Thomas C. Südhof

The nitric oxide (NO)-producing enzyme, nitric oxide synthase was originally identified in the brain at postsynaptic densities (Bredt and Snyder, 1990), where they interact with glutamatergic receptors of the type N-methyl-D-aspartate (NMDA), through the anchoring protein PSD95/SAP90. The present work provides further evidence to previous findings (Alagarsamy et al., 1994, Casado et al., 2000, Sequeira et al., 1997) that a similar complex exists at the presynaptic nerve terminal. NO generated by NO donors or by the stimulation of the endogenous enzyme NO synthase (NOS) modulates neurotransmission in the central nervous system by changing the synaptic efficacy of neuronal signaling (Boulton et al., 1994, Casado et al., 2000, Colasanti et al., 2000, Lu et al., 1999, Sequeira et al, 1997, 1999, 2000, Meffert et al., 1996). The physical proximity of NOS to the NMDA receptor is thought to provide the calcium necessary for the activation of the enzyme, during synaptic activity however calcium channels and non-NMDA glutamate receptors have also been reported to induce NOS-mediated NO production (Alagarsamy et al., 1994, Kendrick et al., 1996) suggesting multiple modes of production of NO. In addition, NMDAR activation plays an essential role in glutamatergic signaling and toxicity in the hippocampus (Carvalho et al., 1999, Stijbos et al., 1996) and NMDA is involved in two common paradigms of memory and learning, long-term potentiation (Shimuzu et al., 2000) and depression (Casado et al., 2000) at glutamatergic synapses.

The distribution and subunit composition of NMDAR at the synapse are critical factors in shaping transmission through the activation of diverse signaling processes (Ito et al., 2000) and may underlie the observations that postsynaptic NMDAR activation is involved in LTP (Arancio et al., 1996) whereas presynaptic NMDAR activation (MacDermott et al., 2000) induces synaptic depression at the parallel fiber-Purkinje cell synapse, where post-synaptic NMDAR are absent (Casado et al., 2000). NMDAR are localized at the plasma membrane by a dynamic network of transporter (Setou et al., 2000), anchoring, and signal transduction proteins (for review see Kim and Huganir, 1999], which potentially modulate NMDAR function and NMDAR subunit variability is observed during development, (Contestabile, 2000) and at mature synapses (Sprengle and Single, 1999) suggesting functional diversity of NMDAR in subsets of neurons. On the other hand, NOS is expressed in nerve terminals (Jones et al., 1995, Sequeira et al., 1997, Sistiaga et al., 1997) as well as in postsynaptic formations and it was previously reported that stimulation of NOS with the substrate L-arginine inhibits the exocytotic release of glutamate and GABA from hippocampal nerve terminals through a cGMP-dependent mechanism (Sequeira et al., 1997, 1999). However, it is not clear whether *in vivo*, presynaptic NOS is stimulated by NMDAR activation, nor have the protein equivalents of PSD-95 been identified, although some potential candidates exist (Butz et al., 1998).

Synaptosomes are a useful preparation for the evaluation of presynaptic events and were thus used in this work, to measure the release of exocytotic glutamate and GABA from hippocampal nerve terminals by a fluorimetric and perfusion-based assay evoked by either 4-aminopyrydine (4-AP, 100 μM) which evokes action-potential like depolarization, or by hypertonic sucrose (0.5 M) which triggers the Ca^{2+}-independent release of docked synaptic vesicles from the readily-releasable pool (Lonart and Südhof, 1998). With this approach, it was investigated whether presynaptic ionotropic glutamate receptor activation modulates the vesicular release of glutamate or GABA from rat hippocampal nerve terminals through the activation of presynaptic NOS. The results show that NMDA causes the inhibition of evoked-glutamate release in a NO-dependent manner, whereas α-amino-3-hydroxy-5-methylisoxazole-4-proprianato (AMPA) and kainate (KA) receptor activation inhibit glutamate release through a NO-independent mechanism: NMDA dose-dependently inhibited the release of glutamate evoked by 4-aminopyridine (4-AP) (IC_{50}= 155 μM), and this effect was reversed by the NMDA receptor antagonist D-AP5 and by the NO synthase inhibitor, L-nitroarginine in synaptosomes isolated from whole hippocampus, CA3 and CA1 areas, but interestingly, not from the dentate gyrus. In contrast, the 4-AP-evoked release of glutamate was reduced by α-amino-3-hydroxy-5-methyl-4-isoxazolepropionic acid (AMPA) or kainite (KA) but it was not blocked by L-nitroarginine, and NMDA, but not AMPA or KA, significantly increased cGMP formation. Presynaptic NMDA receptors are probably involved since removing extracellular NO with the NO scavenger C-PTIO did not block the depression of glutamate release by NMDA. The mechanism underlying this depression involves the direct inhibition of synaptic vesicle exocytosis (Machado et al., 2000) since NMDA/NO inhibited the release of [^{3}H]-glutamate and [^{14}C]-GABA evoked by hypertonic sucrose. The results also suggest that presynaptic NMDAR function as heteroreceptors that modulate the exocytotic release of GABA from hippocampal interneurons. Since glutamate mediates fast excitatory transmission and GABAergic neurons input essential inhibitory tones to glutamatergic circuits, these results may have important implications in the development of therapeutic drugs in brain disease.

References:

1. Alagarsamy, S., Lonart, G. and Johnson, K.M. (1994) Regulation of NO synthase activity in cortical slices by excitatory amino acids and calcium. *J. Neurosci. Res.* **38**:6 648-53.

2. Arancio, O., Kiebler, M., Lee, C.J., Lev-Ram, V., Tsien, R.Y., Kandel, E.R. and Hawkins, R.D. (1996) Nitric oxide acts directly in the presynaptic neuron to produce LTP in cultured hippocampal neurons. *Cell* **87**:6, 1025-1035.

3. Boulton, C.L., Irving, A.J., Southam, E., Potier,B., Garthwaite, J. and Collingridge, G.L. (1994) The nitric oxide-cyclic GMP pathway and synaptic depression in rat hippocampal slices. *Eur.J.Neurosci.* **6**, 1528-1535.

4. Bredt, D.S and Snyder, S.H. (1990) Isolation of nitric oxide synthase, a calmodulin-requiring enzyme. *Proc.Natl.Acad.Sci USA* **86**, 9030-9033.

5. Butz, S., Okamoto, M. and Südhof, T.C. (1998) A tripartite protein complex with the potential to couple synaptic vesicle exocytosis to cell adhesion in brain. *Cell* **94**:6, 773-782.

6. Carvalho, C.M., Sequeira S., Duarte C.B. and Carvalho, A.P. (2000) Modulation of Glutamate Release and Toxicity by Nitric Oxide. In *Free Radicals in Brain Pathophysiology* (G.Poli, E. Cadenas and L. Packer, eds.), Marcel Dekker, N.Y., 157-176.

7. Casado, M., Dieudonné, S. and Ascher, P. (2000) Presynaptic N-methyl-D-aspartate receptors at the parallel fiber-Purkinje cell synapse. *Proc. Natl. Acad. Sci. U S A* **97**:21,11593-97.

8. Colasanti, M. and Suzuki, H. (2000) The dual personality of NO. *TiPS* **21**, 249-252.

9. Contestabile, A. (2000) Roles of NMDA receptor activity and of nitric oxide production in brain development. *Brain Res. Brain Res. Rev.* **33**:2-3, 476-509.

10. Ito, I., Kawakami, R., Sakimura, K., Mishira, M. and Sugiyama, H. (2000) Input-specific targeting of NMDA receptor subtypes at the mouse hippocampal CA3 pyramidal neuron synapses. *Neuropharmacology* **39**:6, 943-951.

11. Jones, N.M., Loiacono, R.E. and Beart, P.M. (1995) Roles for nitric oxide as an intra-and interneuronal messenger at NMDA release-regulating receptors:evidence from studies of the NMDA-evoked release of [^{3}H]noradrenaline and D-[^{3}H]aspartate from rat hippocampal slices. *J. Neurochem.* **64**, 2057-2063.

12. Kendrick, K.M., Guevara-Guzman, R., de la Riva, C., Christensen, J., Ostergaard, K. and Emson, P.C. (1996) NMDA and KA-evoked release of classical transmitters in the rat striatum: evidence that NO may play a neuroprotective role. *Eur. J. Neurosci.* 8 (12): 2619-2634.

13. Kim, J.H, and Huganir, R.L. (1999) Organization and regulation of proteins at synapses. *Curr. Opin. Cell Biol.* **11**(3):248-56.

14. Lonart, G. and Südhof, T.C. (1998) Mechanism of action of rab3A in mossy fiber LTP. *Neuron* **21,** 1141-1150.

15. Lu, Y.F., Kandel, E.R. and Hawkins, R.D. (1999) NO signalling contribute to late-phase LTP and CREB phosphorylation in the hippocampus. *J. Neurosci.* **19**:23, 10250-10261.

16. Machado, J.D., Segura, F., Brioso, M.A. and Borges, R. (2000) Nitric oxide modulates a late step of exocytosis. *J. Biol. Chem.* **275**:27, 20274-20279.

17. MacDermott, A.B., Role, L.W. and Siegelbaum,S.A. (2000) Presynaptic ionotropic glutamate receptors and the control of transmitter release. *Annu. Rev. Neurosci.* **22**, 443-485.

18. Meffert, M.K., Calakos, N.C., Scheller, R.H. and Schulman, H. (1996) Nitric oxide modulates synaptic vesicle docking/fusion reactions. *Neuron* **16**, 1229-1236.

19. Sequeira, S.M., Ambrósio, A.F, Malva, J.O., Carvalho, A.P. and Carvalho, C.M. (1997) Modulation of glutamate release from hippocampal synaptosomes by nitric oxide. *Nitric Oxide: Biol. Chem.* **1**, 315-339.

20. Sequeira, S.M., Carvalho, A.P. and Carvalho, C.M. (1999) Both PKG dependent and independent mechanisms are involved in the modulation of glutamate release by nitric oxide in rat hippocampal nerve terminals. *Neurosci. Lett.* **261**:1-2, 29-33.

21. Setou, M., Nakagawa, T., Seog, D.H., and Hirokawa, N. (2000) Kinesin superfamily motor protein KIF17 and mLin10 in NMDA receptor-containing vesicle transport. *Science* **288**:5472, 1796-1802.

22. Shimuzu, E., Tang, Y.P., Rampon, C. and Tsien, J.Z. (2000) NMDA receptor-dependent synaptic reinforcement as a crucial process for memory consolidation. *Science* **290**:5494, 1170-1174.

23. Sistiaga, A., Miras-Portugal, M.T. and Sánchez-Prieto, J. (1997) Modulation of glutamate release by a nitric oxide/cGMP-dependent pathway. *Eur. J. Pharmacol.*, **331**, 247-257.

24. Sprengel, R. and Singlo, F.N. (1999) Mice with genetically modified NMDA and AMPA-receptors. *Ann. N.Y. Acad. Sci.* **868**, 494-501.

25. Stijbos, P.J., Leach, M.J. and Garthwaite, J. (1996) Vicious cycle involving Na+ channels, glutamate release, and NMDA receptores mediates neurodegeneration through nitric oxide formation. *J. Neurosci.* **16**.16, 5004-13.

Nitric Oxide
R.J. Gryglewski and P. Minuz (Eds.)
IOS Press, 2001

LPS-INDUCED NO PRODUCTION IN HUMAN T84 INTESTINAL EPITHELIAL CELLS. Characterization and pharmacological control.

Mari Lähde and Eeva Moilanen, The Immunopharmacological Research Group, Medical School, 33014 University of Tampere and Tampere University Hospital, Finland

Nitric oxide (NO) production is increased in inflammatory bowel disease (IBD). In the early stages of experimentally induced intestinal inflammation low levels of NO produced by cNOS are protective, whereas later on when NO is produced in high amounts by iNOS, NOS inhibitors have anti-inflammatory effects. The aim of the present study was to characterize the LPS-induced NO production in human intestinal T84 cells and to test if drugs used in the treatment of IBD effect on NO production in these cells.

Lipopolysaccharide (LPS) induced detectable nitrite synthesis in T84 human intestinal epithelial cells measured as accumulated nitrite into the culture medium. This effect was suppressed with NOS inhibitors in the order following of potency 1400W (a novel selective iNOS inhibitor) > L-NIO > L-NMMA > L-NAME. The nitrite production was sensitive to cycloheximide indicating that it was dependent on *de novo* protein synthesis. iNOS protein was detected by Western blot analysis and the accumulation of iNOS mRNA was confirmed by RT-PCR. NF-κB inhibitor PDTC abolished the iNOS mRNA and protein synthesis as well as nitrite production suggesting the critical requirement of NF-κB activation in the iNOS induction by LPS in T84 cells. Combination of cytokines IL-1β, TNFα and IFNγ showed no effect on NO production in T84 cells.

Cyclosporine A (CsA) inhibited LPS-induced NO production in a dose-dependent manner at therapeutically relevant concentrations. Methotrexate and sulfasalazine had a slight inhibitory action when used at high concentrations but did not alter LPS-induced NO production at therapeutic drug concentrations. TNFα antagonists infliximab and etanercept did not alter NO production.

Intestinal epithelial cells express iNOS and produce NO in a NF-κB-dependent manner when exposed to LPS. CsA, a drug used in graft rejection, rheumatoid arthritis and in inflammatory bowel disease, inhibited iNOS expression and NO production but methotrexate, sulfasalazine and TNFα antagonists were practically ineffective. These findings are implicated in the pathogenesis and drug treatment of IBD.

Nitric Oxide
R.J. Gryglewski and P. Minuz (Eds.)
IOS Press, 2001

Modified Method for Capillary Electrophoresis to Determine Nitric Oxide Correlated Nitrate in Tissue Homogenates

Muzaffer Tunçel[†§], Dilek Dogrukol-Ak[†], Nilüfer Erkasap[‡]

[†] University of Anadolu, Faculty of Pharmacy, Department of Analytical Chemistry, 26470;
[‡]University of Osmangazi , Faculty of Medicine, Department of Physiology, 26040, Eskişehir, Türkey.

[§] Correspondence author: Muzaffer Tunçel, Prof.Dr.
Anadolu University, Faculty of Pharmacy, Dept. Anal. Chem. 26470-ESKIŞEHIR/TURKEY
e-mail mtuncel@anadolu.edu.tr

1. Introduction

Nitric oxide (NO) has been accepted as one of the important cellular mediator for almost ten years. Since the endothelium-derived relaxing factor (EDRF) has been agreed by the scientists that it is identical to NO, lots of researches are being performed to elucidate certain physiological or patho-physiological phenomena. It is obviously known that it is absolutely responsible from some physiological mechanisms [1].

Certain studies have been attempted to monitor the level of NO *in vivo* systems, instantly. To achieve these, electroanalytical methods have been employed using microelectrodes [2,3]. Other methods have been reviewed in an article [1] that the researches carried out them based on the measurement of the products of NO. Capillary electrophoresis and certain techniques are also employed to determined the level of nitrite and nitrate in urine, synovial fluid, plasma and several tissue homogenates [4-6].

The aim of the present study is to improve a new and modified CE method for the analysis of nitrate in the body fluids or homogenates and to compare the present methods. The results were evaluated by the methodologies of statistics and application of the method was examined by measuring nitrate level employing certain tissue homogenates.

2. Experimental

All of the chemicals (Merck, Germany) and acetonitrile were of analytical and HPLC grade (Carlo Erba, Italy). CE experiments were conducted using a SpectraPhoresis 100 system capillary electrophoresis equipped with a high-voltage power supply, a modular injector and a model of SpectraFOCUS scanning detector (Thermo Separation Products, CA, USA) cabled to a model Etacomp 486 DX4 100 computer processing the data using PC 1000 (version 2.6) working under OS/2 Warp Program (version 3.0).

In the present study, fused silica capillary was filled with the background electrolyte consisting of 200 mM lithium chloride, 10 mM borate buffer at pH 8.5. All of the solutions were prepared in a nitrate free double distilled water. The dilutions of nitrate were made from the stock solutions of $1.06x10^{-3}$ M KNO_3 and $4.39x10^{-4}$ M KBr as internal standard (IS). The final concentration of IS was always $2.92x10^{-4}$ M in the calibration and sample solutions.

Twelve Spraque-Dawley male rats that have 200-250 g were employed to examine the performance of the method. The tissues were homogenized in a 3-ml isotonic buffer solution at pH 7.4.

3. Results and Discussion

Quantitative measurements of released NO during the *in vivo* studies are very difficult due to the short half-life of mentioned compound. Since it is rapidly oxidized to nitrite and nitrate in body fluids or tissue homogenates, most studies have been achieved to measure nitrite and/or nitrate as markers of NO generation.

Nitrite and nitrate in biological material as indices of NO generation and the measurement of their levels are very complicated by several factors. Most significantly human and animals get these by dietary intake. Therefore, the researchers have to consider there would be an offset for the measurement of NO products in the organism.

Well-resolved and sharp peaks were obtained in the use of 200 mM LiCl and 10 mM borate buffer at pH 8.5, detecting at 214 nm in our study. High efficient separation was observed with their 572 000 plates m^{-1}. The peaks were rather symmetrical with 0.97 and 1.08 CFR asymmetry factor and they appeared at the migration times of (t_M±RSD%) 7.2±1.1 min and 8.9±1.01 min for bromide and nitrate, respectively. The statistical evaluation of integrated area of $7.1x10^{-5}$ M nitrate and $2.92x10^{-4}$ M bromide was made by using the ratio of peak normalization values nitrate to IS for all quantification studies.

Whole validation process including peak precision, linearity, accuracy, specificity, and method precision procedures were applied and reasonable results were obtained.

The linearization studies were performed in the range of $1.41x10^{-5}$-$7.05x10^{-5}$ M nitrate and fix concentration of $2.92x10^{-4}$ M bromide. The calibration equations were computed considering the data of inter-day and intra-day studies using the ratio values of normalized nitrate and bromide areas. Statistical evaluation of the calibration studies was tried to validate the method. The related results covering accuracy of the method in the range mentioned and mean recoveries are at the level of 95% confidence interval are in Table I.

The limit of the detection (LOD) as concentration was calculated to be $5.70x10^{-7}$ M according to the criteria of the division of 3.3 times of standard deviation of intercept to the slope of the calibration curve. The limit of quantification (LOQ) was computed as 1.73x 10^{-6} M via the division of ten times of standard deviation of intercept to slope of the calibration curve.

Table I. Results of inter-day and intra-day calibration studies for linearity and accuracy of the method.

Regression parameters	
Linearity	
r^2	0.9986; 0.9974; 0.9978
intercept (mean±S.D., n=3)	0.048±0.006
slope (mean±S.D., n=3)	16405±584
Accuracy	mean recovery ± CL (%)
50 %	100.5±13.1
100 %	101.3±8.2
150 %	100.6±3.4

CE techniques are very appropriate for the biological analysis. During the processes, only precipitation of proteins is realized and acetonitrile was used for precipitating agent of proteins in this study. Besides, acetonitrile makes the peaks sharp and symmetrical and decrease the electroosmotic mobility, as well.

The applicability of the method was tested using certain tissue homogenates of rats such as kidney, brain, heart, stomach, lung, testis and liver. The groups were consisting of control and group to which is applied 50 mg kg^{-1} L-NAME. Any interference is not observed during the analysis of tissue homogenates, at all. Although, the t_M values of the standard bromide and nitrate shift a little due to the conditioning of the capillary surface. The original electropherogram of stomach tissue homogenate of rat were demonstrated in Figure 2a control group and 2b L-NAME applied group. The brief results for the tissues are demonstrated in Table II. This application was achieved to investigate whether the method was effected by any factor in biological studies. As is seen from the results, L-NAME significantly causes decrease of nitrate levels, which corresponds to NO production.

Table II. Brief results of nitrate levels ± SD in the relevant tissue homogenates of rats as nmol g^{-1} wet weight of tissue (n= 3 for each).

	Nitrate level (mean±SD) (nmol g^{-1}, w/w)		Rate of
Tissue	Control	L-NAME	Decrease (%)
Kidney	21.8± 6.0	16.0± 5.5	26.8
Brain	20.4±4.6	17.6±8.9	13.7
Heart	19.3±5.1	9.2±9.2	52.4
Stomach	43.1±15.1	12.8± 7.7	70.3
Lung	50.7±26.1	8.8±6.6	82.6
Testis	16.4±2.7	4.2±2.6	74.5
Liver	17.7±6.5	7.5±2.4	57.4

As conclusion, the method is worthwhile for the analysis of nitrate especially in the body fluids and tissue homogenates. It is precise, accurate, cheap and a single analysis takes about fifteen minutes. All of the points mentioned above exhibit the advantageous of the study.

Acknowledgements

The authors appreciate the Research Fund of University of Anadolu for their supporting this study.

REFERENCES

[1] G. Ellis, I. Adatia, M. Yazdanpanpanah and S.K. Makela, Clin. Biochem. 31 (1998) 195.

[2] H. Tsukahara, D.V. Gordienko and M.S. Goligorsky, Biochem. Biophys. Res. Commun. 193(2) (1993) 722.

[3] T. Malinski and Z. Taha, Nature 358 (1992) 676.

[4] T. Ueda, T. Maekawa, D. Sadamitsu, S. Oshita, K. Ogino and K. Nakamura, Electrophoresis 16 (1995) 1002.

[5] M.A. Friedberg, M.E. Hinsdale and Z.K. Shihabi, J. Chromatogr. A 781 (1997) 491.

[6] G. Zunic, S. Spasic and Z. Jelic-Ivanovic, J. Chromatogr.B 727 (1999) 73.

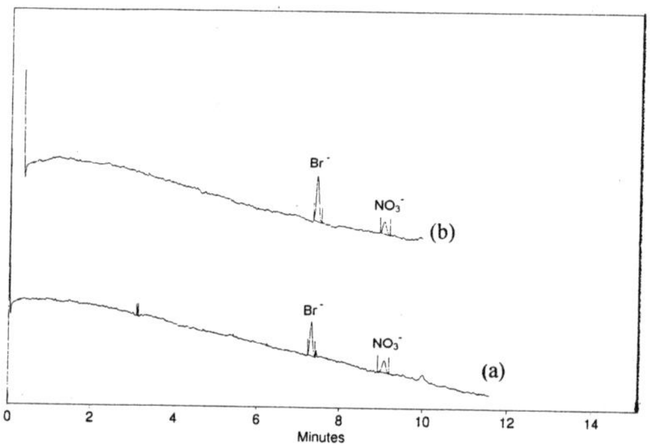

Figure 1.

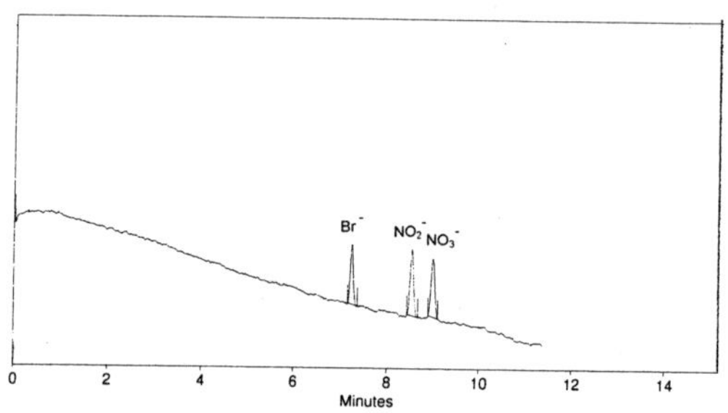

Figure 2.

Figure 1. Electropheregrams of bromide, nitrite and nitrate. Conditions: 200 mM LiCl and 10 mM borate (pH 8.5); injection, hydrodinamically 0.5 s; applied voltage. –15 kV; capillary, uncoated fused silica, 75 μm I.D., 82-cm total and 43-cm effective length; detection wavelength, 214 nm. Sample concentration: 2.92×10^{-4} M bromide, 8.88×10^{-5} M nitrite and 5.64×10^{-5} M nitrate in acetonitrile-water (1:2, v/v).

Figure 2. Electropherograms of stomach tissue homogenates of rat (a) control group and (b) L-NAME applied group. Other conditions as described in Figure 1.

Nitric Oxide
R.J. Gryglewski and P. Minuz (Eds.)
IOS Press, 2001

Role of Nitric Oxide on Exercise-Induced Changes in Active Subjects: Effects of Physical Training and Acute Exercise

Giuseppina Benoni & Laura Cuzzolin

Department of Medicine & Public Health – Section of Pharmacology, University of Verona, Italy

Abstract

The mechanism responsible of the beneficial cardiovascular effects of exercise awards an important role to nitric oxide as vasodilating and antiadhesive agent. We compared, in our experiments, subjects moderately and highly trained to inactive subjects, in order to evaluate the NO production, platelet adhesion and erythropoietin concentrations as parameters of performance. In subjects engaged in a regular physical activity (either moderate or strenuous) we found higher NO levels than in inactives and these values well correlated with a decrease in platelet adhesion and an increase in erythropoietin in hypoxic conditions. So, the increase in NO production induced by long-term physical activity may be protective in athletes and improve their performance in a hypoxic situation.

Introduction

The effects of exercise on the cardiovascular system have been well documented. Various benefits of regular exercise include an increase in coronary flow reserve [1] and a significant decrease in the incidence of cardiovascular diseases [2]. Although the precise mechanism of these beneficial effects of exercise remain unclear, it has been suggested to be multifactorial and involving a modulation of vascular endothelial function, with an important role for nitric oxide (NO) as vasodilating agent [3, 4] released in response to many stimuli such as acetylcholine, catecholamines, shear stress and local hypoxia [5]. Modest increases in NO production have been reported during an acute exercise [6, 7] and regular physical exercise has been linked to an upregulated expression of the endothelial NOS enzyme either in isolated specimens of vascular tissue or *in vivo* [8, 9]. Chronic exercise leads to a greater NO production by the endothelium, most probably as a result of increased levels or activity (or both) of endothelial NOS: sustained increases in muscle blood flow and shear stress occurring during chronic aerobic exercise could lead to adaptive increases in NO synthesis. However, human studies are contraddictory: the level of physical activity required to produce beneficial increases in NO production and the duration of this effect after exercise cessation have not yet been established. Some authors have found that NO plays a significant role in exercise-induced arteriolar vasodilatation [10], whereas others have found that it does not [11]. While many protective effects of adaptation to stress or exercise have been shown to be mediated by their stimulating influence on NO synthesis, the effect of adaptation to hypoxia on NO metabolism remains obscure in many respects [12], probably due to the use of different models of hypoxic exposures [13, 14]. Intermittent short-term hypoxic exposures of moderate intensity seem to be accompanied by transient increases in intracellular Ca^{++} which activates eNOS [15] and adaptation to this situation induces a number of processes in the organism which may underlie the increased production and storage of NO [16].

Since NO is a highly reactive gas with a short half-life [17], in experimental conditions it is possible to detect this molecule in the exhaled air by a chemiluminescence analyzer or to measure nitrite and nitrate, their major stable-end metabolites, in plasma and/or urine.

The cellular source of NO detected in exhaled air is not clear, but may include pulmonary endothelial cells and/or resident macrophages of the lung [18]. Bauer et al. [19] hypothesized that acute changes in cardiovascular status could alter the pulmonary excretion rate of NO: slight increases in expired NO concentrations, ranging from 6-20%, were observed in healthy subjects during a 10 min exercise at a cycle ergometer, but the changes were reversed within 10 min after exercise had stopped; in addition, the well-conditioned athletes had a basal NO excretion rate two to three times higher compared to the other subjects. Also other authors [20, 21] observed in subjects of varying fitness, after a standardized bycicle exercise protocol, an increase in exhaled NO as workload increased. Also Matsumoto [22] found a threefold increase in NO output after 11 min of bycicle exercise test.

Several studies have examined the effects of an acute exercise on NO production evaluated as nitrite/nitrate concentrations, and the maximal aerobic capacity in healthy subjects has been found to correlate with plasma and urinary nitrate concentrations [23]. Transient elevation of these NO metabolites have been observed after an acute exercise in human healthy subjects [6, 24] and it has been hypothesized to be of vascular origin. However, the nitrate levels in plasma and their excretion rate could reflect also other endogenous sources comprised leukocytes [25] and platelets [26].

In a work evaluating the effects on NO production of an intense acute exercise such as 42 km marathon race and 15 km race [27], the urinary excretion of NO metabolites was found to be related to the level of physical activity, with the greatest excretion encountered in subjects highly trained. In another study, the resting levels of plasma nitrate were higher in trained than in sedentary subjects and a period of physical exercise (a graded exercise test on a cycle ergometer) further increased the plasma nitrate levels both in athletic and nonathletic subjects [28].

Experimental studies

In our experiments, we studied the effects of an acute exercise (made in normoxic and hypoxic conditions) in inactive subjects and in moderately or highly trained subjects on NO production (evaluated as nitrite/nitrate, their major stable-end metabolites), platelet adhesion and Epo concentrations.

At this scope, we enrolled the members of the Research Center of Physical Fitness and Sports (age 21 to 25 years) and a group of cross-country skiers (mean age 25.5 years). The first group (group A), considered moderately trained, engaged in regular exercise at least for the last two years and were investigated after an extensive training period of one month consisting of running, swimming and cycling; the second group (group B) comprised highly trained athletes, usually living at 800-1200 mt of altitude. Both groups were compared to control inactive subjects.

The subjects of group A made a 10 min acute exercise on a racing cycle ergometer performing stepwise increases in intensity until reaching, within 5 min, a constant heart rate of 150 beats x min^{-1} which represents an oxygen consumption of about 75% of VO_2 max. The skiers of group B underwent 6 days of intensive training (4 h/day at an exercise intensity corresponding to 70-90% VO_2 max) at 3100 mt of altitude, preceeded by 2 days of acclimatisation.

From the first group and compared controls, blood samples were taken before the acute exercise, immediately after the acute exercise and 1 h later: nitrite/nitrate concentrations were measured in the plasma using a simple two-step procedure employing an enzymatic conversion of nitrate to nitrite (Cayman Chem. Co), platelets were obtained by different fractionated centrifugations and tested for adhesion, as previously described [29].

From the second group and compared controls, red blood cells, hemoglobin, hematocrit, Epo and nitrite/nitrate were measured prior, immediately after the period at altitude and 7 days after returning at home as reported by Schena et al. [30].

Results and Conclusions

In our opinion, the most important result of our experimental investigations was the baseline higher plasma levels of nitrite/nitrate observed in moderately and highly trained subjects, before an acute exercise or a 6-day period of training at altitude. These values were 44 μmol/L in the first group (members of Research Center of Physical Fitness) and 50 μmol/L in the second group (cross-country skiers), compared to 25 μmol/L observed in inactives. Moreover, in the students these values were correlated to the decrease in platelet adhesion of about 50%.

Ten minutes of an acute exercise increased nitrite/nitrate plasma levels (about 80 μmol/L), but this increase did not modify platelet adhesion. Probably this was a too short time to induce NOS enzyme and NO production: so, we explain this high value with the haemoconcentration due to a loss of water with the sweatness. Fortunately, these high nitrite/nitrate concentrations persist for a short time, since it is known that too high levels of NO for long periods could be dangerous for tissues and cells [17]. In the highly trained athletes we correlated NO levels with Epo concentrations in the serum. From our data, it seems that physical exercise could induce an increase in Epo levels only in hypoxic conditions, without affecting nitrite/nitrate plasma levels. So, we hypothesized that in a hypoxic situation about 50 μmol/L could be sufficient to influence Epo production .

Finally, in our experimental studies we did not observe any difference in erythrocytes, hemoglobin and hematocrit, even if other authors [31] referred that intermittent hypoxia is able to stimulate erythropoiesis and improve aerobic capacity.

Final considerations

From our data physical fitness, rather than a single acute exercise, and NO formation are positively linked to each other, as also observed by other authors [28]. In fact, in subjects engaged in a regular physical activity (whether moderate or strenuous) higher nitrite/nitrate concentrations (40-50 μmol/L) have beneficial effects by decreasing platelet adhesion and stimulating Epo production in hypoxic conditions.

The increase in NO production induced by long-term physical activity may be protective in athletes and improve their performance in a hypoxic situation. Moreover, a chronic moderate physical activity could be useful in humans in the prevention of cardiovascular diseases.

References

1) M.H. Laughlin, J.N. Diana, C.M. Tipton. Effects of exercise training on coronary reactive hyperemia and blood flow in the dog. *Journal of Applied Physiology* 45 (1978) 604-610.

2) T.F. Schiable, J. Scheuer. Cardiac adaptations to chronic exercise. *Progress in Cardiovascular Diseases* 27 (1985) 297-324.

3) J. LoScalzo, J.A. Vita. Ischemia, hyperemia, exercise and nitric oxide. Complex physiology and complex molecular adaptations. *Circulation* 90 (1994) 2556-2559. 1994

4) M.D. Delp. Effects of exercise training on endothelium-dependent peripheral vascular responsiveness. *Medicine and Science in Sports and Exercise* 27 (1995) 1152-1157.

5) G.M. Rubanyi, J.C. Romero, P.M. Vanhoutte. Flow-induced release of endothelium-derived relaxing factor. *American Journal of Physiology* 250 (1986) H1145-H1149.

6) S.M. Bode-Boger, R.H. Boger, E.P. Schroeder, J.C. Frolich. Exercise increases systemic nitric oxide production in men. *Journal of Cardiovascular Risk* 1 (1994) 173-178.

7) A. Wennmalm. Metabolism and excretion of nitric oxide in man: basal studies and clinical applications. *Agents Actions* 45 (1995) 129-138.

8) J. Wang, M. Wolin, T.H. Hintze. Chronic exercise enhances endothelium-mediated dilatation of epicardial coronary artery in conscious dogs. *Circulation Research* 73 (1993) 829-838.

9) W.C. Sessa, K. Pritchard, N. Seyedi, J. Wang, T.H. Hintze. Chronic exercise in dogs increases coronary vascular nitric oxide synthase gene expression. *Circulation Research* 4 (1994) 349-353.

10) D.M. Gilligan, J.A. Panza, C.M. Kilcoyne, M.A. Waclawiw, P.R. Casino, A.A. Quyumi. Contribution of endothelium-derived nitric oxide to exercise induced vasodilatation. *Circulation* 90 (1994) 2853-2858.

11) T. Endo, T. Imaizumi, T. Tagawa, M. Shiramoto, S. Ando, A. Takeshita. Role of nitric oxide in exercise-induced vasodilatation of the forearm. *Circulation* 90 (1994) 2886-2890.

12) I.Y. Malyshev, E.B. Manukhina. Stress, adaptation and nitric oxide. *Biochemistry* 63 (1998) 992-1006.

13) C. Xue, A. Rengasamy, T.D. Lecras, P.A. Koberna, G.C. Dailey, R.A. Johns. Distribution of NOS in normoxic vs hypoxic rat lung: upregulation of NOS by chronic hypoxia. *American Journal of Physiology* 11 (1994) L667-L678.

14) P.W. Shaul, A.J. North, T.S. Brannon, K. Ujiie, L.B. Wells, P.A. Nisen, C.J. Lowenstein, S.H. Snyder, R.A. Star. Prolonged in vivo hypoxia enhances nitric oxide synthase type I and type III gene expression in adult rat lung. *American Journal of Respiratory Cell and Molecular Biology* 13 (1995) 167-174.

15) V. Hampl, D.N. Cornfield, N.J. Cowan, S.L. Archer. Hypoxia potentiates nitric oxide synthesis and transiently increases cytosolic calcium level in pulmonary artery endothelial cells. *European Respiratory Journal* 8 (1995) 515-522.

16) E.B. Manukhina, I.Y. Malyshev, B.V. Smirin, S.Y. Mashina, V.A. Saltykova, A.F. Vanin. Production and storage of nitric oxide in adaptation to hypoxia. *Nitric oxide: Biology and Chemistry* 3 (1999) 393-401.

17) S. Moncada, R.M.J. Palmer, E.A. Higgs. Nitric oxide: physiology, pathophysiology and pharmacology. *Pharmacological Reviews* 43 (1991) 109-142.

18) L.E. Gustafsson, A.M. Leone, M.G. Persson, N.P. Wiklund, S. Moncada. Endogenous nitric oxide is present in the exhaled air of rabbits, guinea pigs and humans. *Biochemichal and Biophysical Research Communication* 181 (1991) 852-857.

19) J.A. Bauer, J.A. Wald, S. Doran, D. Soda. Endogenous nitric oxide in expired air: effects of acute exercise in humans. *Life Sciences* 55 (1994) 1903-1909.

20) M.G. Persson, N.P. Wiklund, L.E. Gustafsson. Endogenous nitric oxide in single exhalations and the change during exercise. *American Review of Respiratory Disease* 148 (1993) 1210-1214.

21) J. Koller-Strametz, B. Matulla, M. Woltz, M. Muller, J. Entlicher, H. Eichler, L. Schmetterer. Role of nitric oxide in exercise-induced vasodilatation in man. *Life Sciences* 62 (1998) 1035-1042.

22) A. Matsumoto, Y. Hirata, S. Momomura, H. Fujita, A. Yao, M. Sata, T. Serizawa. Increased nitric oxide production during exercise. *Lancet* 343 (1994) 849-850.

23) L. Jungersten, B. Wall, A. Wennmalm. Physical fitness regulates nitric oxide formation in healthy man. *Circulation* 94 (1996) 497-502.

24) C.D. Leaf, J.S. Wishnok, S.R. Tannenbaum. Nitric oxide: the dark side. In: S. Moncada & E.A. Higgs (eds) Nitric oxide from L-arginine: a bioregulatory system. Elsevier, New York, 1990, pp. 291-299.

25) A.M. Niess, H.H. Dickhuth, H. Northoff, E. Fehrenbach. Free radicals and oxidative stress in exercise. Immunological aspects. *Exercise Immunology Review* 5 (1999) 22-56.

26) Q. Zhou, G.R. Hellermann, L.P. Solomonson. Nitric oxide release from resting human platelets. *Thrombosis Research* 77 (1995) 87-96.

27) L.G. Rodriguez-Plaza, A.B. Alfieri, L.X. Cubeddu. Urinary excretion of nitric oxide metabolites in runners, sedentary individuals and patients with coronary artery disease: effects of 42 km marathon, 15 km race and a cardiac rehabilitation program. *Journal of Cardiovascular Risk* 4 (1997) 367-372.

28) L. Jungersten, A. Ambring, B. Wall, A. Wennmalm. Both physical fitness and acute exercise regulate nitric oxide formation in healthy humans. *Journal of Applied Physiology* 82 (1997) 760-764.

29) L. Cuzzolin, S. Lussignoli, F. Crivellente, A. Adami, F. Schena, P. Bellavite, G. Brocco, G. Benoni. Influence of an acute exercise on neutrophil and platelet adhesion, nitric oxide plasma metabolites in inactive and active subjects. *International Journal of Sports Medicine* 21 (2000) 289-293.

30) F. Schena, L. Cuzzolin, L. Rossi, M. Pasetto, G. Benoni. Plasma nitrite/nitrate and erythropoietin levels in cross-country skiers during altitude training. *Clinical Physiology* (in press).

31) H.C. Gunga, L. Rocker, C. Behn, W. Hildebrandt, E. Koralewski, I. Rich, W. Schobersberger, K. Kirsch. Shift working in the chilean Andes (>3,600 m) and its influence on erythropoietin and the low-pressure system. *Journal of Applied Physiology* 81 (1996) 846-852.

Author Index